Advances in Corrosion and Protection of Materials

Advances in Corrosion and Protection of Materials

Editor

Renato Altobelli Antunes

MDPI • Basel • Beijing • Wuhan • Barcelona • Belgrade • Manchester • Tokyo • Cluj • Tianjin

Editor
Renato Altobelli Antunes
Center for Engineering,
Modeling and Applied Social Sciences,
Federal University of ABC,
Santo André, Brazil

Editorial Office
MDPI
St. Alban-Anlage 66
4052 Basel, Switzerland

This is a reprint of articles from the Special Issue published online in the open access journal *Metals* (ISSN 2075-4701) (available at: https://www.mdpi.com/journal/metals/special_issues/corrosion-protection).

For citation purposes, cite each article independently as indicated on the article page online and as indicated below:

LastName, A.A.; LastName, B.B.; LastName, C.C. Article Title. *Journal Name* **Year**, *Volume Number*, Page Range.

ISBN 978-3-0365-7980-1 (Hbk)
ISBN 978-3-0365-7981-8 (PDF)

Contents

About the Editor

Renato Altobelli Antunes

Renato Altobelli Antunes is a professor at the Federal University of the ABC (UFABC), at the Center for Engineering, Modeling and Applied Social Science. Currently, his research activity is centered on the study of corrosion mechanisms of metallic materials, protection methods, scanning probe techniques, and the correlation between surface chemistry and the corrosion behavior of light alloys and steels. Prof. Antunes is a Guest Editor for the journal Metals. He has been awarded a Research Productivity Grant from the Brazilian agency CNPq (Brazilian National Council for Scientific and Technological Development) since 2015.

metals

MDPI

Editorial

Advances in Corrosion and Protection of Materials

Renato Altobelli Antunes

Center for Engineering, Modeling and Applied Social Sciences, Federal University of ABC, Av. Dos Estados, 5001, Santo André 09210-580, SP, Brazil; renato.antunes@ufabc.edu.br

1. Introduction and Scope

From an engineering standpoint, the corrosion of metallic materials is a challenging phenomenon, responsible for huge costs and serious failures. This situation has driven the quest for increasing scientific initiatives to improve the current knowledge on corrosion mechanisms and mitigation strategies in different areas. Hence, intense research activity is dedicated to develop new alloy compositions, protection methods, and to investigate corrosion mechanisms in depth.

As part of this scenario, this Special Issue was conceived to gather new, innovative contributions towards a deeper understanding of corrosion processes in a variety of applications.

2. Contributions

This volume collected 12 research papers devoted to different aspects of corrosion and its mechanisms. In the first paper, Seikh et al. [1] prepared low-Ni, Co-free maraging steels with varying Mo concentrations using an electroslag remelting process. The effect of Mo addition on the electrochemical behavior of the new maraging steels was investigated in H_2SO_4 and NaCl solutions. The corrosion resistance was improved with respect to the Mo-free steel, depending on the Mo content.

Queiroz et al. [2] compared the corrosion resistance of the AA2024-T3 and AA2524 T3 aeronautic Al alloys, giving emphasis to the interplay between microstructural aspects (presence of intermetallic compounds) and the electrochemical behavior of the alloys. The surface area of intermetallic compounds was associated with the localized corrosion susceptibility of the alloys.

The local and global corrosion behaviors of composite Ni–P-multiwalled carbon nanotube films obtained by electroless deposition on carbon steel substrates were studied by Oliveira et al. [3]. The addition of multiwalled carbon nanotubes increased the adhesion strength and reduced the electrochemical activity on the surface of the coated samples.

As ship hull steel is subject to severe corrosion attack, Zhang et al. [4] submitted a contribution towards understanding the corrosion mechanism of EH47 ship steel in seawater. Electrochemical tests and immersions tests (up to 80 days of immersion) were conducted. A mixed pitting corrosion and crevice corrosion mechanism was proposed, depending on the immersion time.

Li et al. [5] obtained a copper-rich surface on copper-bearing steels by wet/dry cyclic corrosion tests, thus forming an in situ composite material. The new copper-rich surface significantly increased the corrosion resistance of the copper-bearing steels.

Anodization is a traditional surface treatment for magnesium alloys. The effect of anodization parameters on the corrosion-fatigue behavior of structural Mg alloys is yet not well understood. The paper by Oliveira et al. [6] aimed to shed light on this topic. They investigated the corrosion-fatigue behavior of an anodized AZ31B Mg alloy in phosphate-buffered solution. The effect of the anodized layer on the fatigue resistance of the alloy is discussed.

Wang et al. [7] evaluated the effect of pH on the localized corrosion behavior of the L360 pipeline steel with and without S8 coating in a typical sour-simulated environ-

Citation: Antunes, R.A. Advances in Corrosion and Protection of Materials. *Metals* **2023**, *13*, 1059. https://doi.org/10.3390/met13061059

Received: 8 February 2023
Accepted: 14 February 2023
Published: 1 June 2023

ment. An innovative wire beam electrode (WBE) was employed for studying the local corrosion processes.

The oxidation behavior of Fe-W-Cr-Zr steel with dispersed Fe_2Zr particles was studied by Chen and Rong [8]. This material is a candidate for advanced generation IV nuclear reactors. In this work, the authors characterized the composition of the oxide layer formed after oxidation tests in stagnant air and oxygen-saturated stagnant liquid Pb-Bi eutectic, as well as the influence of microstructural features on the scale growth rate.

Microbiologically induced corrosion (MIC) is a serious concern in oilfield production. The contribution by Welikala et al. [9] was focused on the study of biofilm growth by iron-reducing bacteria *Shewanella putrefaciens* on carbon steel and its effect on the corrosion behavior of the material in different culture media. They provided an in-depth evaluation of the effect of the IRB biofilms on the corrosion behavior of the carbon steel.

Aperador et al. [10] investigated the corrosion resistance of structural rebar embedded in non-Portland cement alkali-activated slag concrete. The effect of immersion time in chloride solution was investigated. The correlation between concrete porosity and corrosion resistance was discussed for the new cement composition.

Ductile cast iron used for water transmission pipelines are subject to corrosion. In the paper by Wang et al. [11], the effect of surface cracks on the corrosion resistance of a ductile iron casting was studied. They identified the main environmental factors that influence the corrosion behavior, such as oxygen solubility, temperature, pH, chloride concentration, and water hardness.

Howyan et al. [12] developed polystyrene/organoclay nanocomposites to investigate their corrosion protection ability as coatings for carbon steel parts. They used commercial Indian clay and Khulays clay as nanofillers. A corrosion protection efficiency of up to 81.4% was reached using the commercial clay at a filler loading of 1 wt.%. Dispersion of the nanoclays in the polymer matrix was the essential factor influencing the corrosion resistance of the coated samples.

Acknowledgments: As Guest Editor, I would like to thank all the reviewers and authors who contributed to this Special Issue. Special thanks are given to all MDPI staff for the professional, polite, and timely support in all publication steps.

Conflicts of Interest: The author declares no conflict of interest.

References

1. Seikh, A.H.; Halfa, H.; Soliman, M.S. Effect of molybdenum content on the corrosion and microstructure of low-Ni, Co-free maraging steels. *Metals* **2021**, *11*, 852. [CrossRef]
2. Queiroz, F.M.; Terada, M.; Bugarin, A.F.S.; de Melo, H.G.; Costa, I. Comparison of corrosion resistance of the AA2524-T3 and the AA2024-T3. *Metals* **2021**, *11*, 980. [CrossRef]
3. Oliveira, M.C.L.; Correa, O.V.; Silva, R.M.P.; Lima, N.B.; Oliveira, J.T.D.; Oliveira, L.A.; Antunes, R.A. Structural characterization, global and local electrochemical activity of electroless Ni-P-multiwalled carbon nanotube composite coatings on pipeline steel. *Metals* **2021**, *11*, 982. [CrossRef]
4. Zhang, H.; Yan, L.; Zhu, Y.; Ai, F.; Li, H.; Li, Y.; Jiang, Z. The effect of immersion corrosion time on electrochemical corrosion behavior and the corrosion mechanism of EH47 ship steel in seawater. *Metals* **2021**, *11*, 1317. [CrossRef]
5. Li, N.; Yan, L.; Wang, S.; Wang, C.; Zhang, H.; Ai, F.; Jiang, Z. Corrosion behavior of copper bearing steels and the derived in-situ coating. *Metals* **2021**, *11*, 1462. [CrossRef]
6. Oliveira, L.A.; Santos, S.L.; Oliveira, V.A.; Antunes, R.A. Influence of anodization on the fatigue and corrosion-fatigue behaviors of the AZ31B magnesium alloy. *Metals* **2021**, *11*, 1573. [CrossRef]
7. Wang, F.; Li, J.; Qu, C.; Yu, T.; Li, Y.; Zhu, S.; Yang, B.; Cheng, F. Corrosion mechanism of L360 pipeline steel coated with S8 in CO_2-Cl^- system at different pH values. *Metals* **2021**, *11*, 1975. [CrossRef]
8. Chen, S.; Rong, L. Oxidation behavior of intermetallic phase and its contribution to the oxidation resistance in Fe-Cr-Zr ferritic alloy. *Metals* **2022**, *12*, 827. [CrossRef]
9. Welikala, S.; Al-Saadi, S.; Gates, W.P.; Panter, C.; Raman, R.K.S. Biofilm development on carbono steel by iron reducing bacterium *Shewanella putrefaciens* and their role in corrosion. *Metals* **2022**, *12*, 1005. [CrossRef]
10. Aperador, W.; Bautista-Ruiz, J.; Sánchez-Molina, J. Effect of immersion time in chloride solution on the properties of structural rebar embedded in álcali-activated slag concrete. *Metals* **2022**, *12*, 1952. [CrossRef]

11. Wang, B.; Liu, T.; Tao, K.; Zhu, L.; Liu, C.; Yong, X.; Cheng, X. A study of the mechanisms and kinetics of the localized corrosion aggravation of ductile iron in a harsh water quality environment. *Metals* **2022**, *12*, 2103. [CrossRef]
12. Howyan, H.A.; Al Juhaiman, L.A.; Mekhamer, W.K.; Altilasi, H.H. Comparative study of protection efficiency of C-steel using polystyrene clay nanocomposite coating prepared from commercial Indian clay and local Khulays clay. *Metals* **2023**, *13*, 879. [CrossRef]

metals

Article

Comparative Study of Protection Efficiency of C-Steel Using Polystyrene Clay Nanocomposite Coating Prepared from Commercial Indian Clay and Local Khulays Clay

Nashwa A. Howyan [1], Layla A. Al Juhaiman [1,*], Waffa K. Mekhamer [1,2] and Hissah H. Altilasi [1]

[1] Chemistry Department, King Saud University, Riyadh 145111, Saudi Arabia
[2] Material Science Department, Institute of Graduate Studies, Alexandria University, Alexandria 21526, Egypt
* Correspondence: ljuhiman@ksu.edu.sa

Abstract: This work aimed to compare the coating protection efficiency of C-steel using two kinds of clay: a local Khulays clay (RC_{Kh}) from Saudi Arabia and a commercial clay (CC_{In}) from India. Clay-based polymer nanocomposites have a unique layered structure, rich intercalation chemistry, and availability at low cost. They are promising reinforcements for polymers. The raw clay for both clay types was washed before being treated with NaCl to produce sodium clay (NaC). The cationic surfactant cetylpyridinium chloride (CPC) was then used to convert the NaC into the organoclay (OC) form. Polystyrene/organoclay nanocomposites (PCNs) were prepared by combining different concentrations of organoclay (1%, 3%, and 5% OC) in toluene solvent and polystyrene (PS) as the matrix. To ensure the success of the PCN modification process, the organoclay and PCN films were characterized using a variety of techniques, including Fourier transform infrared spectroscopy (FT-IR), X-ray diffraction (XRD), and transmission electron microscopy (TEM). The shifts in the FT-IR spectra after the CPC treatment of NaC confirmed the presence of CPC in the organoclay samples and the presence of OC in the PCNs. The exfoliated structure was obtained from the XRD spectrum for low clay loading (1–3% PCN), while the intercalated structure was the dominant form for the 5% PCN. The XRD results were confirmed by TEM images. To calculate the coating efficiency of the PCNs, various electrochemical methods were used. The electrochemical measurements included electrochemical impedance spectroscopy (EIS), the electrochemical frequency modulation (EFM) method, and Tafel plots. The PCN with a concentration of 1 wt.% OC has a fully exfoliated structure and higher coating efficiency than the PCNs with partially exfoliated structures (3 wt.% and 5 wt.%). It was found from the Tafel plots that commercial Indian clay has better corrosion protection (81.4%) than local Khulays clay (60.2%). A comparison with other studies using current density values shows that our results are superior to those of many studies.

Keywords: clay; polystyrene clay nanocomposite; C-steel; protection efficiency; electrochemical measurements

Citation: Howyan, N.A.; Al Juhaiman, L.A.; Mekhamer, W.K.; Altilasi, H.H. Comparative Study of Protection Efficiency of C-Steel Using Polystyrene Clay Nanocomposite Coating Prepared from Commercial Indian Clay and Local Khulays Clay. *Metals* **2023**, *13*, 879. https://doi.org/10.3390/met13050879

Academic Editors: Changdong Gu and Renato Altobelli Antunes

Received: 15 February 2023
Revised: 12 April 2023
Accepted: 18 April 2023
Published: 2 May 2023

1. Introduction

Many metals are widely used in construction and other industrial fields. Corrosion problems, which limit the long-term use of metals such as iron, zinc, and their alloys, are one of the greatest challenges in the metals industry. Carbon steel (C-steel) is a common steel alloy that is used extensively in a variety of applications, such as installation, transportation, mining, and construction [1–7]. C-steel has a carbon content ranging from 0.12% to 2.0% (w/w), which increases the strength and hardness of steel objects, as well as their corrosion resistance [1–7]. Metal corrosion protection is the most difficult problem that many engineers and chemists face [1–4]. Many studies have been conducted to protect steel bodies and other metal alloys from damage caused by aggressive corrosion [1–7]. Organic coatings are one of many coating applications used to protect various steel bodies and other metallic substrates from corrosion due to their good barrier properties. Recently, scientists have attempted to prepare nanocomposite materials by interacting polymer coating materials with nanosized particles (e.g., metallic, organic, and inorganic nanofillers, and many nanoscale additives) that are added to improve the barrier, mechanical, and thermal properties of these polymer coatings [7–12]. Some filler nanoparticles are SiO_2, TiO_2, ZrO_2, Fe_2O_3, Al_2O_3 [10,11], and clay derivatives (e.g., organic and inorganic clay) [11–19].

A nanocomposite is described as a mixture of two or more different materials, at least one of which has nanostructural dimensions between 1 and 100 nm. Clay and polymer or monomer molecules can be combined to create nanocomposite materials, or they can be made via a variety of other methods [20–23]. According to the degree of silicate-layer separation and the strength of the interfacial tension between the polymer matrix and clay layers, polymer organoclay nanocomposites are divided into three categories: intercalated nanocomposites, exfoliated nanocomposites, and conventional composites [21]. Due to their superior properties as compared with pure polymers, polymer organoclay nanocomposites (PCNs) have been used in several applications in numerous fields. These characteristics include flame resistance, thermal stability, barrier properties, mechanical properties, chemical resistance, and optical qualities [13,14]. Additionally, many studies have examined the coating characteristics of polymers and assessed the impact of adding clay to various polymers and epoxy coating [13,16–19,23,24].

Montmorillonite clay (MMT), which belongs to the Smectite group, is used for intercalation or exfoliation with the polymer matrix in PCN application [12,21]. MMT is a very soft layered silicate with the following chemical formula:

$$Na_{0.67}K_{0.13}Ca_{0.02}Ba_{0.04}(Si_{7.47}Al_{0.53})(Al_{2.59}Fe_{0.78}Ti_{0.14}Mg_{0.44}Cr_{0.04})O_{20}(OH)_4$$

MMT possesses a high cationic-exchange capacity (CEC), which ranges between 80 and 150 meq./100 g, as well as high reactivity and a large surface area. MMT clay's crystal structure consists of nanometer-thick layers or plates (approximately 1 nm) of aluminum octahedron sheets sandwiched between two silicon tetrahedron sheets. When the layers are arranged and stacked, a gap between them is created that is known as d-spacing or gallery spacing [12,20,21]. MMT and other layered silicate clays are hydrophilic by nature. This makes them unsuitable for interacting and mixing with most hydrophobic polymer matrices. The surface of the inorganic clay should be treated with an organic surfactant to make it compatible with the organic polymer. The conventional ion-exchange method is a simple way to modify the clay surface. The most common organic surfactants used for clay modification are phosphonium or ammonium ions in the primary, secondary, tertiary, and quaternary states. Organic cations can be exchanged for inorganic cations (K^+, Na^+, and Ca^{++}) that are not strongly bound to the clay surface [12–19,21–24]. This organic modification causes the d-spacing to increase in proportion to the length of the alkyl group in the surfactant. A wide range of matrix polymers are compatible with modified organic clay (OC) [12–19,21–24]. The organic polymer can diffuse into the clay galleries after reducing the electrostatic interaction between the clay layers, which helps to separate the clay platelets so they can be more easily intercalated and exfoliated. Polymer nanocomposites represent an exciting and promising alternative to conventional composites due to the dispersion of nanometer clay platelets and their improved performance in mechanical, thermal, barrier, optical, electrical, and other physical and chemical properties [8,17–19,21–26]. In the present study, proceeding from our previous findings, we aimed to enhance the coating protection of C-steel using low clay loading and two types of clay.

Because of the amount of sea salt deposited on metallic and nonmetallic bodies, the marine system is one of the most important aggressive-corrosion environments. To simulate the seawater system, the corrosion behavior of steel bodies is frequently studied in a solution containing 3.5% w/v % sodium chloride. The difference in the coating efficiency of C- steel differs according to the clay type or nanoadditive intercalated with the polymer. In the first stage, the goal of this research was to modify and prepare polystyrene/organoclay nanocomposites (PS/OC PCNs). In the second stage, we aimed to characterize these PCN formulations using FT-IR, XRD, and TEM. The final stage was to investigate the corrosion behavior of C-steel rods coated with PCN using commercial Indian clay (CC_{In}) and local Khulays clay (RC_{Kh}) at various concentrations (1, 3, 5% PCN). The coating efficiency of the RC_{Kh} and CC_{In} after preparing the anticorrosive PCN coatings was then calculated using various electrochemical methods, such as electrochemical impedance spectroscopy (EIS), the electrochemical frequency modulation (EFM) method, and Tafel plots.

2. Materials and Methods

2.1. Materials

To compare the coating efficiency, two types of raw clay were studied: a commercial clay from India (CC_{In}) provided from the Clay Craft India, Jaipur [25] and a local clay (RC_{Kh}) from Saudi Arabia's Khulays region, Jeddah. The RC_{Kh} was prepared by grinding raw rocks with a mortar and pestle to achieve a fine grain size in the micrometer range. X-ray Florescence (XRF, from Hitachi Company, Tokyo, Japan) was used to characterize both types of clay. The dominant CC_{In} components were montmorillonite (64.9%), kaolinite (10.6%), goethite (9.5%), hematite (9.4%), and boehmite (5.6%), as found earlier [25]. The XRD patterns of the RC_{Kh} show that the dominant components were montmorillonite (35.22%), mica (22.8%), kaolinite (13.33%), quartz (8.57%), feldspar (6.66%), and ilmenite (5.71%). Dolomite and gypsum (3.81%) and calcite (3.81%) were the remaining components [26].

Polystyrene (PS), with a molecular weight of 259,000 g/mol, was used to prepare the polymer clay nanocomposite in this study, and it was provided by the Sabic Company, Riyadh, Saudi Arabia.

Cetylpyridinium chloride (CPC) was purchased from the BDH Co., Istanbul, Turkey. It was used as a cationic surfactant in the production of organoclay (OC). The general formula is $C_{21}H_{38}ClN$, and its molecular weight is 358.01 g/mol. The Avonchem Company, Cheshire, UK, provided toluene ($C_6H_5CH_3$), with a high purity percentage of 99.5%, which was used as a solvent for the polymer matrix PS. The Win lab Company, Queensland, Australia, provided analytical-grade sodium chloride AR.

The carbon steel (C-steel) rods were provided by the ODS Co., Schleswig-Holstein, Germany. C-steel has a chemical composition of 98.468% Fe and 0.46% C [17].

2.2. Methods

2.2.1. Modification Methods of Clay

The scheme of synthesis of PCN starting from raw clay using present study technique is shown in Figure 1. The local Khulays clay was prepared in our laboratory from raw rocks by grinding to a fine grain size in the micrometer range [26], and the CC_{In} was supplied as a fine powder. An amount of 30 g of raw clay (RC) was dispersed in 500 mL of distilled water and shaken overnight to remove undesirable salts and contaminants from both clay types. After setting the solution aside for a while, the supernatant solution was removed. Three days were spent repeating this action. A 250 mL solution of 0.25 M NaCl was added to the raw clay suspension of both types of clay (RC_{Kh} and CC_{In}), which was then agitated overnight. This process was repeated for three days. This is referred to as the inorganic cationic-exchange modification that creates a homo ionic clay surface. Following this process, the solution suspension was separated by centrifuging at 3800 rpm for roughly 10 min. When the $AgNO_3$ test for chloride ions was negative, the produced homo ionic clay (NaC) was washed and separated by centrifuging. After this alteration, the NaC was ground to fine powder and stored in a desiccator after being dried in an oven. The organic modification of the NaCl of both clay types was then achieved through a cationic-exchange reaction with cetylpyridinium chloride (CPC), in which organic cations replaced sodium ions to prepare the organoclay (OC).

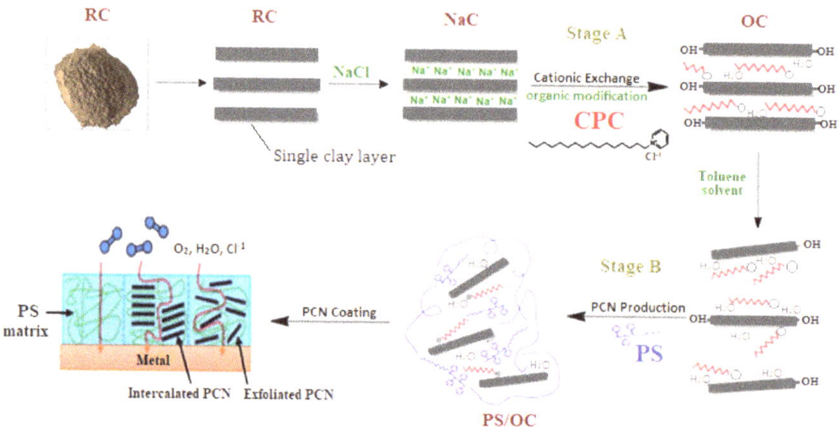

Figure 1. Scheme of synthesis of PCN starting from raw clay using present study technique.

This procedure was carried out by dispersing 5 g of NaC from both clay types in 500 mL of distilled water, stirring the solution, and leaving it overnight. CPC solution was made by dissolving 3.58 g of cetylpyridinium chloride (two times the CEC of clay) in 100 mL distilled water. The CPC surfactant solution was added dropwise to the NaC suspension while vigorously stirring, and the mixture was stirred for 24 h at room temperature to ensure CPC insertion within the clay layers. The OC was collected and separated by centrifugation at 3800 rpm for about 10 min, followed by several washes with distilled water until no chloride ions were found in the solutions. The organoclay was oven-dried overnight before being ground with a mortar and pestle to a fine grain size in the micrometer range and stored in a desiccator [17].

2.2.2. PCN Preparation Methods

Polystyrene/organoclay (PS/OC) composites, or PCNs, were prepared utilizing solution techniques, with the OC contents ranging from 1% to 5%. A specific amount of organoclay (0.02, 0.06, or 0.1 g) was added to 10 mL of toluene solvent in 50 mL glass bottles, and the mixture was magnetically agitated for 24 h at room temperature. The next day, two grams of PS was added to the organoclay suspension, which was magnetically stirred for six hours at room temperature. The PS/OC (PCN) mixture was then subjected to 10 min of sonication. For the characterization methods, the PCN solutions were cast into Petri dishes and allowed to dry at room temperature for 24 h before being delicately removed from the glass surface [26]. Figure 1 shows the process of preparing a PS/OC nanocomposite from raw clay to the final goal of preparing the coating protecting the C-steel using PCN.

2.2.3. Characterization Methods

For the characterization, the methods FT-IR and XRD were performed at King Saud University's College of Science, the Department of Chemistry. In the absorbance range from 400 to 4000 cm^{-1}, Perkin Elmer FT-IR (Waltham, MA, USA) was used to characterize the raw clay, Na clay, organoclay, and all prepared polystyrene nanocomposites. Wide-angle X-ray diffraction (XRD) patterns were used to examine all the PCN samples and modified clays using an XRD (Bruker model D8 ADVANCE from Hamburg, Germany). The operating conditions were as follows: a Cu anode radiation source with a wavelength of $\lambda = 0.154$ nm, and a current and voltage generator of 40 mA and 40 kV, respectively. The experiments were conducted at a scanning rate of 0.30/sec in the $2\theta = 3–50^{\circ}$ range. Transmission electron microscope (TEM) (JOEL-1400 from Japan) measurements were taken in the central laboratory for the Center of Science and Medical Studies for Female Students at KSU. All prepared films were immersed in epoxy resin and cured at room temperature overnight. They were then divided into segments with a thickness of roughly 70–100 nm using a microtome. The nanocomposite samples were placed on carbon-coated 200-mesh Cu grade, and the ultrathin segments were cut with a diamond knife to create clear images. TEM was used to examine the morphology of the pure PS, OC, and polystyrene/organoclay nanocomposite (1–5% wt.) films.

2.2.4. Electrochemical Methods

All electrochemical measurements were performed using a three-electrode electrochemical cell, with the C-steel as the working electrode, having a surface area of 9.95 cm^2, which was calculated using a Mitutoyo gauging tool (Kanagawa, Japan). The standard Calomel electrode (SCE) served as the reference electrode, and rigid platinum foil served as the auxiliary/counter electrode, having a surface area of 100 mm^2. The working electrode was polished to a mirror image by the polishing machine (Metaserve 2000, Buehler, London, UK) using emery papers of various grades (80, 220, 600, 1000). The electrodes were then washed several times with distilled water and cleaned with an ultrasonic cleaner in acetone for 2 min. The PCN solutions were cast dropwise onto the polished C-steel rods to completely cover their surfaces. The coated C-steel samples were then dried at room temperature for 1 h before being dried at 50 °C overnight. Another layer was added to achieve a final thickness of 100 ± 10 μm, as measured by a coating-thickness gauge (Elcometer 465, Manchester, UK) [8]. All of the electrodes were immersed in a 3.5 wt.% NaCl aqueous solution at room temperature (30 ± 0.5 °C).

The electrochemical measurements were performed using a Gamry Potentiostat/Galvanostat ZRA (model 3000, Warminster, PA, USA). Three electrochemical methods were used in the following order to assess the protection efficiencies of the prepared PCN coatings. The first step was to attain equilibrium. The open-circuit potential (OCP) was activated for 90 min to stabilize the system and achieve steady-state potential (E_{SS}). The first method was electrochemical impedance spectroscopy (EIS) using Nyquist plots, with a sweep frequency from 10^5 to 10^{-1} Hz and at an AC amplitude of 10 mV. The perturbation amplitude for the second approach, electrochemical frequency modulation (EFM), was 10 mV with a base frequency of 0.1 Hz. The potential 250 mV from the E_{SS} was scanned using the fourth method, potentiodynamic polarization (Tafel plot), at a rate of 1 mV/sec. Each experiment was conducted three times, and the reported results are the averages of three similar results.

3. Results and Discussion

3.1. *Characterization Methods*

The following parts will exhibit the FT-IR spectra, XRD, and TEM analyses of the different types of clay: commercial Indian clay (CC_{In}), local Khulays clay (RC_{Kh}), and their derivatives (NaC, OC), in addition to all recorded from 1–5 wt.% PCN.

3.1.1. FT-IR Analysis

Clay surface changes are identified using FT-IR analysis [27,28]. The FT-IR spectra of the unmodified, Na-modified, and organically modified clay from the CC_{In} and RC_{Kh} are shown in Figures 2 and 3, respectively. From Figure 2, the infrared spectrum of commercial clay, the band in the OH-stretching region at 3628 cm^{-1} is attributed to the hydroxyl groups coordinated to Al^{+3} octahedral cations (Al-OH) [28]. At 3439 cm^{-1} and 1646 cm^{-1}, respectively, the OH-stretching and -bending vibrations of the adsorbed water molecules were seen [29]. Additionally, a band at 522 cm^{-1} is for Al–O–Si deformation, and a band width at 1033 cm^{-1} is for the clay's Si–O stretching vibrations. Additionally, a wide weak band at 462 cm^{-1} has been linked to Si-O-Si deformation [28,30,31]. Absorption bands for water molecules were seen at 3440 and 1650 cm^{-1}. The bands for the deformations of Al–O–Al and Si–O–Si are situated at 523 and 464 cm^{-1}, respectively.

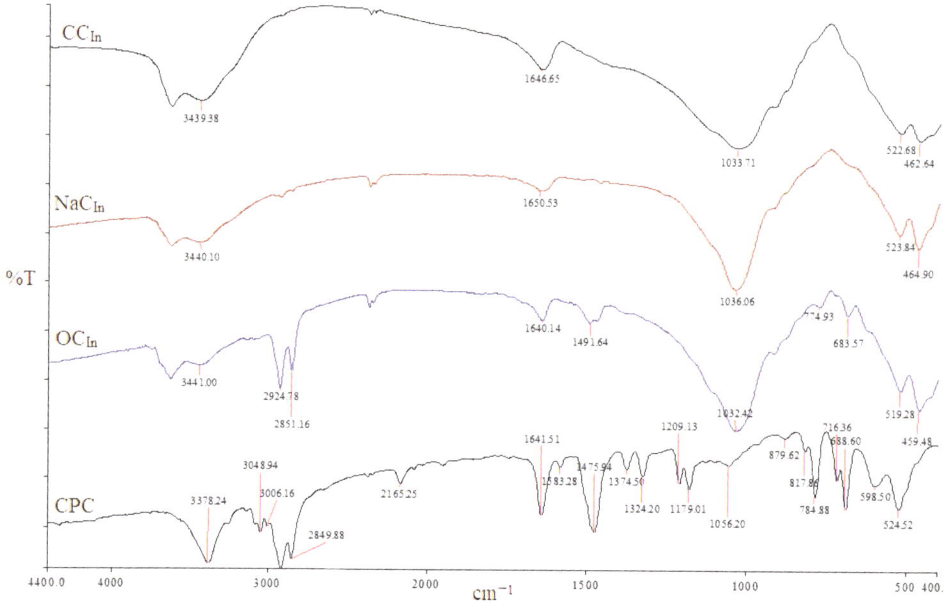

Figure 2. FT-IR spectra of CC_{In}, NaC_{In}, OC_{In}, and CPC from CC_{In} modification.

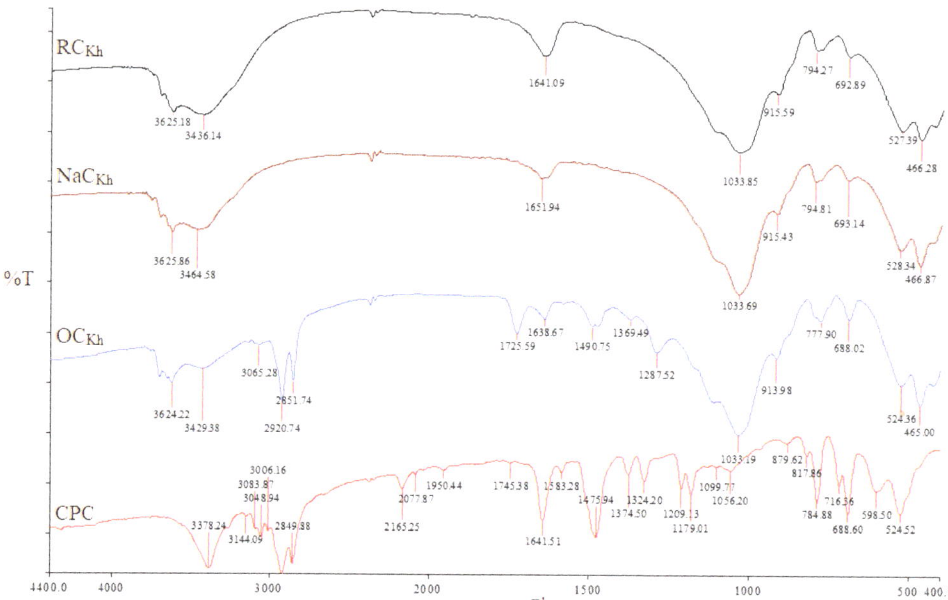

Figure 3. FT-IR spectra of RC_{Kh}, NaC_{Kh}, OC_{Kh}, and CPC from RC_{Kh} modification.

The asymmetric and symmetric stretching vibrations of the (C–H) in the alkyl group of CPC have significant sharp peaks at 2924 and 2851 cm^{-1}, respectively [32]. These two distinct, strong bands are unmistakably an indication of the intermolecular interactions that take place between neighboring alkyl chains of the organic surfactant (CPC) in clay galleries that have been treated with sodium. Additionally, a new peak for the aromatic (C = C) stretching vibrations of CPC molecules has been identified at 1491 cm^{-1} [33]. In addition to the Si–O vibrations' strong broadband properties, which were seen at 1032 cm^{-1}, the OH groups of water's asymmetric stretching and bending vibrations were visible at 3441 cm^{-1} and 1640 cm^{-1}. The success of the CPC cation intercalation between silicate layers and the organic alteration of NaC by CPC were both validated by all of these shifts and modifications in the OC spectrum [34]. Figures 4 and 5 for the CC_{In} and RC_{Kh}, respectively, display the FT-IR spectra for all synthesized PS/OC nanocomposites (PCNs) and their derivatives. These spectra show the presence of OC in the polymer matrix. The absorption bands emerged in these spectrums at the same places of the polystyrene matrix with little shifting, while some novel band alterations were attributed to the OC structure [34,35]. Two absorption bands related to the aromatic stretching vibration of sp^2 can be seen in the PS spectrum at 3059 and 3027 cm^{-1} (C–H). Additionally, at a wavelength from 2923 to 2852 cm^{-1}, there are two powerful absorption bands that are connected to the asymmetric/symmetric vibrations of the aliphatic (C–H) stretching of (–CH_2).

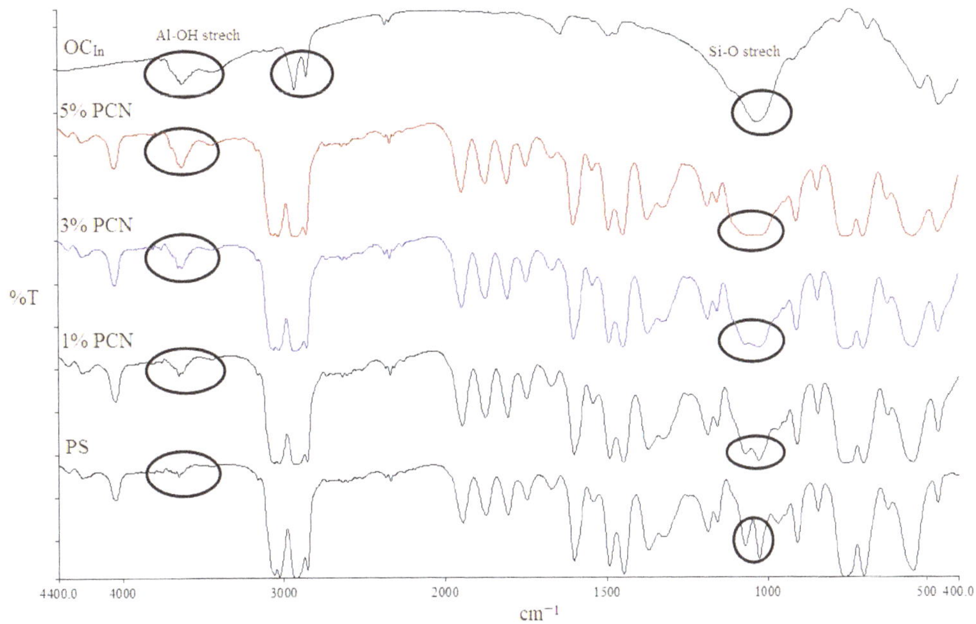

Figure 4. FT-IR spectra of PS, 1–5% wt.% PCN, and OC_{In} using CC_{In}.

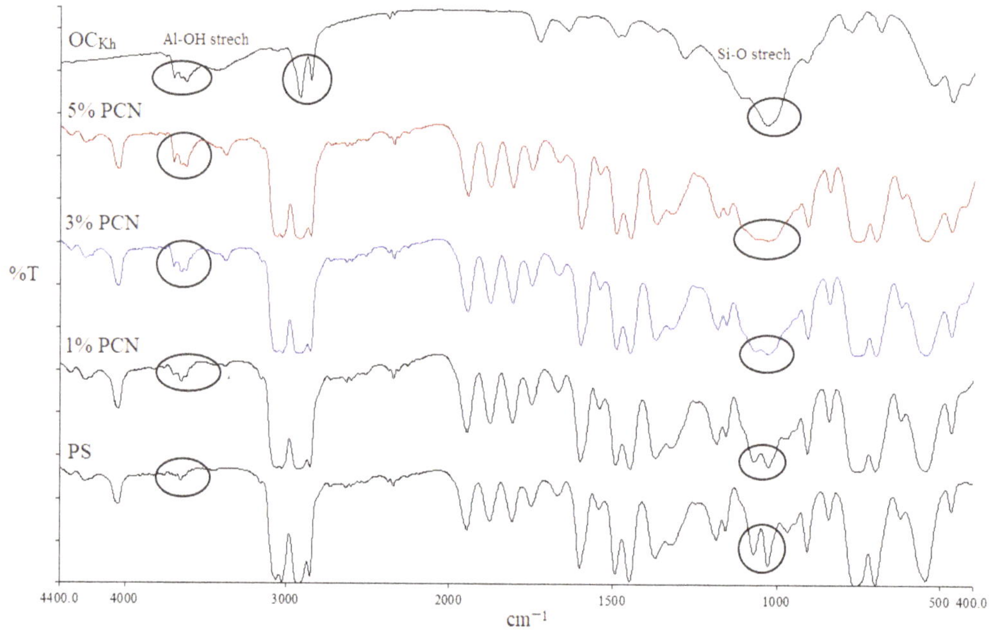

Figure 5. FT-IR spectra of PS, 1–5% wt.% PCN, and OC_{Kh} using RC_{Kh}.

Additionally, it was discovered that in all the PCN spectra, the aromatic ring stretching vibrations of (C = C–C) had absorptions between 1492 and 1600 cm^{-1}, and their overtone bands were seen between 1945 and 1746 cm^{-1}. For the Indian and Khulays clay types, as well as for all manufactured nanocomposites (1–5% wt.% PCNs), the evident change in the weak band of pure PS at 3652 cm^{-1} was noted, especially in the case of the 5% PCN. Additionally, for all the nanocomposites, the overlap with the aliphatic and aromatic C–H stretching of PS is to blame for the removal of the CH_2 stretch peak for OC. Additionally, it was noted that the form of the CH_2 bending peak for PS at 1027 cm^{-1} changed, and it was particularly obvious in the case of the 5 wt.% PCN. These alterations all pointed to the successful intercalation between the OC and PS matrices. Additionally, the overlap with the aliphatic and aromatic C–H stretching of PS is to blame for all the nanocomposites' elimination of the CH_2 stretch peak for OC.

3.1.2. X-ray Diffraction

The distances between the clay layers for the raw clay, NaC, OC, and PCN films were obtained from the peak position of the XRD pattern using Bragg's equation:

$$2d \sin\theta = n \lambda \quad (1)$$

where d is the basal spacing between the clay layers; θ is the angle from the diffraction beam to the atomic plane; n (equal to one here) relates to the order of the reflection; λ is the wavelength of the X-ray radiation employed in the experiment (λ = 1.54060 Å).

X-ray Diffraction of RC, NaC, and OC

In Figure 6 the spectra of the raw Indian clay (CC_{In}) showed a characteristic peak at a 2θ value of 7.15°, corresponding to a basal spacing of 12.35 Å. The small peak in NaC appeared at a 2θ value of 9.17°, with a d-spacing of 9.64 Å. However, the NaC_{In} peak after modification with a CPC organic surfactant to obtain in the OC_{In} shifted to a lower 2θ value at 4.21°, with a d-spacing equal to 21 Å due to the loss of water molecules. The XRD patterns for the local Khulays clay (RC_{Kh}) in Figure 7 show the characteristic peak of d-spacing at a 2θ equal to 7.00° (d = 12.62 Å). This peak in the NaC_{Kh} and OC_{Kh} was observed at a 2θ equal to 7.39° (d = 11.95 Å) and 2θ equal to 4.30° (d = 20.55 Å), respectively. The changes and shifting of the OC peaks (for the Indian and Khulays clay) to a lower 2θ and the increase in the d-spacing refer to the increase in the gap between the clay platelets. Thus, the large increase between the NaC platelets provides a good interlayer exchange of sodium cations by CPC cations [34], which means that the CPC surfactant was successfully intercalated in these types of clay [34,36].

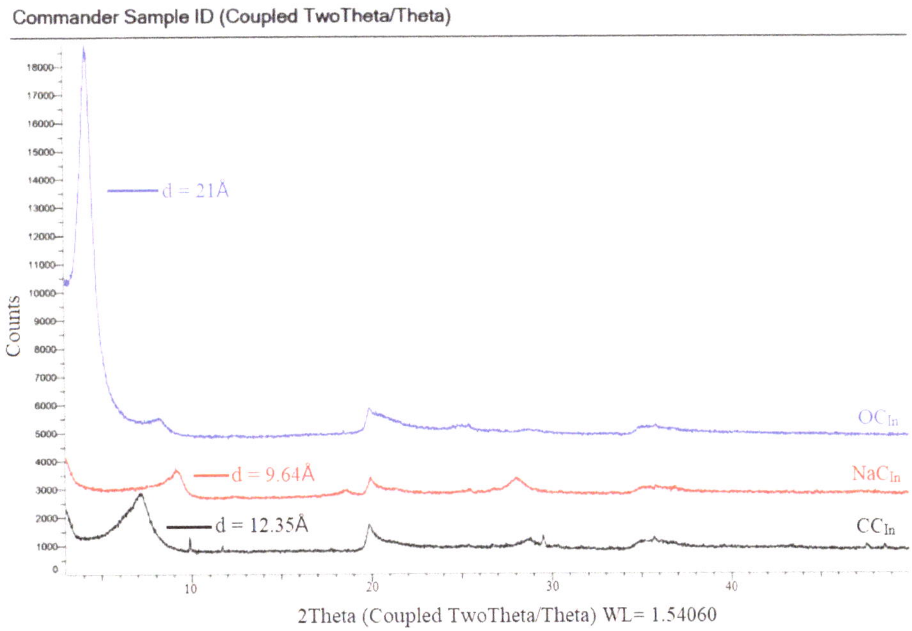

Figure 6. XRD patterns for RC, NaC, and OC of commercial Indian clay.

X-ray Diffraction of PS, OC, and PCN

The XRD of the PCN samples from both clay types are shown in Figures 8 and 9. Both figures show the amorphous PS humps. Low crystallinity is the factor behind the broad diffraction peaks in PS. In the 1–5% PCN patterns, the diffraction peak of polystyrene appeared in the same position in these PCNs (1–5% PCNs). Importantly, the diffraction peak of the OC disappeared in the prepared PCNs, which indicated the amorphous PCN structure. The elimination of the OC peak denotes a significant opening of the clay layers, which is consistent with a good intercalation between the OC and PS matrix. The produced PCNs thus acquired an exfoliated structure. Our findings are consistent with numerous earlier investigations that found polymer clay nanocomposites to have an exfoliated structure [34,36]. The intercalated organic cations can be distinguished based on their different configurations using the d_{001} spacing value that Chen et al. [37] previously published. The change in the basal spacing for both types of OC is attributed to the increase in the interlayer space distance, indicating a successful intercalation of surfactant molecules between clay layers for both CC_{In} and RC_{Kh} [38,39].

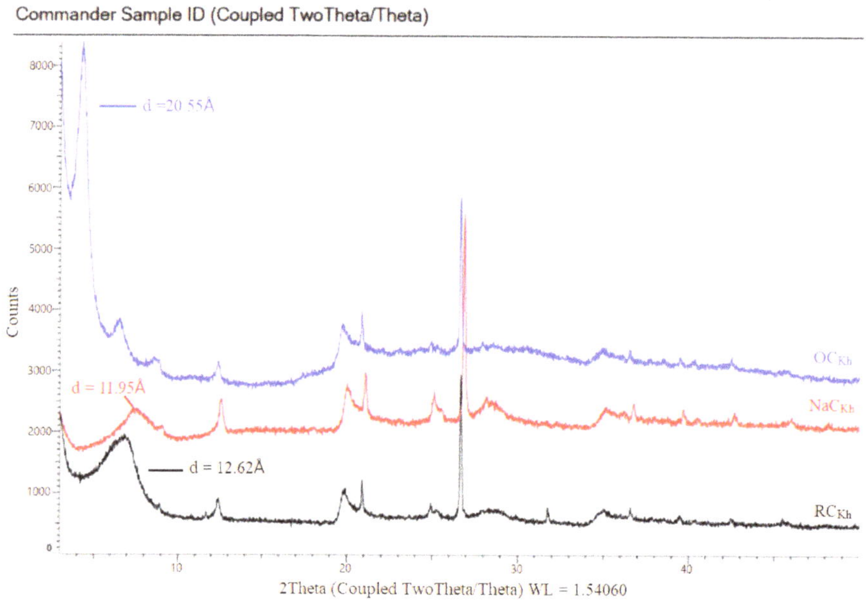

Figure 7. XRD patterns for RC, NaC, and OC of Khulays clay.

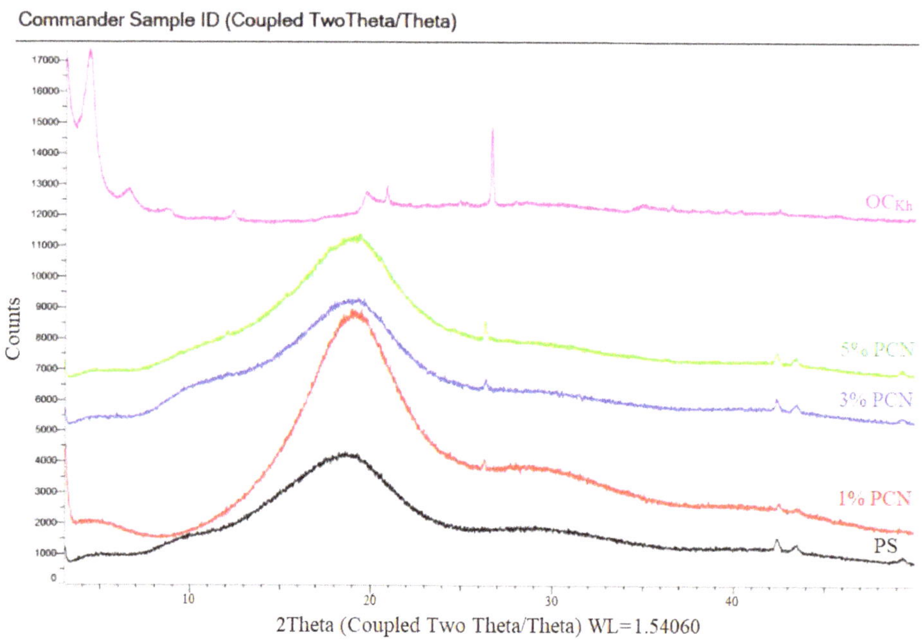

Figure 8. XRD patterns of pure OC, pure PS, and 1–5% wt. PCN using Indian organoclay.

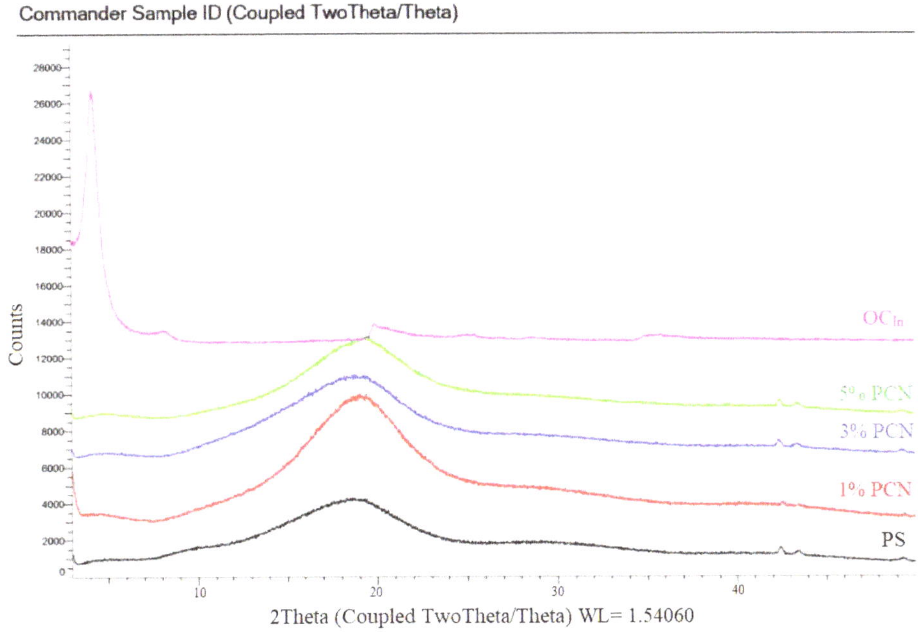

Figure 9. XRD patterns of pure OC, pure PS, and 1–5% PCN using Khulays organoclay.

3.1.3. TEM Analysis

In this investigation, TEM was employed to confirm and explore the dispersion states of polystyrene chains into the organoclay (OC) clay galleries [36]. Figures 10 and 11 show the TEM micrographs of pure organoclay (OC) from the Indian and Khulays clay, as well as all prepared nanocomposites (1–5 wt.% PCNs). The bright areas in these photos repre-sent the PS matrix, and the dark lines represent the OC sheets [17]. Images of the 1% PCN from both types of clay revealed that the OC sheets are entirely exfoliated, whereas at 3% and 5% PCN, the OC sheets are still well ordered in certain tiny regions. This finding demonstrates categorically that the structure of the PS/OC nanocomposite at 1 wt.% is totally exfoliated, while the structure at 3 and 5 wt.% PS/OC nanocomposites are only partially exfoliated. In the case of the 1% PCN, the results are consistent based on the available XRD data and TEM micrographs.

However, the XRD results for the 3 and 5% PCNs revealed the formation of an exfoliated structure, while the TEM data revealed the presence of an intercalated structure in localized locations. The same outcomes were attained for all the PS/OC nanocomposites made from the local Khulays and Indian clay. As a result, the TEM technique is regarded as an important tool for evaluating the dispersion degree of nanoparticles in polymer matrices [11,12,21]. When the TEM pictures are compared, it is clear that the CC_{In} exhibited greater exfoliation than the RC_{Kh} at all the formulations tested.

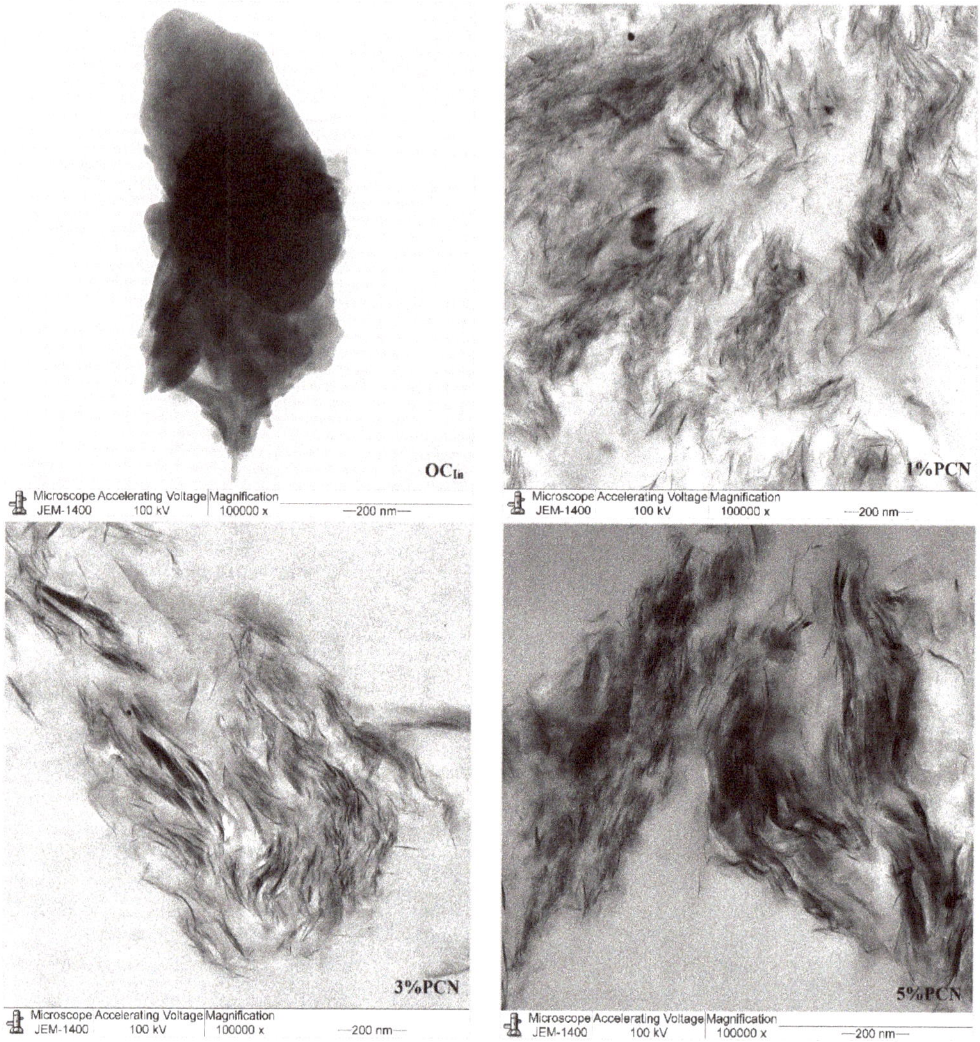

Figure 10. TEM micrographs of pure OC and 1–5 wt.% PCN for CC_{In} at high magnifications.

3.2. *Electrochemical Methods*

3.2.1. Electrochemical Impedance Spectroscopy (EIS)

The coating efficiencies of pure PS and all PS nanocomposite coatings on C-steel rods were investigated using electrochemical impedance spectroscopy (EIS), which is a vital tool in electrochemistry research. It is a strong nondestructive tool for examining and evaluating various electrical characteristics of materials [2–4]. Figures 12 and 13 show Nyquist plots of pure PS and all samples of commercial and local clay PCNs, as well as similar circuits. Table 1 shows the EIS parameters as corrosion resistance (R_{Corr} or R_{ct}), electrical double-layer capacitance (C_{dl} or C_{Corr}), coating resistance (R_{po}) (pore resistance), and coating capacitance (C_C). As illustrated in Figures 12 and 13, these parameters were derived by fitting the Nyquist plots of all the coated C-steel samples that had undergone testing to an analogous circuit using Gamry software.

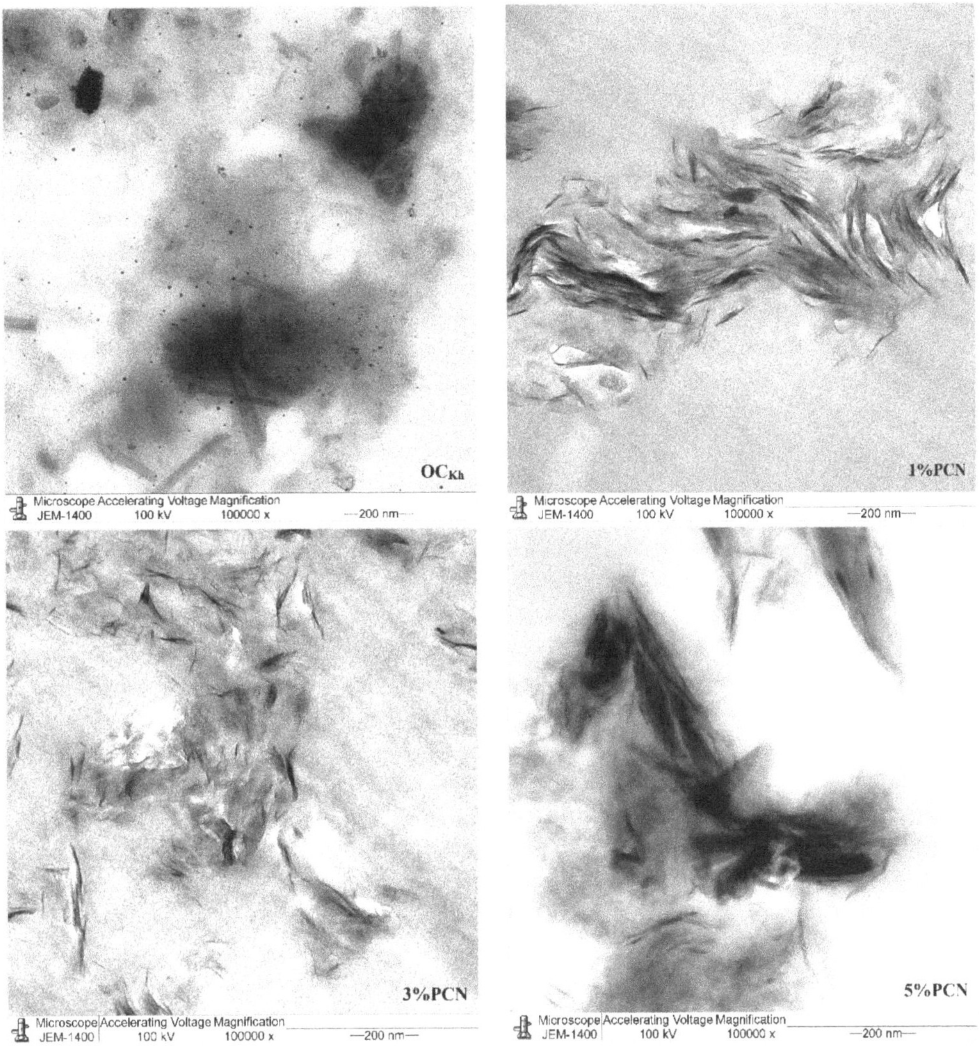

Figure 11. TEM micrographs of pure OC and 1–5 wt.% PCN for RC_{Kh} at high magnifications.

Table 1. EIS data of bare C-steel, pure PS, and 1–5 wt.% PCN coating from commercial Indian and local Khulays clay in 3.5% NaCl solution at 30 °C.

Sample Code		EIS Parameters			
		R_{ct} (Ω.cm^2)	C_{Corr} (F/cm^2)	R_{po} (Ω.cm^2)	C_c (F/cm^2)
Bare C-steel		1.29×10^2	8.61×10^{-3}	-	-
Pure PS		5.24×10^5	4.79×10^{-6}	7.59×10^5	6.81×10^{-10}
CC_{In}	**1% PCN**	6.26×10^6	1.73×10^{-7}	1.21×10^5	2.72×10^{-10}
	3% PCN	5.25×10^6	9.81×10^{-7}	2.21×10^5	2.89×10^{-10}
	5% PCN	4.30×10^6	1.99×10^{-6}	2.59×10^4	3.60×10^{-10}
RC_{Kh}	**1% PCN**	2.46×10^6	2.71×10^{-8}	4.84×10^5	2.98×10^{-10}
	3% PCN	1.31×10^6	6.01×10^{-8}	1.88×10^5	2.83×10^{-10}
	5% PCN	8.83×10^5	1.48×10^{-7}	8.43×10^4	2.84×10^{-10}

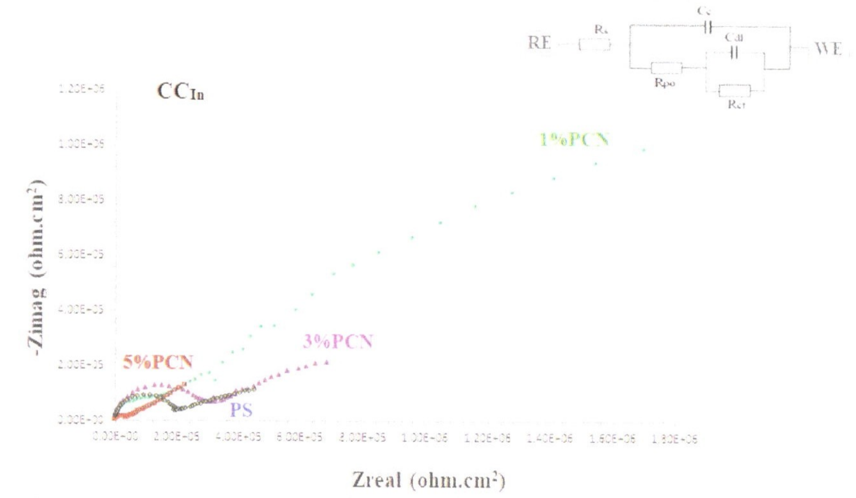

Figure 12. EIS plots of coated C-steel by pure PS and all prepared 1–5% PCNs from Indian clay (CC_{In}) in 3.5% NaCl solution at 30 °C.

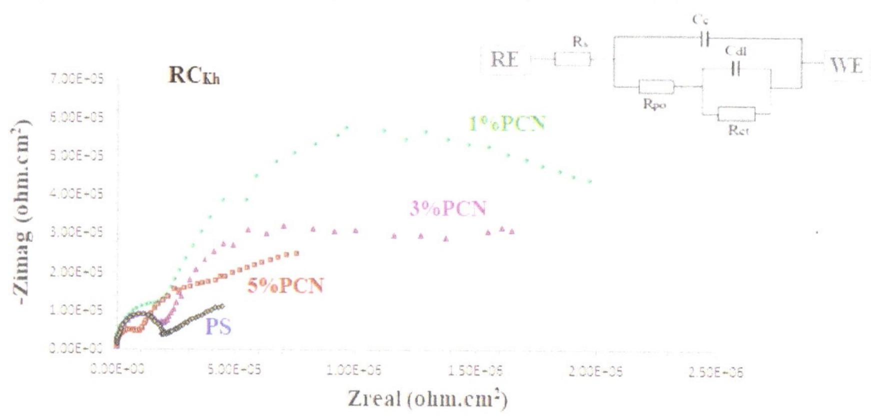

Figure 13. EIS plots of coated C-steel by pure PS and all prepared (1–5% PCNs for RC_{Kh} in 3.5% NaCl solution at 30 °C.

The fitted EIS data of bare C-steel and PS and the 1–5% PCN data for both types of clay are presented in Table 1. It is noticed that the coating of C-steel with PS greatly increased the R_{Corr} values and decreased the C_{Corr} values relative to bare C-steel. As reported in the literature for the coated substrates, the first semicircle in the high-frequency region was related to the resistance and capacitance of the protective coating and its properties [13–15,22]. The second semicircle in the low-frequency region was attributed to the electrochemical reactions on the C-steel surface. The findings demonstrated that, in comparison with pure PS coating, the introduction of a tiny amount of OC to the PS polymer enhanced the corrosion resistance (R_{Corr}) and pore resistance (R_{po}) of the prepared PCNs. It was realized that, as the diameter of the second semicircle increased, so did the values of the R_{Corr} and R_{po}. By adding the small percentage of OC, the values of the corrosion and coating capacitance (C_{Corr}, C_C) decreased. These observations for the R_{Corr}, R_{po}, C_{Corr}, and C_C values indicate that the PCN coatings have higher corrosion protection than pure PS coating. Generally, this indicates that the protective properties of pure PS coating were improved by adding OC content. This behavior was observed for all the prepared PCNs from commercial Indian clay and Khulays clay. In the present study, our impedance results are in agreement with many previous studies [2,3,19,23,24]. Moreover, adding the organic form of the commercial Indian clay at 1% PCN increased the corrosion resistance value (R_{Corr}) of the coating by about three times the value of the local Khulays clay at 1% PCN. The value of the R_{Corr} in the case of the commercial Indian clay was 6.26 MΩ·cm^2, while in the case of the local Khulays clay, the value was 2.46 MΩ·cm^2 This was probably due to the high montmorillonite content in the CC_{In}, which could be easily exfoliated and allows for maximum protection.

3.2.2. EFM Method

The electrochemical frequency modulation (EFM) method is considered a new nondestructive way of electrochemical corrosion monitoring. Consequently, it is a good choice for many metals and metal alloys in different aqueous corrosion systems. The corrosion parameters of the EFM test are good for comparison with the results from the Tafel method and linear polarization resistance [40,41]. The results from the EFM are included in Table 2, with the potentiodynamic polarization results prepared from 1–5 wt.% PCNs for commercial Indian (CC_{In}) and local Khulays (RC_{Kh}) clay in 3.5% NaCl solution at 30 °C. The percentage of coating, or the protection efficiency, was calculated from the following equation:

$$\%PE = \frac{(Icorr(PS) - Icorr(PCN))}{Icorr(PS)} \quad (2)$$

Table 2. EFM data of bare C-steel, pure PS, and 1–5% PCN coating from commercial Indian and local Khulays clay in 3.5% NaCl solution at 30 °C, and relative protection efficiencies (%PE) calculated using I_{Corr}.

Sample Code		I_{Corr} (µA/cm^2)	CR (mpy)	CF2	CF3	%PE
Bare C-steel		3.65×10^2	17.17	2.75	-	-
PS		8.43×10^{-2}	3.87×10^{-3}	1.30	-	-
CC_{In}	**1% PCN**	1.29×10^{-2}	5.92×10^{-4}	2.36	1.34	84.69
	3% PCN	1.96×10^{-2}	8.98×10^{-4}	1.23	1.75	76.75
	5% PCN	1.97×10^{-2}	9.07×10^{-4}	4.84	1.14	76.63
RC_{Kh}	**1% PCN**	1.56×10^{-2}	7.16×10^{-4}	1.86	1.53	81.49
	3% PCN	1.56×10^{-2}	7.17×10^{-3}	2.01	1.70	81.49
	5% PCN	2.61×10^{-2}	1.20×10^{-3}	1.16	1.34	69.04

The EFM parameters (the corrosion current (I_{Corr}), corrosion potential (E_{Corr}), corrosion rate (CR), and calculated relative coating efficiency from the I_{Corr} values) are summarized in Tables 2 and 3. It can be noticed that the coating of C-steel with PS greatly decreased the CR and I_{Corr} values relative to the bare C-steel. Moreover, the PCN coating further decreased the corrosion rate and increased the coating efficiency for both clay types. For the EFM in Table 2, the parameters are the current density in µA/cm^2 and the corrosion rate in MPY (milli-inch per year). The goodness of fit for the EFM method is presented in causality factors 2 and 3. Almost all of our results fit this criterion. It is shown in Table 2 that for the 1% PCN of CC_{In}, the CR is 5.920×10^{-4} mpy, which is the lowest CR of all the formulations.

3.2.3. Potentiodynamic Polarization (Tafel Plots)

Tafel plots are represented in Figures 14 and 15 for the coated C-steel samples prepared from PCNs prepared from CC_{In} and RC_{Kh}, respectively, in 3.5 wt.% NaCl solution at 30 °C. The cathodic and anodic curves were analyzed directly from a Gamry potentiostat/galvanostat to determine the Tafel corrosion parameters.

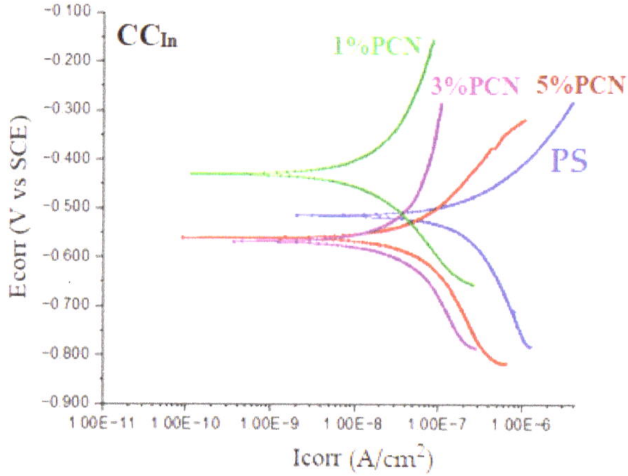

Figure 14. Tafel plots of coated C-steel by pure PS and from 1–5 wt.% PCN prepared from commercial Indian clay (CC_{In}) in 3.5% NaCl solution at 30 °C.

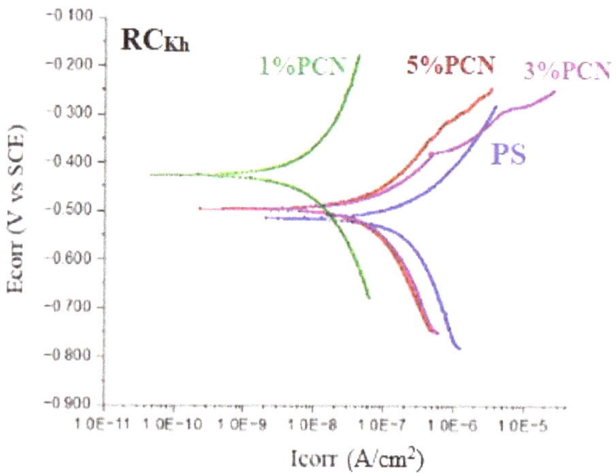

Figure 15. Tafel plots of coated C-steel by pure PS and 1–5 wt.% PCN prepared from local Khulays clay (RC_{Kh}) in 3.5% NaCl solution at 30 °C.

Table 3. Tafel data of bare C-steel, pure PS, and 1–5% PCN coating from CC_{In} and RC_{Kh} in 3.5% NaCl solution at 30 °C, and the relative protection efficiencies (PE%) calculated using I_{Corr}.

Sample Code		E_{Corr} (mV)	I_{Corr} (μA/cm²)	CR (mpy)	Chi.sq.	%PE
Bare C-steel		−631	1.46×10^{2}	6.87	2.55×10^{-2}	-
PS		−515	3.89×10^{-1}	1.79×10^{-2}	1.20	-
CC_{In}	**1% PCN**	−431	7.23×10^{-2}	3.32×10^{-3}	2.38	81.41
	3% PCN	−568	1.34×10^{-1}	6.12×10^{-3}	4.48	65.55
	5% PCN	−561	9.20×10^{-2}	4.23×10^{-3}	2.29	76.35
RC_{Kh}	**1% PCN**	−425	1.52×10^{-1}	6.99×10^{-3}	3.93×10^{-2}	60.93
	3% PCN	−496	1.71×10^{-1}	1.06×10^{-2}	3.83	56.04
	5% PCN	−496	2.90×10^{-1}	1.33×10^{-2}	16.18	25.44

In Table 3 for the Tafel plots, the parameters are the current density in $\mu A/cm^2$ and the corrosion rate in mpy. For Tafel, the values of the chi-square values are included, which reflect the goodness of fit of the present data, which lies in the accepted range. As in the EFM, the best coating efficiency is shown for the 1% PCN of CC_{In}.

The Tafel plot results showed that the PCN-coated C-steel samples had smaller I_{Corr} and CR values than the PS-coated sample. This indicates that an incorporation of OC in the PS matrix enhances the protective properties of this polymer as a coating for C-steel. This behavior is consistent with previous studies that evaluated the effect of adding small amounts of clay to different types of polymers, such as polyurethane [13], polyaniline [16,23], polypyrrole [22], polystyrene [17,18], and epoxy coating [9,24]. Moreover, the trend in the Tafel results was in agreement with the EFM investigation. According to the EIS, EFM, and Tafel results, the PCN-coated C-steel samples exhibited better corrosion resistance and coating efficiency than the pure PS coating. The best electrochemical protection efficacy was observed at 1% PCN for both types of clay. This is most likely due to the good dispersion of the 1 wt.% organoclay layers in the PS chains, as demonstrated by the XRD and TEM results. Other research groups [13,17] have obtained greater protection efficiencies at 3 wt.% organoclay with polymers. In the case of 1% PCN, however, the fully exfoliated structure was obtained by comparing the electrochemical results with the characterization method results (XRD, TEM). However, the corrosion protection performances of the 3% and 5% PCNs were worse than that of the 1% PCN. This suggests that the results of the characterization methods are consistent with the electrochemical investigations (EIS, EFM, and Tafel). The order of the coating efficiency for both clay types from Tables 2 and 3 is as follows:

$$1\%\ PCN > 3\%\ PCN > 5\%\ PCN$$

3.2.4. Comparison Studies

Finally, Figure 16 compares the results of the three methods used in the present study to assess the corrosion protection efficiency. It was found that the CC_{In} had a superior corrosion resistance to the local clay in all of these procedures. The corrosion resistance value of the 1% PCN of the commercial Indian clay is about three times higher than the value of the 1% PCN of the local Khulays clay, as illustrated in Figure 16. Furthermore, the corrosion resistance of these polymer clay nanocomposites decreased gradually from 1–5% PCN. When comparing the corrosion resistance (Rct) obtained from the EIS results (Figure 16), it was found that they coincide with the corrosion current density (I_{Corr}) values derived from the EFM and Tafel methods (Figure 17a,b). This main conclusion may be attributed to the chemical composition of the commercial Indian clay and the high percentage of montmorillonite content compared with the local Khulays clay. The high percentage of montmorillonite content (64.9%) [25] increased the ability of this clay to swell more than the local Khulays clay, which contains 35.22% montmorillonite. This resulted in the improved CPC intercalation and greater dispersion of the commercial Indian clay layers inside the polystyrene chains, which increased the tortuous course of the corrosive ions and resulted in a higher coating efficiency. Fully exfoliated organoclay layers within the PS matrix improve the protective qualities of the PS coating more than partially exfoliated nanocomposite and pure PS coatings. Thus, the permeating molecules are forced to take a wiggling path and diffuse via the convoluted channels around them, especially at low OC content, which increases the tortuosity of the oxygen, water, and chloride ion diffusion pathways and increases the coating efficiency compared with using a pure PS coating [4,9,17].

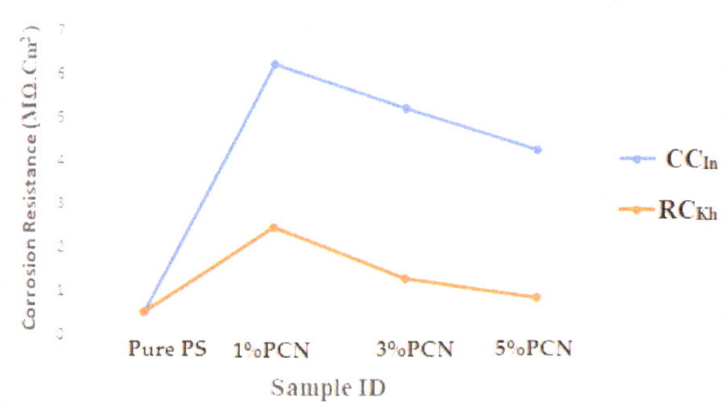

Figure 16. Comparison of electrochemical results from EIS method of PCN coating of present study for both clay types.

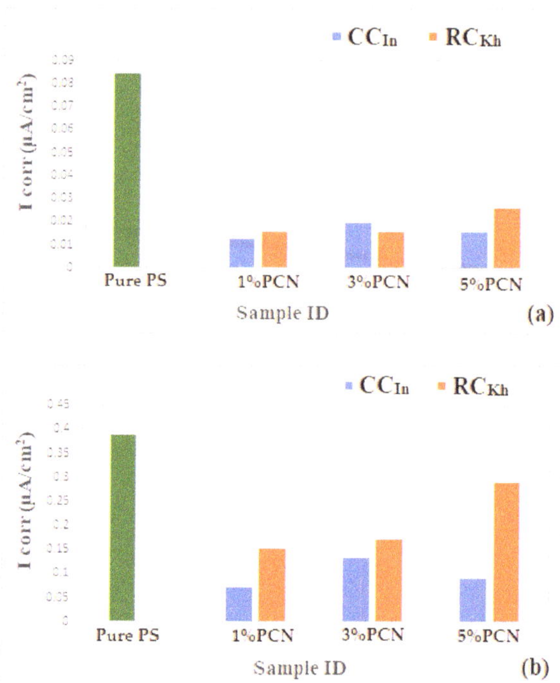

Figure 17. (**a**,**b**) Comparison of electrochemical results from (**a**) EFM and (**b**) Tafel for PCN coating of present study for both clay types.

A comparison with other studies is shown in Table 4 using current density values. It is noticed that the current density values of our study are much better than those of the best study, which is Chen-Yang et al. [42] using polyurethan. The current density value of PCN (CC_{In}) 7.23×10^{-8} A/cm^2 is better than any other formulation, as shown in Table 4. However, it is hard to make such a comparison with other studies, as there are many parameters that need to be taken into consideration, such as the adhesion of the polymer and PCN on the metal, the coating thickness, the metal, the temperature of the study, and the exfoliation of the organoclay layers within the polymer matrix.

Table 4. A comparison of our results with previous ones.

Code	Ref. No.	Polymer	I_{Corr} (A/cm^2)
Previous studies	Chen-Yang et al. [42]	Polyurethan	3.12×10^{-7}
	Navarchian et al. [16]	Polyaniline	4.67×10^{-7}
	Motlatle et al. [23]	Polyaniline	2.10×10^{-4}
	Piromruen et al. [4]	Polyaniline	1.69×10^{-5}
	Olad and Rashidzadeh [43]	Polyaniline	2.58×10^{-6}
	Olad et al. [44]	Polypyrrole	6.30×10^{-5}
	Raju et al. [18]	Polystyrene	2.40×10^{-6}
	Yeh et al. [45]	Polystyrene	1.40×10^{-7}
	Nematollahi et al. [46]	Epoxy	4.96×10^{-7}
	Huang et al. [47]	Polyimide	1.09×10^{-6}
	Chang et al. [48]	PMMA	1.28×10^{-6}
Present study	PCN (RC_{Kh})	Polystyrene	1.52×10^{-7}
	PCN (CC_{In})	Polystyrene	7.23×10^{-8}

4. Conclusions

Polymer clay nanocomposite (PCN) samples were generated in this study by combining different concentrations of organic clay (1%, 3%, and 5% OC) in toluene solvent and polystyrene as the matrix (OC/PS). Many techniques, including FT-IR, XRD, and TEM, were used to characterize the organoclay and PCN films to ensure the success of the modification procedure. After treating sodium clay (NaC) with CPC, the shifts in the FT-IR spectra confirmed the presence of CPC in the organic clay samples. An exfoliated structure was obtained from the XRD spectrum for low clay loading (1–3% PCN), whereas an intercalated structure was the dominant form for the 5% PCN. The XRD results were confirmed by TEM images. Exfoliated structure preparation is crucial, as it might be employed in polymer clay nanocomposite coating. Furthermore, the electrochemical measurements (EIS, EFM, and Tafel plots) coincide with the morphological data from the XRD and TEM. According to the results of the electrochemical experiments (EIS, EFM, and Tafel plots), commercial Indian clay provides greater corrosion protection than native Khulays clay. This is clearly demonstrated by comparing the corrosion resistance values obtained from the EIS technique, as well as the relative coating efficiencies obtained from the EFM test and Tafel plots of both types of PCNs. The nanocomposite with a concentration of 1% has a fully exfoliated structure and showed stronger protective characteristics than the nanocomposites with partially exfoliated structures (3 wt.% PCN and 5% PCN). This is due to the increased tortuosity of the diffusion paths of oxygen, water molecules, and chloride ions when compared with pure PS coating. Even with a modest amount of clay added, polymer clay nanocomposites exhibit improved coating characteristics. This is due to the clay particles' nanoscale size, which results in a wide contact area between the polymer matrix and filler. The structure of clay, with layers of a high aspect ratio, provides superior barrier characteristics as well as enhanced anticorrosive capabilities. This principal result can be attributed to the chemical composition of commercial Indian clay and the high percentage of the montmorillonite concentration when compared with the local Khulays clay. The high montmorillonite content (64.9%) boosted the ability of this clay to swell more than Khulays clay, which has 35.22% montmorillonite, which increases the tortuous track of corrosive ions. It was discovered that commercial Indian clay has better corrosion protection (81.4%) than local Khulays clay (60.2%). A comparison with other studies using current density values proved that the current density values of our study are much better than those of the best study of Chen-Yang et al. using polyurethan. The current density value of PCN (RC_{In}) 7.23×10^{-8} A/cm^2 is better than any other formulation.

Author Contributions: Methodology and supervision, L.A.A.J. and W.K.M.; writing—original draft, N.A.H.; reviewing, L.A.A.J. and W.K.M.; revision, H.H.A. All authors have read and agreed to the published version of the manuscript.

Funding: The authors extend their appreciation to the Deputyship for Research & Innovation, the Ministry of Education, in Saudi Arabia for funding this research work through project no. IFKSURG-2-480.

Data Availability Statement: Not applicable.

Acknowledgments: This work is based on the master's thesis of Nashwa A. Howyan, who worked under the supervision of Layla A. Al Juhaiman (corrosion) as the major supervisor, and Waffa K. Mekhamer (polymers) as the assistant supervisor.

Conflicts of Interest: The authors declare no conflict of interest.

Abbreviations

Symbol	Meaning	Symbol	Meaning
RC_{Kh}	Raw Khulays clay	CEC	Cation-exchange Capacity
CC_{In}	Commercial Indian clay	OCP	Open-circuit potential
NaC	Sodium clay	E_{SS}	Steady-state potential
CPC	Cetylpyridinium clay	R_{Corr}	Corrosion resistance
OC	Organoclay	C_{dl}	Electrical double-layer capacitance
PS/OC	Polystyrene organoclay nanocomposite	C_{Corr}	Corrosion capacitance
MMT	Montmorillonite	R_{po}	Pore resistance
PS	Polystyrene	C_C	Coating capacitance
EIS	Electrochemical impedance spectroscopy	%PE	% Protection efficiency
EFM	Electrochemical frequency modulation	I_{Corr}	Corrosion current density
E_{Corr}	Corrosion potential	CR	Corrosion rate

References

1. Zarras, P.; Stenger-Smith, J. *Intelligent Coatings for Corrosion Control*, 1st ed.; Butterworth Heinemann: Oxford, UK, 2015; pp. 59–91.
2. Zaarei, D.; Sarabi, A.A.; Sharif, F.; Kassiriha, S.M. Structure, properties and corrosion resistivity of polymeric nanocomposite coatings based on layered silicates. *J. Coat. Technol. Res.* **2008**, *5*, 241–249. [CrossRef]
3. Hang, T.; Truc, T.X.; Olivier, T.A.; Vandermiers, M.G.; Guérit, C.; Pébre, N. Corrosion protection mechanisms of carbon steel by an epoxy resin containing indole-3 butyric acid modified clay. *Prog. Org. Coats.* **2010**, *69*, 410–416. [CrossRef]
4. Piromruen, P.; Kongparakul, S.; Prasassarakich, P. Synthesis of polyaniline/montmorillonite nanocomposites with an enhanced anticorrosive performance. *Prog. Org. Coat.* **2014**, *77*, 691–700. [CrossRef]
5. Speight, J.G. *Oil and Gas Corrosion Prevention*; Gulfe Professional: Waltham, NY, USA, 2014; pp. 5–147.
6. Emich, F. *Engineering Chemistry*, 1st ed.; Scientific International PVT. Ltd.: New Delhi, India, 2014; pp. 20–66.
7. El-Sherik, A.M. *Trends in Oil and Gas Corrosion Research and Technologies*, 1st ed.; Woodhead Publishing Series in Energy; Elsevier Inc.: Amsterdam, The Netherlands, 2017; pp. 1–65.
8. Niu, L.; Guo, R.; Tang, C.; Guo, H.; Chen, J. Surface characterization and corrosion resistance of fluoferrite conversion coating on carbon steel. *Surf. Coat. Technol.* **2016**, *300*, 110–117. [CrossRef]
9. Sari, M.G.; Ramezanzadeh, B.; Shahbazi, M.; Pakdel, A.S. Influence of nanoclay particles modification by polyester-amide hyperbranched polymer on the corrosion protective performance of the epoxy nanocomposite. *Corros. Sci.* **2016**, *92*, 162–172. [CrossRef]
10. Zhou, C.; Lu, X.; Xin, Z.; Liu, J.; Zhang, Y. Polybenzoxazine/SiO2 nanocomposite coatings for corrosion protection of mild steel. *Corros. Sci.* **2014**, *80*, 269–275. [CrossRef]
11. Kotal, M.; Bhowmick, K.A. Polymer nanocomposites from modified clays: Recent advances and challenges. *Prog. Polym. Sci.* **2015**, *51*, 127–187. [CrossRef]
12. Sattler, K.D. *Handbook of Nanophysics: Functional Nanomaterials*, 1st ed.; CRC Press: Boca Raton, FL, USA, 2010; pp. 3–31.
13. Heidarian, M.; Shishesaz, M.R.; Kassiriha, S.M.; Nematollahi, M. Characterization of structure and corrosion resistivity of polyurethane/organoclay nanocomposite coatings prepared through an ultrasonication assisted process. *Prog. Org. Coats.* **2010**, *68*, 180–188. [CrossRef]
14. Shreir, L.L.; Jarman, R.A.; Burstein, G.T. *Corrosion2: Corrosion Control*, 3rd ed.; Newnws-Butterworths: London, UK, 1994; pp. 9:41–9:57.
15. Roberge, P. *Corrosion Engineering: Principles and Practice*, 1st ed.; McGraw-Hill: New York, NY, USA, 2008; pp. 370–383.
16. Navarchian, A.H.; Joulazadeh, M.; Karimi, F. Investigation of corrosion protection performance of epoxy coatings modified by polyaniline/clay nanocomposites on steel surfaces. *Prog. Org. Coat.* **2014**, *77*, 347–353. [CrossRef]

17. Al Juhaiman, L.A.; Al-Enezi, D.A.; Mekhamer, W.K. Polystyrene/Organoclay Nanocomposites as Anticorrosive Coatings of C-Steel. *Int. J. Electrochem. Sci.* **2016**, *11*, 5618–5630. [CrossRef]
18. Raju, A.; Lakshmi, V.; Prataap, R.K.V.; Resmi, V.G. Adduct modified nano-clay mineral dispersed polystyrene nanocomposites as advanced corrosion resistance coatings for aluminum alloys. *Appl. Clay Sci.* **2016**, *126*, 81–88. [CrossRef]
19. Madhup, M.K.; Shah, N.K.; Parekh, N.R. Investigation and improvement of abrasion resistance, water vapor barrier and anticorrosion properties of mixed clay epoxy nanocomposite coating. *Prog. Org. Coat.* **2017**, *102*, 186–193. [CrossRef]
20. Fontana, M. *Corrosion Engineering*, 3rd ed.; McGraw-Hill: New York, NY, USA, 1987; pp. 1–30.
21. Bergaya, F.; Theng, B.; Lagaly, G. *Handbook of Clay Science Part A: Fundamental*, 2nd ed.; Developments in Clay Science; Elsevier Ltd.: Amsterdam, The Netherlands, 2006; pp. 100–427.
22. Yeh, J.-M.; Chin, C.-P.; Chang, S. Enhanced corrosion protection coatings prepared from soluble electronically conductive polypyrrole-clay nanocomposite materials. *J. Appl. Polym. Sci.* **2003**, *88*, 3264–3272. [CrossRef]
23. Motlatle, A.M.; Sinha, S.; Scriba, M. Polyaniline-clay composite-containing epoxy coating with enhanced corrosion protection and mechanical properties. *Synth. Met.* **2018**, *245*, 102–110. [CrossRef]
24. Truc, T.A.; Thuy, T.T.; Oanh, V.K.; Xuan Hang, T.T.; Nguyen, A.S.; Caussé, N.; Pébère, N. 8-hydroxyquinoline-modified clay incorporated in an epoxy coating for the corrosion protection of carbon steel. *Surf. Interfaces* **2019**, *14*, 26–33. [CrossRef]
25. Alandis, N.M.; Aldayel, O.A.; Mekhemer, W.K.; Hefne, J.A.; Jokhab, H.A. Thermodynamic and Kinetic Studies for the Adsorption of Fe (III) and Ni (II) Ions From Aqueous Solution Using Natural Bentonite. *J. Dispers. Sci. Technol.* **2001**, *31*, 1–9. [CrossRef]
26. Al Juhaiman, L.A.; Al-Enezi, D.A.; Mekhamer, W.K. Preparation and Characterization of Polystyrene/Organoclay Nanocomposites from Raw Clay. *Dig. J. Nanomater. Biostructures* **2016**, *11*, 105–114.
27. Djomgoue, P.; Njopwouo, D. FT-IR Spectroscopy Applied for Surface Clays Characterization. *J. Surf. Eng. Mater. Adv. Technol.* **2013**, *3*, 275–282. [CrossRef]
28. Madejova, J.; Gates, W.P.; Petit, S. IR Spectra of Clay Minerals. In *Developments in Clay Science: Infrared and Raman Spectroscopies of Clay Minerals*; Gates, W.P., Kloprogge, J.T., Madejova, J.T., Bergaya, F., Eds.; Elsevier: Amsterdam, The Netherlands; Holland, MI, USA, 2017; Volume 8, pp. 107–149.
29. Yuehong, M.; Jianxi, Z.; Hongping, H.; Peng, Y.; Wei, S.; Dong, L. Infrared investigation of organo-montmorillonites prepared from different surfactants. *Spectrochim. Acta Part A Mol. Biomol. Spectrosc.* **2010**, *76*, 122–129. [CrossRef]
30. Madejova, J.; Bujdak, J.; Janek, M.; Komadel, P. Comparative FT-IR study of structural modifications during acid treatment of dioctahedral smectites and hectorite. *Spectrochim. Acta Part A* **1998**, *54*, 1397–1406. [CrossRef]
31. Tabak, A.; Afsin, B.; Aygun, S.F.; Koksal, E. Structural Characteristics of Organo-Modified Bentoonites of Different Origin. *J. Thermal Anal. Calorim.* **2007**, *87*, 375–381. [CrossRef]
32. Cervantes-Uc, J.M.; Cauich-Rodríguez, J.V.; Vázquez-Torres, H.; Garfias-Mesías, L.F.; Paul, D.R. Thermal degradation of commercially available organoclays studied by TGA–FTIR. *Thermochim. Acta* **2007**, *457*, 92–102. [CrossRef]
33. Krishna, S.V.; Pugazhenthi, G. Properties and thermal degradation kinetics of polystyrene/organoclaynanocomposites synthesized by solvent blending method: Effect of processing conditions and organoclay loading. *J. Appl. Polym. Sci.* **2011**, *120*, 1322–1336. [CrossRef]
34. Alshabanat, M.; Al-arrash, A.; Mekhamer, W. Polystyrene/Montmorillonite Nanocomposites: Study of the Morphology and Effects of Sonication Time on Thermal Stability. *J. Nanomater.* **2013**, *12*, 2–12. [CrossRef]
35. Tyagi, B.; Chudasama, C.D.; Jasra, R.V. Determination of structural modification in acid activated montmorillonite clay by FT-IR spectroscopy. *Spectrochim. Acta Part A* **2006**, *64*, 273–278. [CrossRef]
36. Li, X.; Kang, T.; Cho, W.; Lee, J.; Ha, C. Preparation and Characterization of Poly (butyleneterephthalate)/Organoclay Nanocomposites. *Macromol. Mater. Eng.* **2001**, *22*, 1306–1312. [CrossRef]
37. Chen, B.; Zhu, L.; Zhu, J.; Xing, B. Configurations of the bentonite-sorbed myristylpyridinium cation and their influences on the uptake of organic compounds. *Environ. Sci. Technol.* **2005**, *39*, 6093–6100. [CrossRef]
38. Xi, Y.; Ding, Z.; Frost, R.L. Structure of organo-clays an X-ray diffraction and thermogravimetric analysis study. *J. Colloid Interf. Sci.* **2004**, *277*, 116–120. [CrossRef]
39. Giannakas, A.; Spanos, C.G.; Kourkoumelis, N.; Vaimakis, T.; Ladavos, A. Preparation, characterization and water barrier properties of PS/organo-montmorillonite nanocomposites. *Eur. Polym. J.* **2008**, *44*, 3915–3921. [CrossRef]
40. Obot, I.B.; Onyeachu, I.B. Electrochemical frequency modulation (EFM) technique: Theory and recent practical applications in corrosion research. *J. Mol. Liq.* **2018**, *249*, 83–96. [CrossRef]
41. Abdel-rehim, S.S.; Khaled, K.F.; Abd-elshafi, N.S. Electrochemical frequency modulation as a new technique for monitoring corrosion inhibition of iron in acid media by new thiourea derivative. *Electrochim. Acta* **2006**, *51*, 3269–3277. [CrossRef]
42. Chen-Yang, Y.W.; Yang, H.C.; Li, G.J.; Li, Y.K. Thermal and anticorrosive properties of polyurethane/clay nanocomposites. *J. Polym. Res.* **2005**, *11*, 275–283. [CrossRef]
43. Olad, A.; Rashidzadeh, A. Preparation and anticorrosive properties of PANI/Na-MMT and PANI/O-MMT nanocomposites. *Prog. Org. Coat.* **2008**, *62*, 293–298. [CrossRef]
44. Olad, A.; Rashidzadeh, A.; Amini, M. Preparation of polypyrrole nanocomposites with organophilic and hydrophilic montmorillonite and investigation of their corrosion protection on iron. *Adv. Polym. Technol.* **2013**, *32*, 21337. [CrossRef]
45. Yeh, J.; Liou, S.; Lin, C.; Chang, Y.; Yu, Y.; Cheng, C. Effective Enhancement of Anticorrosive Properties of Polystyrene by Polystyrene–Clay Nanocomposite Materials. *J. Appl. Polym. Sci.* **2003**, *92*, 1970–1976. [CrossRef]

46. Nematollahi, M.; Heidarian, M.; Peikari, M.; Kassiriha, S.M.; Arianpouya, N.; Esmaeilpour, M. Comparison between the effect of nanoglass flake and montmorillonite organoclay on corrosion performance of epoxy coating. *Corros. Sci.* **2010**, *52*, 1809–1817. [CrossRef]
47. Huang, H.Y.; Huang, T.C.; Yeh, T.C.; Tsai, C.Y.; Lai, C.L.; Tsai, M.H.; Chou, Y.C. Advanced anticorrosive materials prepared from amine-capped aniline trimer-based electroactive polyimide-clay nanocomposite materials with synergistic effects of redox catalytic capability and gas barrier properties. *Polymer* **2011**, *52*, 2391–2400. [CrossRef]
48. Chang, K.C.; Chen, S.T.; Lin, H.F.; Lin, C.Y.; Huang, H.H.; Yeh, J.M.; Yu, Y.H. Effect of clay on the corrosion protection efficiency of PMMA/Na+-MMT clay nanocomposite coatings evaluated by electrochemical measurements. *Eur. Polym. J.* **2008**, *44*, 13–23. [CrossRef]

metals

MDPI

Article

A Study of the Mechanisms and Kinetics of the Localized Corrosion Aggravation of Ductile Iron in a Harsh Water Quality Environment

Bingqin Wang [1,2,†], Tao Liu [3,†], Kai Tao [4], Lingsheng Zhu [5], Chao Liu [1,2,*], Xingyue Yong [6] and Xuequn Cheng [1,2,*]

1 Institute for Advanced Materials and Technology, University of Science and Technology Beijing, Beijing 100083, China
2 School of Metallurgical and Ecological Engineering, University of Science and Technology Beijing, Beijing 100083, China
3 National Materials Corrosion and Protection Data Center, University of Science and Technology Beijing, Beijing 100083, China
4 School of Materials Science and Engineering, Henan Institute of Technology, Xinxiang 453011, China
5 R&D Center, National, Xinxing Ductile Iron Pipes Co., Ltd., Handan 056300, China
6 State Key Laboratory of Organic and Inorganic, Beijing University of Chemical Technology, Beijing 100029, China
* Correspondence: liuchao@ustb.edu.cn (C.L.); chengxuequn@ustb.edu.cn (X.C.); Tel.: +86-010-62333931(C.L.)
† These authors contributed equally to this work.

Abstract: Ductile iron is a high-strength cast iron material. The spherical graphite obtained by inoculation treatment effectively improves the mechanical properties of cast iron, resulting in higher strength than carbon steel. However, severe corrosion may occur under specific circumstances, especially in thermal water pipelines. In this paper, the corrosion mechanisms at the main defective points of ductile iron were investigated using microscopic morphological characterization after accelerated tests combined with numerical simulations. The influence law of each environmental factor on the corrosion kinetics of ductile iron in a complex water quality environment was studied using dynamic potential polarization tests. The results showed that the main causative factors leading to the increased corrosion of ductile iron were the presence of tail-like gaps on its surface, and the crescent-shaped shrinkage and loosening organization around the graphite spheres. After mechanical treatment was applied to eliminate the obvious defects, the number of corrosion pits was reduced by 41.6%, and the depth of the pits was slowed down by 40% after five days. By comparison, after ten days, the number of pits was reduced by 51%, and the depth of the pits was slowed down by 50%. The dynamic potential polarization test results show that the dissolved oxygen concentration has the greatest influence on the corrosion of ductile iron in the simulated water environment; meanwhile, the water hardness can slow down the corrosion of ductile iron. The relative influence of each environmental factor is as follows: dissolved oxygen concentration > temperature > immersion time > water hardness > pH > Cl^-.

Keywords: ductile iron; harsh water; mechanical treatment; localized corrosion

Citation: Wang, B.; Liu, T.; Tao, K.; Zhu, L.; Liu, C.; Yong, X.; Cheng, X. A Study of the Mechanisms and Kinetics of the Localized Corrosion Aggravation of Ductile Iron in a Harsh Water Quality Environment. *Metals* **2022**, *12*, 2103. https://doi.org/10.3390/met12122103

Academic Editor: Renato Altobelli Antunes

Received: 31 October 2022
Accepted: 5 December 2022
Published: 7 December 2022

1. Introduction

With increasing demand for water resources, the laying of water transmission pipelines has been gradually accelerated [1]. Additionally, the demand for water transmission pipelines such as steel pipes, ductile iron pipes, pre-stressed steel cylinder concrete pipes (PCCP), and sandwich glass steel pipes has increased sharply. In the application of cast pipes with cement lining, harmful ions such as Cr^{6+} often enter into the water from contaminated cement and endanger human health. With the increased use of cement-free lined cast iron pipes in plumbing lines, the fluid in the pipeline comes into direct contact with the line material. In these circumstances, the flow rate and shear stress destroy the resulting corrosion film on the material's surface [2], and elevate the potential risk of the corrosion-based deterioration of pipelines [2–8]. This may endanger the safe operation of the entire water supply system and lead to a series of problems, such as water quality deterioration, and ecological and economic losses.

In non-cement-lined water supply pipelines, the water temperature, pH [1,9], oxygen content, and chloride ion concentration are important factors that lead to the immersion corrosion of the transmission pipeline; these factors exist in addition to the flow immersion corrosion effect. Stefano [3] described the effect of scaling ions (Ca^{2+} and Mg^{2+}) in water quality on the corrosion process of geothermal galvanized steel pipes. The concentration of Zn^{2+} and OH^- has a clear effect on the formation of the nuclei of corrosion products in a solution with a higher concentration of scaling ions, which results in the presence of spherical corrosion products and scales on the surface of the pipe. This corrosion product mainly comprises zinc oxide, calcium carbonate, and magnesium carbonate, which could decrease the corrosion rate. Georgii [6] showed the effect of different flow rates on the corrosion process of mild steel, whereby, at higher flow rates (0.19–0.45 m/s), the protective properties of the metal surface layer were brought about by enhanced mass transport. In this condition, crystals of calcite and iron oxide could be formed in the cathodic and anodic regions, respectively, leading to the formation of a self-healing barrier film that blocked the oxygen supply and prevented the release of iron into the water.

Ductile iron, as the material most widely used in aqueducts, can lead to the accelerated damage of aqueducts due to microstructural defects such as microscopic shrinkage loosening on the material's surface, which can increase the corrosion area of castings. These defects are also more likely to cause the formation of pits. However, these defects could be improved by suitable surface mechanical treatment techniques [10,11], thus enhancing the corrosion resistance of ductile iron castings in aqueduct services.

Studies of the corrosion resistance of ductile iron are currently dominated by single influencing factors [12–16]. The corrosion of ductile iron [14] under cyclic salt spray exposure consists of three processes: (i) the pitting of ferrite, (ii) the inward oxidation of the austenitic ferrite matrix, associated with the formation of a dense inner layer, and (iii) the mixed oxidation of the matrix, associated with the formation of a rust layer on the sample surface, where chloride ions are transferred and become involved in the corrosion reaction by changing the properties and composition of the rust layer. The levels of corrosion resistance of carbon steel and ductile iron castings show significant differences at different levels of chloride ion content [15]. An increase in the chloride content leads to an increase in the corrosion rate, and carbon steel is more susceptible to damage from chloride than ductile iron is. However, the differences in the mechanisms of the carbon steel and ductile iron castings in a multi-factor coupled environment remain unclear.

This study investigated the influence of surface cracks on the corrosion resistance of ductile iron castings. In this work, the main causative factors for the acceleration of the corrosion of ductile iron materials were investigated, and a variety of water quality environments that may exist in water pipelines were simulated. We offer a quantitative ranking of the degree of influence of each environmental factor on the corrosion kinetics of ductile iron, and screen out the environmental factors that have the greatest influence on the acceleration of the corrosion of ductile iron at each corrosive stage. The main environmental factors that influence the corrosion resistance of ductile iron castings were identified.

2. Materials and Methods

2.1. Materials

The chemical composition of the ductile iron castings used in this work is 3.75% C, 1.8% Si, 0.45% Mn, 0.07% P, 0.012% S, 0.06% Mg, and Bal. Fe. Defects including shrinkage and holes can be easily formed on the material's surface in the manufacturing process. Two different specimens were produced for this research. One type of specimen was produced by being slightly mechanically ground to 2000#, with surface defects maintained on the specimen surface. Another type of specimen was mechanically ground to 2000#

until the defect layer disappeared. All of the specimens were polished to a mirror finish with diamond abrasives with a particle size of 1.0 μm.

2.2. Environmentally Simulated Solution

Before the immersion experiment, nitrogen was introduced into the environment; this required deoxygenation to simulate an anoxic environment, in which the simulated liquid was deoxygenated at a constant speed of every 250 mL for 3 h. The concentration was then measured and controlled at 0.3 mg$*$L^{-1}, and then start the immersion experiment. A total of 8 different environments were configured for the experiments, as shown in Table 1. The water hardness was adjusted with $CaCO_3$, and the Cl^- concentration was adjusted with NaCl. The same batch was immersed in a large 8-hole water bath during the immersion process. A small water bath was used for the electrochemical measurements to ensure a constant temperature during the test.

Table 1. Characteristic parameters of the simulated solution.

No.	1	2	3	4	5	6	7	8
Hardness/ppm	20	20	20	20	80	80	80	80
pH	7	7	10	10	7	7	10	10
Temperature/°C	60	90	90	60	90	60	60	90
Cl^-/ppm	20	120	20	120	20	120	20	120
Oxygen concentration/mg$*$L^{-1}	0.3	0.3	6.5	6.5	6.5	0.3	6.5	0.3

2.3. Autoclave Immersion Experiment

Solution environment no. 8 was selected for the experiment. Ductile iron samples of 50 mm × 25 mm × 4 mm, both with and without surface mechanical treatment, were polished with silicon carbide paper (2000#) and then cleaned with ethanol and dried using cold air. After determining the original weight of the specimen, the specimen was suspended from the inner wall of the beaker using a nylon fishing line. During the immersion process, the autoclave was pressurized to 1 MPa and the water temperature was controlled at 90 °C; the test cycles were 1220 h and 2440 h. Four parallel specimens were prepared for each material, and three of them were used to determine the loss in thickness. The last sample was used for scanning electron microscopy (SEM) and laser scanning confocal microscopy (CLSM) characterization. The rust on the specimen was ultrasonically cleaned in a descaling solution that consisted of 98%HCl (500 mL), H_2O (500 mL), and hexamethylenetetramine.

2.4. Electrochemical Test

Electrochemical measurements were performed using the PARSTAT4000A electrochemical workstation and electrochemical system using the classic three-electrode system with a platinum sheet as the counter electrode, a saturated calomel electrode (SSCE) as the reference electrode, and ductile iron samples as the working electrodes. Electrochemical samples were sealed with high-temperature-resistant epoxy resin. The exposure area was 1 cm^2. The open circuit potential (OCP) was measured for at least 30 min until a steady state was reached; then, electrochemical experiments were conducted. Three sets of parallel experiments were set up for each environment to minimize experimental errors [17].

Potentiodynamic polarization curves were obtained by performing kinetic polarization curve measurements with a scan rate of 0.333 mV/s and a scan potential range of −0.5 V/OCP to 0.8 V/OCP.

The test conditions for electrochemical impedance spectroscopy (EIS) in each solution were the same as the working electrode used for the kinetic potential polarization test. The frequency range for EIS was from 100 kHz to 10 mHz with an amplitude of 5 mV (rms) at the open circuit potential.

2.5. Simulation Model Calculations

The model uses the same solution environment as the autoclave immersion experiment. Finite element simulation was performed using Comsol software [18–21]. Irrespective of insoluble substances, the modeled electrolyte species are H^+, OH^-, Cl^-, Na^+, and Fe^{2+}. Fixed concentrations and electrolyte phase potential were set at the top horizontal electrolyte boundary, facing the air. The iron dissolves at the electrode surface:

$$Fe^{2+} + 2e^- \Leftrightarrow Fe\ (s) \tag{1}$$

Additionally, the kinetics depend on pH (i_0 proportional to H^+):

$$i_{loc} = i_0(\exp(\frac{\partial_a F\eta}{RT}) - \exp(-\frac{\partial_c F\eta}{RT})) \tag{2}$$

$$i_{loc} = i_0(c_R \exp(\frac{\alpha_a F\eta_{ref}}{RT}) - c_0 \exp(-\frac{-\alpha_c F\eta_{ref}}{RT})) \tag{3}$$

Equation (2) is a Butler–Volmer expression [22], where α_c denotes the cathodic charge transfer coefficient, α_a is the anodic charge transfer coefficient, and i_0 is the exchange current density. The kinetic equation is then coupled with the solution environment. Equation (3) is a Butler–Volmer expression with concentration dependence. This type of expression allows for more freedom in defining the concentration-dependent Butler–Volmer type of expressions, where the anodic and cathodic terms of the current density expression typically depend on the local concentrations of the electroactive species at the electrode's surface. c_R and c_0 are dimensionless expressions, describing the dependence on the reduced and oxidized species in the reaction.

η is the overpotential given by

$$\eta = E_m - E_{eq} \tag{4}$$

E_m is the electrode potential and E_{eq} is the equilibrium potential. The metal potential is set to a fixed value, resulting in a mixed potential not affected by the local pit corrosion. Ions such as Cl^- may also be transported in order to maintain electroneutrality. The transport of ions, in combination with the pit shape, determine the local pH. If the iron oxidation reaction is catalyzed by H^+, a lower pH within the pit results in faster metal dissolution compared to the metal surface outside the pit. The tertiary current distribution and Nernst–Planck interface defines the mass and ion transport. The water-based charge balance model with electroneutrality defines the H^+ concentration and the OH^- concentration as built-in variables, and automatically defines the water autoprotolysis equilibrium. The separator node is used to define the pit as a porous structure. Deformed geometry handles the deformation of the pit. The multiphysics nodes couple the electrochemistry to the deformation.

3. Results and Discussion

3.1. Exploration of the Corrosion Mechanisms of Ductile Iron Shrinkage Holes

3.1.1. Microstructure Characteristics

As shown in Figure 1a–f, a large number of shrinkage holes, which are visible to the naked eye, can be seen on the surface of the ductile iron pipe that did not receive surface mechanical treatment. Additionally, these holes are deep and cover a wide area. These holes are highly likely to become areas of aggressive ion accumulation in a harsh environment. This would accelerate the formation of a local acidified environment, in which the dissolution of the iron matrix would be promoted. After the surface mechanical treatment, these obvious defects on the surface of the ductile iron pipe are removed; the microscopic morphology is shown in Figure 1g–k. It can be seen that the shrinkage holes on the surface of this specimen are significantly reduced. The surface uniformity of this specimen is better, and the graphite balls are uniformly distributed on the surface.

3.1.2. Corrosion Morphology

To investigate the corrosion process of the ductile iron pipe in the real service environment of thermal pipeline, an immersion corrosion test was conducted in the autoclave. The corrosion morphology of the two different specimens after different periods of immersion is shown in Figure 2. As the immersion time increases, there was a marked increase in the corrosion product (Figure 2a,b,e,f). As shown in Figure 2c,d, the surface of the ductile iron without surface mechanical treatment featured large pits after the rust removal, indicating that serious localized corrosion was initiated on the specimen's surface. As the acceleration time increases, the number of large-sized holes increases, and local corrosion intensifies [23,24]. The whole interface between the graphite spheres and the matrix was dissolved on the ductile iron specimen that had not received surface mechanical treatment. This is primarily the result of the holes and shrinkages formed around the graphite spheres. Aggressive ions such as Cl^- would be preferentially enriched at these defect sites, resulting in the local acidification of the solution environment. The interface with high electrochemical activity would be dissolved easily. The mechanically treated ductile iron sample shows slight corrosion morphology (Figure 2g). Serval shallow pits were randomly distributed on the specimen's surface after 120 h of immersion. The corrosion is relatively mild, mainly comprising the pits left by the dislodging of graphite balls after the spread of corrosion. As the immersion time increased to 240 h, larger sized pits could be seen on the surface of the mechanically treated ductile iron specimen (Figure 2h). This is mainly due to the integration of the pits formed after the dislodging of the graphite balls. However, the number of pits was significantly smaller than that of the ductile iron samples that had not received surface mechanical treatment (Figure 2d). As the upper graphite sphere is detached, the graphite sphere buried at the bottom is also exposed.

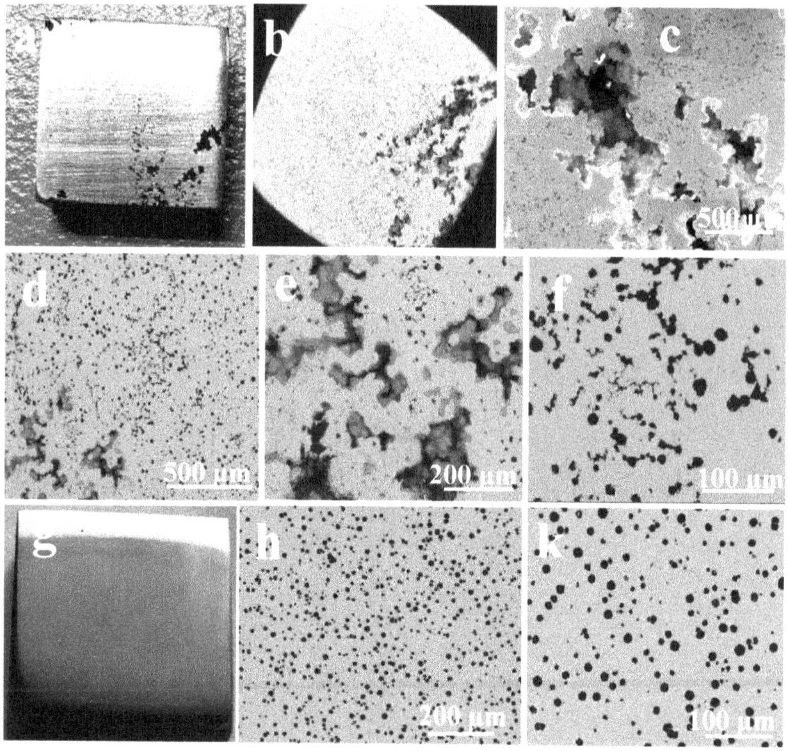

Figure 1. The microscopic morphology of the two ductile iron samples before the experiment. (**a**–**f**) ductile iron pipe with no surface mechanical treatment, (**g**–**k**) ductile iron pipe with surface mechanical treatment.

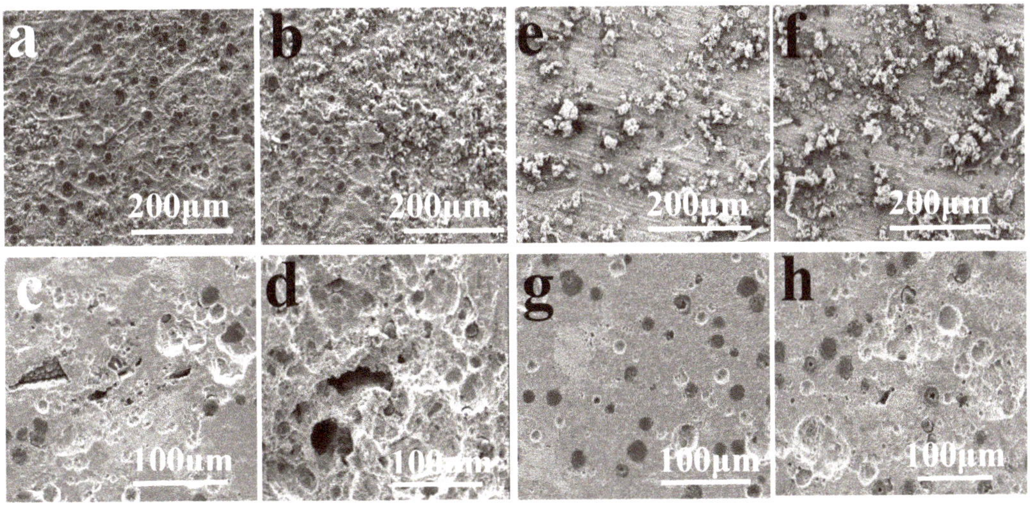

Figure 2. Surface corrosion morphology of different specimens after the immersion test. (**a**,**b**,**e**,**f**) Before and (**c**,**d**,**g**,**h**) after the corrosion rust removal. (**a**,**c**,**e**,**g**) Immersion at 120 h, and (**b**,**d**,**f**,**h**) immersion at 240 h. (**a–d**) Ductile iron with no surface mechanical treatment, (**e–h**) Ductile iron with surface mechanical treatment.

The SEM observation results show that there are many shrinkage holes and tail-like crevices on the surface of the ductile iron. After mechanical polishing to remove these obvious defects and applying a high-pressure immersion corrosion test, the degree of corrosion on the specimen's surface is significantly reduced, as is the number of local corrosion holes. It can initially be concluded that the corrosion resistance of ductile iron is significantly improved by the use of suitable surface mechanical treatments.

In order to analyze the development of the different pits in a quantitative manner, the 3D corrosion pit morphology of the specimen after rust removal was observed using confocal laser scanning microscopy (CLSM). The results are shown in Figure 3. Due to the presence of more tail-like defects with the shrinkage and loosening of the tissue, the corrosion on the surface of the non-mechanically treated ductile iron samples increased after the immersion test (Figure 3a,c). There was also an increase in the depth of the localized corrosion pits and an increase in the corrosion area. In contrast, the surface-treated ductile iron samples corroded slightly (Figure 3b,d); this is consistent with the results observed using SEM.

The pits on the surfaces of both samples were analyzed quantitatively and the results are shown in Figures 4 and 5. After 120 h of immersion corrosion, the number of pits (124/15 μm) and the maximum depth of the pits in the surface of the mechanically treated ductile iron samples were significantly lower than in the untreated specimens (220/25 μm). Equations (5) and (6) are used to describe the effect of the surface treatment on improving the pitting resistance of the ductile iron material:

$$\text{pren}(\text{number}) = \frac{n - n_{st}}{n} \times 100\% \tag{5}$$

$$\text{pren}(\text{depth}) = \frac{d - d_{st}}{d} \times 100\% \tag{6}$$

where n is the number of etch pits on the ductile iron without surface treatment, and n_{st} is the number of etch pits for the ductile iron with surface treatment. Additionally, d is the maximum pit depth for the ductile iron without surface treatment, and d_{st} is the maximum pit depth for the ductile iron with surface treatment. The results show that the number of pits decreased by 43.6%, and the maximum pit depth reduction rate reached 40%. The K-value distribution results show that, after the 120 h immersion corrosion test, the volume of etch pits on the surface of the ductile iron samples without surface mechanical treatment reached 2500–10,000 μm^3. The K value distribution is not completely concentrated in the small corrosion pit area; it is evenly distributed between the medium and large corrosion pits. This indicates that the large pits observed by SEM are not coincidental. As shown in Figure 4b, the volume of corrosion pits on the surface of the sample is distributed below 2500 μm^3 after the surface mechanical treatment.

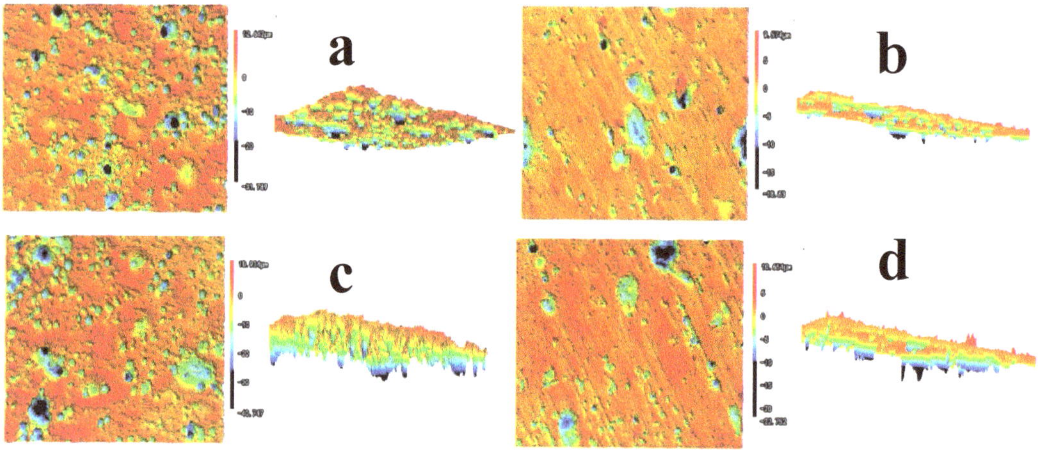

Figure 3. 3D corrosion morphology of different specimens after the immersion test. (**a**,**c**) represent the ductile iron that had not received surface mechanical treatment; (**b**,**d**) represent the ductile iron that had received surface mechanical treatment.

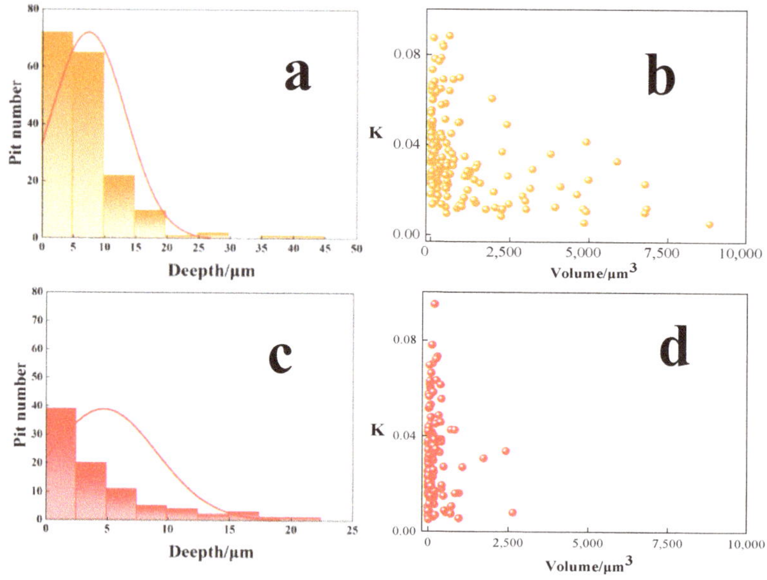

Figure 4. The number of etching pits and the K-value statistics of the CLSM results after the 120 h accelerated test. (**a**,**b**) Ductile iron with no surface mechanical treatment, (**c**,**d**) ductile iron with surface mechanical treatment.

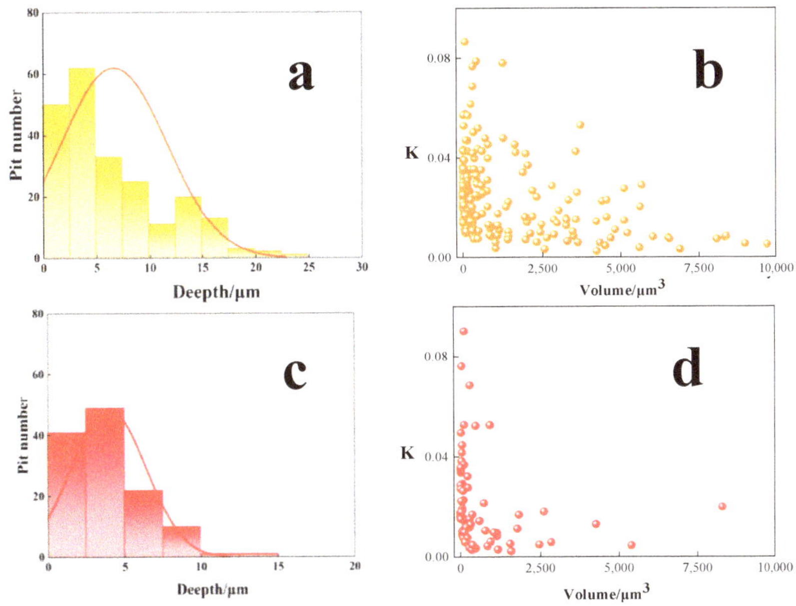

Figure 5. Distribution of the number of etch holes as a function of their size and the statistics of the K value of the CLSM results after the 240 h accelerated test. (**a**,**b**) Ductile iron with no surface mechanical treatment, (**c**,**d**) ductile iron with surface mechanical treatment.

After continuing the immersion in the autoclave up to 240 h, the corrosion of both ductile iron samples increased to different degrees. Specifically, the number, volume, and depth of pits increased, especially for specimens that had not received surface mechanical treatment. Although the surface corrosion pits on the surface of the mechanically treated ductile iron samples remained small, medium and large volume corrosion pits with volumes greater than 2500 μm^3 began to appear, and reached a maximum volume of 8000 μm^3. These occasional large pits are formed by the fusion of several small pits during the corrosion evolution process. Compared to the state of the surface at 120 h, the number of corrosion pits with a volume distribution of 2500 to 10,000 μm^3 was, predictably, increased for the samples without surface mechanical treatment. Calculations show that, in the surface-treated ductile iron material, the number of corrosion pits can be reduced by 51% and the maximum pit depth is slowed by 50% after 240 h of immersion.

3.1.3. Mechanisms of the Localized Corrosion Initiation

The CLSM results showed that the corrosion of the ductile iron was effectively slowed down after the mechanical treatment was used to eliminate the obvious defects on the surface of the ductile iron. Combined with the SEM observations, these finding suggest that the main reason that the increased corrosion of the ductile iron occurs in the simulated water quality is due to localized corrosion [23,25,26]. After the ductile iron that had not had its surface defects removed was soaked in the autoclave, there were many large corrosion pits on the surface of the sample. These local corrosion pits are large, deep, and numerous. Additionally, when these surface defects are removed, the surface of the ductile iron is basically flat, except for a very small number of small holes. However, as the corrosion proceeds, the surface will continue to exhibit shrinkage holes and defects, and the rate of corrosion will gradually increase.

Observation results by SEM, the localized corrosion process can be described as follows: (1) With the aggressive ions accumulated in the defects at the matrix and the graphite nodule, the initiation of localized corrosion is triggered (Figure 6a). (2) With the evolution of the localized corrosion, the galvanic effect [27–29] between the matrix and graphite promotes the development of localized corrosion (Figure 6b). (3) With the growth of smaller pits, larger pits would be formed, resulting from the consolidation of these smaller pits (Figure 6c). The depth of the pits increases as the small graphite spheres continue to fall off during the corrosion process.

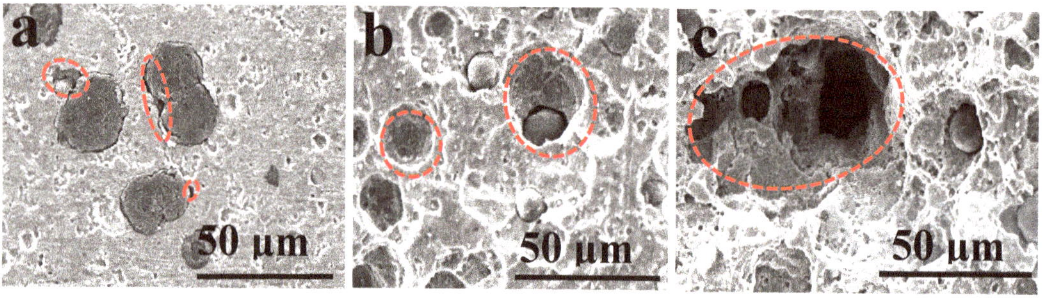

Figure 6. The localized corrosion initiation process in ductile iron (**a–c**).

3.1.4. Simulation of the Corrosion Process

In order to further confirm the conclusion offered in 3.1.3, and to describe clearly the dynamic process of the increased corrosion of ductile iron by shrinkage holes, a finite element simulation [30–33] was used to assess the ductile iron's surface at the shrinkage gap. The model defines the kinetic characteristics of the actual electrolyte environment and the cast iron substrate. The planar geometry and mesh division of the model are shown in Figure 7. The whole model consists of spherical graphite, an electrolyte, and a cast iron electrode. Marking points 1,2,3 facilitates the observation of shape changes due to corrosion in the later simulation studies.

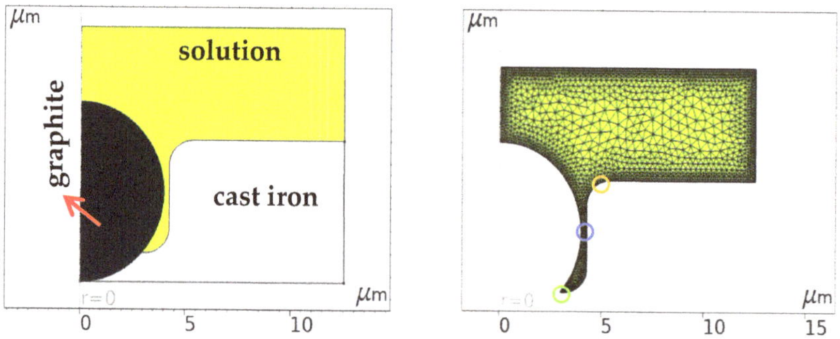

Figure 7. Schematic diagram of the geometry and meshing of the simulation model.

In the early stages of corrosion, when uniform electrochemical corrosion occurs over the entire area, the crescent-shaped gaps around the graphite were filled with the electrolyte solution (Figure 7b). This is because the shape of the gap was not extended when corrosion occurred for 1d (Figure 8a,e). When corrosion occurs for 3 d (Figure 8b,f), the shape of the gap changes due to corrosion. Additionally, because of charge conservation, as Fe^{2+} dissolves and oxygen are consumed, higher amounts of H^{+} and Cl^{-} in the solution diffuse into the gap. The pH and Cl^{-} concentrations also changed. The highest Cl^{-} concentration was found at the bottom of the gap (point 3), reaching 190 ppm, and the pH decreased to 9. This also resulted in a concentration difference within the gap from the overall solution environment [25,34].The gap kept expanding. After 7 d of corrosion (Figure 8c,g), the gap gradually evolved into a teardrop-shaped etch pit [35]. The Cl^{-} concentration in the gap reached a maximum of 260 ppm and the pH dropped to a minimum of 8.9. At this point, the gap expanded at an increasingly rapid rate. This continuous anodic dissolution also continuously produced Fe^{2+} and consumed oxygen, prompting more migration of Cl ions and transport of H into the crescentic gap. further acidifying the solution in the interstitial space. Because of this vicious cycle, the initial gap gradually expanded into a teardrop-shaped corrosion pit at 10 d (Figure 8d,h). This is consistent with the statistics provided by the CLSM. The pH of the corrosion pits (point 3) was only 8.32 at this time, and the Cl^- concentration increased to 700 ppm.

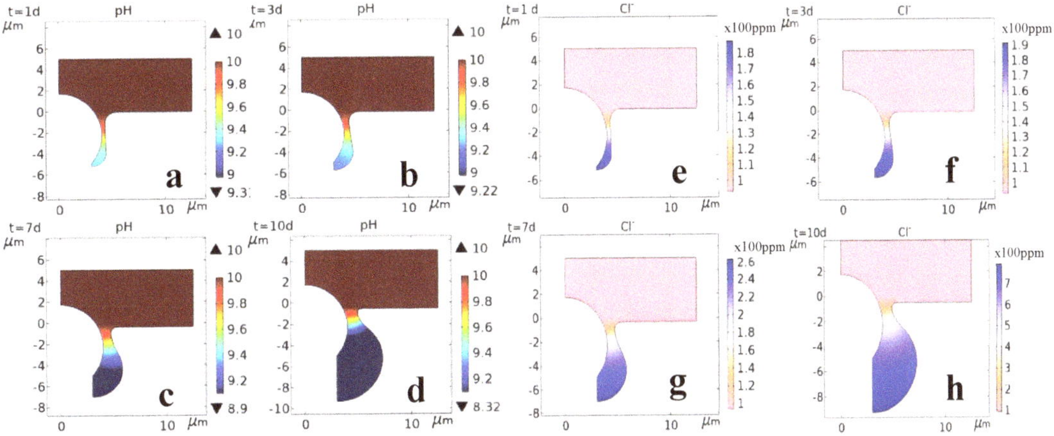

Figure 8. The pH and Cl^- diffusion at different times. (**a–d**) culated results of pH at 24 h, 72 h, 168 h, and 240 h, (**e–h**) alculated results of Cl^- diffusion at 24 h, 72 h, 168 h, and 240 h. On this scale, 1 represents 100 ppm concentration.

According to the guidelines of the physical model adopted in this work, as corrosion occurs, the variation in the anode's surface potential in the crescent-shaped gap is shown in Figure 9a~d. Inside the gap, the local potential difference in the interface iron/near solution gradually becomes more negative from the top to the bottom of the gap. The electrode potential at the top of the gap (point 1) was −0.62 V at 10 h and 90 h. The anode's metal corrosion potential at the bottom of the slit (point 3) decreased to −0.65 V. A more negative electrode surface potential also leads to a higher corrosion rate and more severe corrosion at the bottom of the crevice (Figure 9c). This is because, with the negative shift of the electrode potential, a higher overpotential is generated in the interface region between the electrode and the electrolyte (Equation 4), which results in an enhanced corrosion kinetic process. At 240 h, the surface corrosion potential of the electrodes inside the gap continues to shift negatively. The potential at point 1 decreases to −0.65 V and the surface potential at point 3 decreases to −0.73 V. The corrosion process still gradually increases.

3.2. The Effect of the Water Environment on the Corrosion Kinetics of Ductile Iron

3.2.1. Corrosion Rate Analysis

Figure 10 shows the corrosion rate of the two ductile iron materials after the autoclave immersion experiment. After 120 h of immersion, the corrosion rate of the ductile iron samples without surface treatment was much higher than that of the surface-treated samples. With the immersion time increased to 240 h, the corrosion rate of the surface-treated samples was still lower than the specimen without surface treatment. This corrosion kinetic law indicates that the corrosion rate of ductile iron can be substantially slowed down after surface treatment to eliminate obvious defects on the surface. It also shows that the corrosion damage of ductile iron in a harsh water quality environment is caused by very severe local corrosion [36,37] due to the formation of dense differential cells [38] by shrinkage holes.

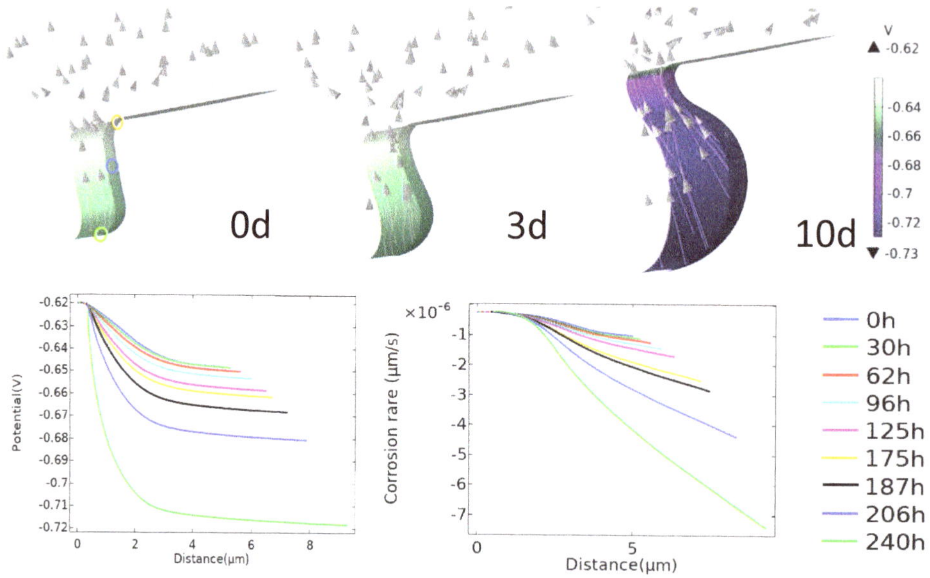

Figure 9. Calculated relative changes of local metal potential at different times. (**a**–**c**) potential distribution clouds at 10 h, 90 h, and 240 h, respectively, (**d**) relative changes of local potential of the metal surface along the y-direction, and (**e**) the corrosion rate distribution along the y-direction.

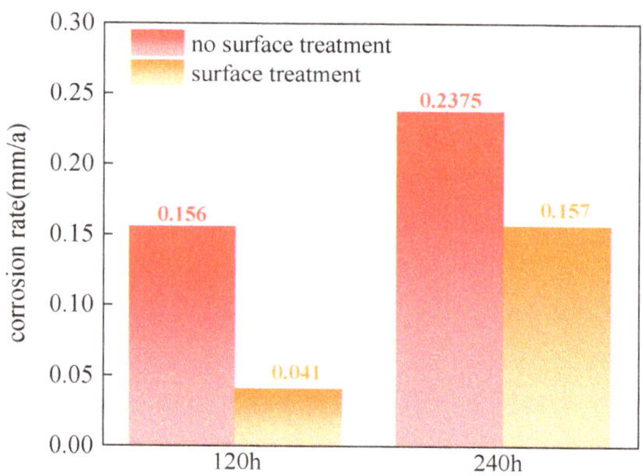

Figure 10. Corrosion rates of different ductile iron samples after different periods of immersion.

3.2.2. Electrochemical Test

The polarization curve can accurately reflect the corrosion kinetics of the electrode. It is able to express both the cathodic oxygen consumption and the anodic electrode dissolution. It is widely used in metal corrosion studies [23,39–42]. The results are shown in Figure 11. The figure shows that the two ductile iron samples exhibit the same electrochemical reaction mechanism in different solution environments. The cathodic process primarily concerns the consumption of oxygen, and the anodic process concerns the dissolution of the metal. Under low oxygen dissolution conditions, the cathodic process is limited. It is worth noting that the surface-treated ductile iron samples in solution no. 4 are always passivated after a certain anodic potential polarization. The two materials also exhibit significant differences in their corrosion kinetics at different immersion cycles when the water environment changes.

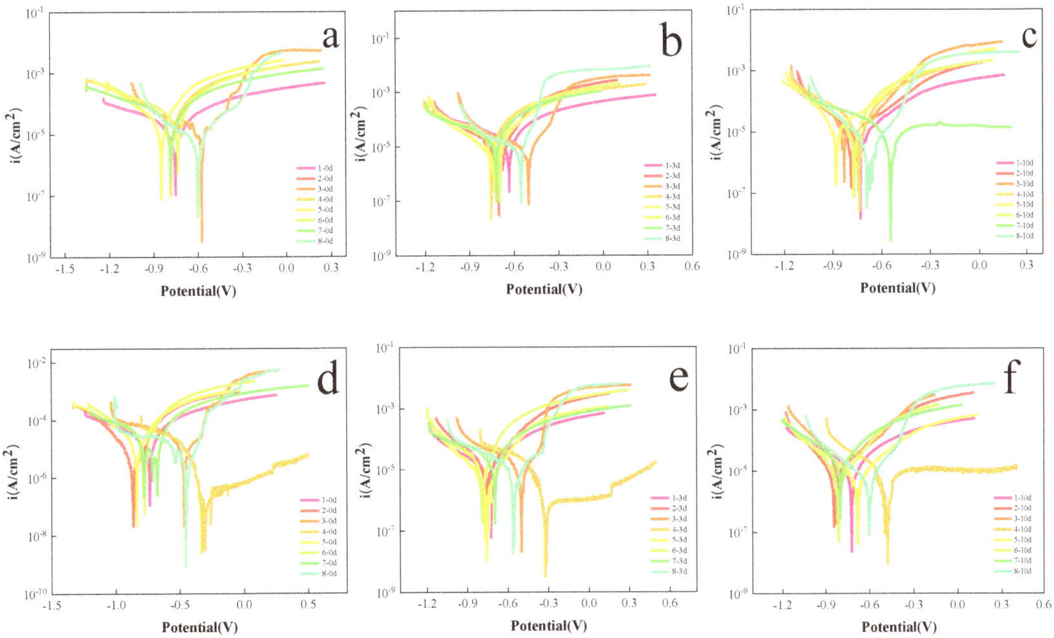

Figure 11. Dynamic potential polarization curves of two types of ductile iron during different periods of immersion in eight simulated water conditions, (**a–c**) are ductile iron without mechanical treatment; (**d–f**) are ductile iron with mechanical treatment.

The polarization curve data in the range of ±(100 to 125) mV, relative to the open circuit potential, were intercepted to fit the i_{corr}, and the results of the fit are shown in Table 2. At 0 d, the i_{corr} of the unsurfaced samples was higher than that of the surface-treated samples in all simulated environments (Figure 12a). However, the i_{corr} of the surface-treated samples started to improve as the immersion time increased. By 10 d of immersion, the i_{corr} of the surface-treated samples in the No. 3, 5, 7, and 8 solutions had surpassed that of the unsurfaced samples (Figure 12c). This verifies the hypothesis previously made in the autoclave immersion experiment. The mechanical treatment eliminates the obvious defects on the ductile iron surface; however, as the corrosion proceeds, uniform corrosion will cause defects to gradually be exposed on the sample surface, accelerating the occurrence of local corrosion. Overall, compared to the i_{corr} average, the surface-treated ductile iron samples showed significantly higher corrosion resistance in harsh water conditions than the non-surface-treated samples.

To confirm the findings of the kinetic potential test, an electrochemical workstation was used to perform EIS tests on both materials under the same conditions in solutions no. 2, 4, 5, and 8. The nyquist plot (Figure 13a) shows that the EIS in solutions no. 8 and 7 appears to be characterized by high frequency capacitive arcs and low frequency Warburg impedance. It shows that there is a diffusion impedance for the electrochemical reaction under this condition. The equivalent circuit in Figure 13h was used to fit the EIS data. Additionally, only one capacitive arc exists for the EIS in solutions no. 3 and 5. The equivalent circuit in Figure 13g was used to fit the EIS data [43]. R1 is the solution resistance, and R2 is the equivalent resistance of the charge transfer impedance in the interface region. CPE1 is the equivalent capacitance of the bilayer in the interface region, and W1 is the Warburg impedance [44,45].

Table 2. Fitting results of i_{corr} by potentiodynamic linear polarization scanning.

Times(d)	i_{corr} without Mechanical Treatment(A/cm^2)							
	1	2	3	4	5	6	7	8
0	8.3	16.8	8.7	15.6	15.4	14.8	17.4	4.6
3	6.8	5.9	5.1	6.7	5.0	9.6	11.1	6.3
10	6.9	7.6	9.7	9.2	5.2	9.1	4.2	3.1
Average	7.3	10.1	7.8	10.5	8.5	11.1	10.9	4.6
	i_{corr} with mechanical treatment(A/cm^2)							
	1	2	3	4	5	6	7	8
0	6.2	2.9	7.2	0.2	13.3	11.8	7.7	2.6
3	3.9	5.8	6.2	0.6	4.2	10.8	4.8	4.2
10	5.9	4.7	10.2	2.1	6.7	3.6	19.2	4.5
Average	5.3	4.5	7.9	1.0	8.1	8.7	10.6	3.8

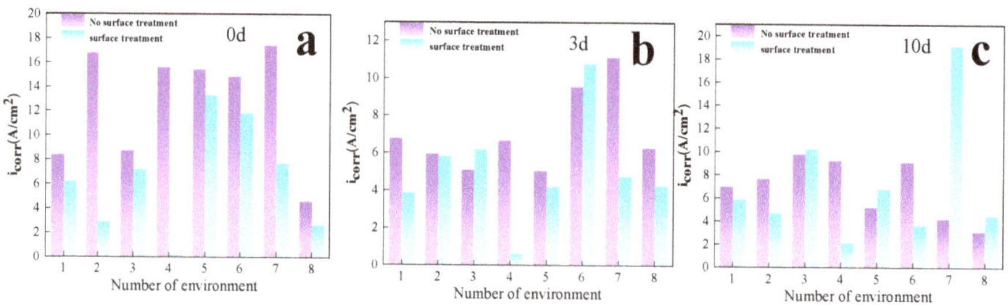

Figure 12. Fitting results of the i_{corr} values of different ductile iron samples (**a**–**c**).

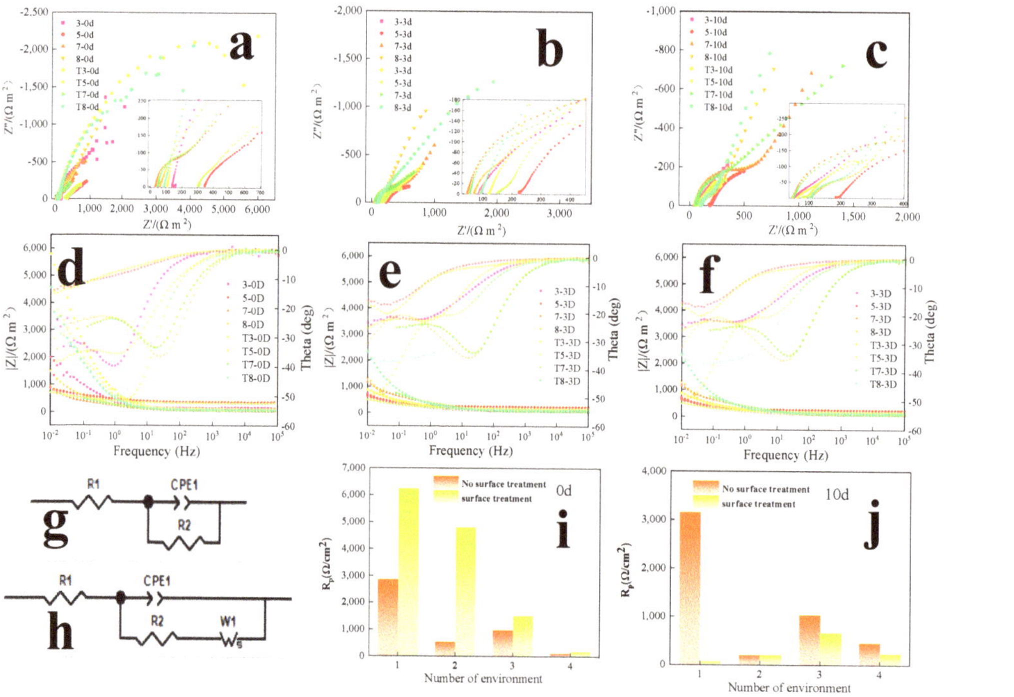

Figure 13. EIS test results. (**a**–**c**) are nyquist plots, (**d**–**f**) are bode plots and phase angles, (**g**,**h**) are the equivalent circuits, (**i**,**j**) are the comparison fitting results of R_P at 0 d and 10 d.

The fitted results are shown in Table 3. It can be seen, the charge transfer resistance of the surface-treated ductile iron samples is generally higher at the beginning of the immersion process (Figure 13i). As the immersion time increases, the Rp of the surface-treated ductile iron material starts to decrease at 10 d of corrosion. A weaker corrosion inhibition effect was exhibited. This also explains the sudden increase in the i_{corr} of the surface-treated ductile iron at 10 d (Figure 12c).

Table 3. Fitting results of R_s, R_p, and W_R by EIS.

Times(d)	R_s without Mechanical Treatment (Ω/cm^2)				R_p/W_R without Mechanical Treatment (Ω/cm^2)			
	3	8	5	7	3	8	5	7
0	112	58	327	31	2848	526/ 3	967	100 /2
3	84	69	226	46.	989	139.4/0.5	821	163 /4
10	49	108	187	45	3149	193/	1020	445 /0.2
Times(d)	R_s with mechanical treatment (Ω/cm^2)				R_p/W_R with mechanical treatment (Ω/cm^2)			
	3	8	5	7	3	8	5	7
0	102	83	297	52	6222	4785	1493	171/ 36
3	118	91	154	35	1409	366.1 /62	547	182 /14
10	58	109	133	47	58.26	150/1.1	650	220 /4.7

3.2.3. The Influence of Environmental Factors

In order to further quantify the influence of different water quality environmental factors on the corrosion process of ductile iron with different treatments, the xgboost algorithm was used to compare the influence of environmental factors (pH, Cl^- concentration, hardness, oxygen content, and temperature) on the corrosion ductile iron with different surface states. The calculation results are shown in Figure 14.

As can be seen from Figure 14a, for the ductile iron material without surface mechanical treatment, the presence of dissolved oxygen in the solution plays a major controlling role among all environmental factors, contributing to 68.5% of the increase in i_{corr}. The remaining environmental factors were evenly distributed below 15% in terms of the degree of influence on i_{corr} [46,47]. It is noteworthy that the contribution of Cl^- concentration to i_{corr} was only 2%. This is because the restricted oxygen levels caused a significant slowing of local corrosion. At the same time, temperature also reduces the corrosion acceleration effect of Cl^- [39]. The contribution of each environmental factor to the elevated i_{corr} under fully oxygenated conditions is shown in Figure 14b. At this point, the Cl^- [25] concentration played a control role and contributed up to 85% of the increase of i_{corr}. This indicates that the acceleration effect brought about by Cl^- can only occur under conditions of sufficient oxygen. The study in 3.1 found that the corrosion of ductile iron intensified because of the severe localized corrosion caused by the concentration cell effect, and the results of the kinetic law analysis here also confirmed this. The continued dissolution of the anode under conditions of sufficient oxygen causes Cl^- to continue to diffuse into the slit between the graphite and the cast iron, exacerbating the corrosion of the ductile iron. this is why the i_{corr} contribution is so high.

Figure 14c shows the contribution of environmental factors to the i_{corr} of ductile iron after eliminating obvious defects on the surface. It can be deduced that the influence effect of each environmental factor becomes uniform. This means that, under this condition, the effect of the deterioration of the water quality environment on the corrosion acceleration of ductile iron is obviously weakened; moreover, the oxygen concentration is not a highly influential environmental factor, but the contribution rate is still high. This may be the reason for the weakening of the local corrosion effect. Under fully oxygenated conditions, the contribution of Cl^- concentration to i_{corr} suddenly decreases. This indicates that, as the degree of local corrosion decreases, Cl^- concentration is no longer the most important environmental factor affecting i_{corr}. However, pH and immersion time contributed to the increase in i_{corr}, which may be due to the fact, that the originally flat surface of the mechanically treated ductile iron samples gradually exhibited more defects as the corrosion progressed, leading to increased localized corrosion. However, it still corrodes less than ductile iron without any surface treatment. For temperature and water hardness, the i_{corr} contribution did not change significantly before and after mechanical treatment under fully oxygenated conditions. This indicates that an increase in temperature from 60 °C to 90 °C does not bring about a significant increase in localized corrosion.

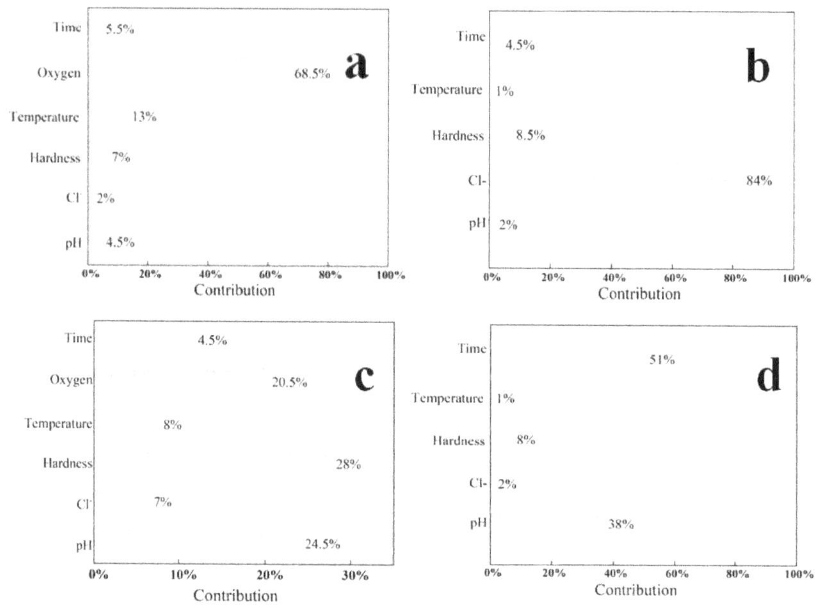

Figure 14. Analysis of the contribution of environmental factors to i_{corr}. (**a**,**c**) show the contribution of non-surface-treated ductile iron in all environments and in fully oxygenated conditions, respectively; (**b**,**d**) show the contribution of the surface-treated ductile iron in all environments and in fully oxygenated conditions, respectively.

4. Conclusions

(1) Defects such as pores and shrinkage generated in the process of producing ductile iron easily become a gathering place for aggressive ions, which induce the dissolution of graphite spheres and the matrix interface, and then induce serious local corrosion. After surface mechanical treatment to eliminate holes and shrinkage, SEM observations and CLSM statistics of the pits show that localized corrosion pits on the material surface are significantly suppressed. The average i_{corr} in the different solutions is lower and the corrosion kinetic processes are inhibited. All of the results prove that surface treatment techniques can improve the corrosion resistance of ductile iron, and inhibit localized corrosion.

(2) Simulations revealed that the solution around graphite spheres for defects such as pores and shrinkage loosening is highly susceptible to acidification, whereby an acidification autocatalytic cell effect is formed, accelerating the dissolution of the substrate and reducing the corrosion resistance of the material.

(3) Electrochemical test results showed that changes in water quality have a greater impact on the corrosion kinetic process of ductile iron. For unsurfaced ductile iron, the

order of the relative weight of the environmental factors that affect corrosion resistance in all environments is oxygen solubility > temperature > water hardness > immersion time > pH > Cl^- concentration. Under oxyfuel conditions, the order of the relative weight of the environmental factors that affect corrosion resistance is Cl^- concentration > water hardness > immersion time > pH > temperature.

(4) For surface-treated ductile iron, with the removal of obvious defects from the ductile iron's surface, oxygen concentration is no longer the main factor in determining the i_{corr} value. Under oxyfuel conditions, the immersion time becomes the largest factor affecting the i_{corr} contribution, not Cl^- or pH. This indicates that uniform corrosion emerges as the dominant type of erosion as immersion time increases.

Author Contributions: Methodology, T.L., K.T. and X.Y.; Software, X.Y.; Validation, T.L.; Formal analysis, T.L. and K.T.; Investigation, L.Z.; Resources, C.L.; Data curation, B.W. and C.L.; Writing—original draft, B.W.; Writing—review and editing, C.L. and X.C.; Project administration, L.Z. and X.C. Funding acquisition, C.L. and X.C. All authors have read and agreed to the published version of the manuscript.

Funding: The National Natural Science Foundation of China (No. 52104319); the National Science and Technology Resources Investigation Program of China (No. 2019FY101400).

Data Availability Statement: The raw/processed data required to reproduce these findings cannot be shared at this time as the data are related to an ongoing study.

Acknowledgments: The authors acknowledge Qinglin Li and Yunyang Wu in University of Science and Technology Beijing for their assistance in the experiment.

Conflicts of Interest: The authors declare no conflict of interest.

References

1. Abubakar, S.A.; Mori, S.; Sumner, J. A Review of Factors Affecting SCC Initiation and Propagation in Pipeline Carbon Steels. *Metals* **2022**, *12*, 1397. [CrossRef]
2. Zhang, G.A.; Cheng, Y.F. Electrochemical characterization and computational fluid dynamics simulation of flow-accelerated corrosion of X65 steel in a CO_2-saturated oilfield formation water. *Corros. Sci.* **2010**, *52*, 2716–2724. [CrossRef]
3. Cassineri, S.; Duff, J.; Cioncolini, A.; Curioni, M.; Banks, A.; Scenini, F. Deposition of corrosion products under pressurised water nuclear reactor conditions: The effect of flow velocity and dissolved hydrogen. *Corros. Sci.* **2019**, *159*, 108113. [CrossRef]
4. Koike, M.H. Erosion corrosion in stainless steel pipe under water vapour two-phase flow conditions. *Corros. Sci.* **2006**, *48*, 617–624. [CrossRef]
5. Ma, W.L.; Wang, H.X.; Barker, R.; Kapur, N.; Hua, Y.; Neville, A. Corrosion behaviour of X65 carbon steel under the intermittent oil/water wetting: A synergic effect of flow velocity and alternate immersion period. *Corros. Sci.* **2021**, *187*, 109507. [CrossRef]
6. Vasyliev, G.S. The influence of flow rate on corrosion of mild steel in hot tap water. *Corros. Sci.* **2015**, *98*, 33–39. [CrossRef]
7. Wu, K.-h.; Zhu, L.-q.; Li, W.-p.; Liu, H.-c. Effect of Ca^{2+} and Mg^{2+} on corrosion and scaling of galvanized steel pipe in simulated geothermal water. *Corros. Sci.* **2010**, *52*, 2244–2249. [CrossRef]
8. Zhu, Y.; Free, M.L.; Cho, J.-H. Integrated evaluation of mixed surfactant distribution in water-oil-steel pipe environments and associated corrosion inhibition efficiency. *Corros. Sci.* **2016**, *110*, 213–227. [CrossRef]
9. Choo, H.; Choi, Y.; Lee, W.; Lee, C. Effect of pH Variations on the Yield Stress of Calcium Bentonite Slurry Treated with pH-Responsive Polymer. *Materials* **2020**, *13*, 2525. [CrossRef]
10. Ceschini, L.; Campana, G.; Pagano, N.; Angelini, V. Effect of laser surface treatment on the dry sliding behaviour of the EN-GJS400–12 ductile cast iron. *Tribol. Int.* **2016**, *104*, 342–351. [CrossRef]
11. Mojisola, T.; Seidu, S.O.; Olubambi, P.A.; Adediran, A.A. Effect of preconditioning on the microstructure and mechanical properties of ductile cast iron. *Mater. Today Proc.* **2022**, *62*, S23–S29. [CrossRef]
12. Hsu, C.-H.; Chen, M.-L. Corrosion behavior of nickel alloyed and austempered ductile irons in 3.5% sodium chloride. *Corros. Sci.* **2010**, *52*, 2945–2949. [CrossRef]
13. Krawiec, H.; Stypuła, B.; Stoch, J.; Mikołajczyk, M. Corrosion behaviour and structure of the surface layer formed on austempered ductile iron in concentrated sulphuric acid. *Corros. Sci.* **2006**, *48*, 595–607. [CrossRef]
14. Krawiec, H.; Vignal, V.; Lelito, J.; Krystianiak, A.; Ozga, P. In-situ monitoring of the corrosion behaviour of austempered ductile iron (ADI) under cyclic salt spray exposure. *Corros. Sci.* **2021**, *185*, 109437. [CrossRef]
15. Song, Y.; Jiang, G.; Chen, Y.; Zhao, P.; Tian, Y. Effects of chloride ions on corrosion of ductile iron and carbon steel in soil environments. *Sci. Rep.* **2017**, *7*, 6865. [CrossRef] [PubMed]

16. Zhou, Y.; Lu, Z.; Zhan, M. An investigation of the erosion–corrosion characteristics of ductile cast iron. *Mater. Des.* **2007**, *28*, 260–265. [CrossRef]
17. Wang, P.J.; Ma, L.W.; Cheng, X.Q.; Li, X.G. Influence of grain refinement on the corrosion behavior of metallic materials: A review. *Int. J. Miner. Metall.* **2021**, *28*, 1112–1126. [CrossRef]
18. Chen, Y.; Huang, H.; Zhang, Y.; Wang, C.; Fan, W. A method of atmospheric corrosion prediction for aircraft structure. *Mater. Corros.* **2018**, *70*, 79–90. [CrossRef]
19. Salleh, S. Modelling Pitting Corrosion in Carbon Steel Materials. Ph.D. Thesis, University of Manchester, Manchester, UK, 2013.
20. Zhao, J.; Liu, Y.; Yang, X.; Xin, H.E.; Wang, L.; Xiong, D.; Yanhong, G.U. Corrosion Behavior of Pipeline Steel in Oilfield Produced Water under Dynamic Corrosion System. *J. Wuhan Univ. Technol.* **2022**, *37*, 15. [CrossRef]
21. Valor, A.; Caleyo, F.; Alfonso, L.; Rivas, D.; Hallen, J.M. Stochastic modeling of pitting corrosion: A new model for initiation and growth of multiple corrosion pits. *Corros. Sci.* **2007**, *49*, 559–579. [CrossRef]
22. Van den Steen, N.; Gonzalez-Garcia, Y.; Mol, J.M.C.; Terryn, H.; Van Ingelgem, Y. Predicting the effect of droplet geometry and size distribution on atmospheric corrosion. *Corros. Sci.* **2022**, *202*, 11308. [CrossRef]
23. Messinese, E.; Casanova, L.; Paterlini, L.; Capelli, F.; Bolzoni, F.; Ormellese, M.; Brenna, A. A Comprehensive Investigation on the Effects of Surface Finishing on the Resistance of Stainless Steel to Localized Corrosion. *Metals* **2022**, *12*, 1751. [CrossRef]
24. Szklarska-Smialowska, Z. Pitting Corrosion of Metals. *Corros. Sci.* **1986**, *41*, 9.
25. Dastgerdi, A.A.; Brenna, A.; Ormellese, M.; Pedeferri, M.P.; Bolzoni, F. Experimental design to study the influence of temperature, pH, and chloride concentration on the pitting and crevice corrosion of UNS S30403 stainless steel. *Corros. Sci.* **2019**, *159*, 108160.1–108160.9. [CrossRef]
26. Dawson, J.L.; Ferreira, M. Crevice corrosion on 316 stainless steel in 3% sodium chloride solution. *Corros. Sci.* **1987**, *26*, 1027–1040. [CrossRef]
27. Kuhn, R.J. Galvanic Corrosion on Cast Iron Pipes. *Ind. Eng. Chem* **2002**, *22*, 335–341. [CrossRef]
28. Tavakkolizadeh, M. Galvanic Corrosion of Carbon and Steel in Aggressive Environments. *J. Compos. Constr.* **2001**, *5*, 200–210. [CrossRef]
29. Varela, F.E.; Kurata, Y.; Sanada, N. The influence of temperature on the galvanic corrosion of a cast iron-stainless steel couple (prediction by boundary element method). *Corros. Sci.* **1997**, *39*, 775–788. [CrossRef]
30. Lu, S.; Zhou, L.; Wang, C.; Wang, Z. Finite Element Analysis of Multipoint Counter Electrode Sensor in Steel Corrosion Rate Measurement. *IEEE Sens. J.* **2014**, *14*, 790–792. [CrossRef]
31. Shafei, B.; Alipour, A.; Shinozuka, M. Prediction of corrosion initiation in reinforced concrete members subjected to environmental stressors: A finite-element framework. *Cem. Concr. Res.* **2012**, *42*, 365–376. [CrossRef]
32. Sharland, S.M.; Jackson, C.P.; Diver, A.J. A finite-element model of the propagation of corrosion crevices and pits. *Corros. Sci.* **1989**, *29*, 1149–1166. [CrossRef]
33. Yang, X.; Yang, Y.; Sun, M.; Jia, J.; Cheng, X.; Pei, Z.; Li, Q.; Xu, D.; Xiao, K.; Li, X. A new understanding of the effect of Cr on the corrosion resistance evolution of weathering steel based on big data technology. *J. Mater. Sci. Technol.* **2022**, *104*, 67–80. [CrossRef]
34. Sung-Ho, T.; Takafumin, N.; Takumi, U. Corrosion Resistance of Cr-bearing Rebar in Macrocell Corrosion Environments Due to Different Concentrations of Chloride Ions. *ISIJ Int.* **2006**, *46*, 1467–1472.
35. Jeong, J.-Y.; Jeong, C.; Kim, Y.-J.; Jang, C. Effect of Stress Magnitude on Pit Growth Rate of 304 Austenitic Stainless Steel in Chloride Environments. *Metals* **2021**, *11*, 1415. [CrossRef]
36. Chao, L.; Rir, C.; Xuan, L.A.; Zj, A.; Sy, D.; Zc, E.; Dza, B.; Hta, C.; Xla, B. Technology: New insights into the mechanism of localised corrosion induced by TiN-containing inclusions in high strength low alloy steel. *J. Mater. Sci.* **2022**, *124*, 141–149.
37. Liu, C.; Li, X.; Revilla, R.I.; Sun, T.; Li, X. Towards a better understanding of localised corrosion induced by typical non-metallic inclusions in low-alloy steels. *Corros. Sci.* **2021**, *179*, 109150. [CrossRef]
38. Qiao, L.J.; Hsiao, C.M.; Chu, W.Y.; Chen, L.; Zou, J.J. The concentration of hydrogen at crack tip of austenitic stainless steel after stress corrosion and polarization. *Scr. Metall.* **1988**, *22*, 627–630. [CrossRef]
39. Eškinja, M.; Moshtaghi, M.; Hönig, S.; Zehethofer, G.; Mori, G. Investigation of the effects of temperature and exposure time on the corrosion behavior of a ferritic steel in CO_2 environment using the optimized linear polarization resistance method. *Results Mater.* **2022**, *14*, 100282. [CrossRef]
40. Choi, Y.S.; Shim, J.J.; Kim, J.G. Effects of Cr, Cu, Ni and Ca on the corrosion behavior of low carbon steel in synthetic tap water. *J. Alloys Compd.* **2005**, *391*, 162–169. [CrossRef]
41. García, I.; Damborenea, J. Corrosion properties of tin prepared by laser gas alloying of ti and ti6al4v. *Corros. Sci.* **1998**, *40*, 01419. [CrossRef]
42. Parapurath, S.; Jacob, L.; Gunister, E.; Vahdati, N. Effect of Microstructure on Electrochemical Properties of the EN S275 Mild Steel under Chlorine-Rich and Chlorine-Free Media at Different pHs. *Metals* **2022**, *12*, 1386. [CrossRef]
43. Jia, S.; Gao, J.; Guo, H. Influence of Water Quality on Corrosion of Cast Iron Pipe in Reclaimed Water. *J. Chin. Soc. Corros. Prot.* **2020**, *40*, 569–576.
44. Aziz, I.; Qi, Z.; Min, X. Using EIS to evaluate anti-corrosion properties of the SiC_p/5A06 aluminium MMC treated by cerium conversion coatings. *J. Rare Earths* **2010**, *28*, 109–116. [CrossRef]
45. Zhang, H.; Yan, L.; Zhu, Y.; Ai, F.; Li, H.; Li, Y.; Jiang, Z. The Effect of Immersion Corrosion Time on Electrochemical Corrosion Behavior and the Corrosion Mechanism of EH47 Ship Steel in Seawater. *Metals* **2021**, *11*, 1317. [CrossRef]

46. Liang, J.; Hu, L.; Hao, J. Improvement of Corrosion Properties of Microarc Oxidation Coating on Magnesium Alloy by Optimizing Current Density Parameters. *Appl. Surf. Sci.* **2007**, *253*, 6939–6945. [CrossRef]
47. Tian, Z.L.; Zhang, T.; Wei, C.J.; Lai, Y.Q.; Li, J. Effect of current density on corrosion of $NiFe_2O_4$ based cermet inert anode for aluminum electrolysis. *Chin. J. Nonferrous Met.* **2014**, *24*, 2360–2365.

metals MDPI

Article

Effect of Immersion Time in Chloride Solution on the Properties of Structural Rebar Embedded in Alkali-Activated Slag Concrete

Willian Aperador [1], Jorge Bautista-Ruiz [2,*] and Jorge Sánchez-Molina [2]

1 Department of Engineering, Universidad Militar Nueva Granada, Bogotá 110111, Colombia
2 Centro de Investigación de Materiales Cerámicos, Universidad Francisco de Paula Santander, San José de Cúcuta 540003, Colombia
* Correspondence: jorgebautista@ufps.edu.co

Abstract: The electrochemical impedance spectroscopy (EIS) technique is widely used in the study of the corrosion of metallic materials. This method also allows for the electrical characterization at the ceramic–metal interface in contact with an aqueous solution composed of chloride ions. EIS makes it possible to discriminate the contribution of the phenomena that occur in the interface to determine the porosity of the cementitious material. The porosity determines the degree of corrosion of the metallic material and the diffusion processes on the electrode surface. In this study, the degradation of a type of non-Portland cement obtained from blast furnace steel slag and activated alkali was evaluated. This type of cement is of great interest because it avoids the emission of CO_2 during its manufacture. Estimating the porosity determined the degree of deterioration suffered by the steel embedded in the concrete as a function of the evaluation time. The hydrated samples were also characterized by ^{29}Si magic angle spinning nuclear magnetic resonance (MAS-NMR) to determine the structure of the formed calcium silicate hydrate (C-S-H) gel. This mixture formed a C-S-H gel, constituted mainly of silicon in the middle groups, in chains in the disilicates. The effect of the slag was remarkable in improving the other evaluated characteristics, i.e., in the porous matrix, the concrete was found to significantly reduce the current passing through as a function of time, showing a reduction in porosity and an increase in impedance because of the generated pozzolanic reaction.

Keywords: porosity; EIS; corrosion; alkali-activated slag; diffusion

Citation: Aperador, W.; Bautista-Ruiz, J.; Sánchez-Molina, J. Effect of Immersion Time in Chloride Solution on the Properties of Structural Rebar Embedded in Alkali-Activated Slag Concrete. *Metals* **2022**, *12*, 1952. https://doi.org/10.3390/met12111952

Academic Editor: Renato Altobelli Antunes

Received: 11 October 2022
Accepted: 11 November 2022
Published: 15 November 2022

1. Introduction

There is extensive scientific information on the potential of alkali-activated concrete as a sustainable material to replace ordinary Portland cement due to its low energy cost, high compressive strength, rapid setting and hardening, and its resistance to fire, acid, and saline solutions compared to ordinary Portland cement [1]. However, the electrochemical characteristics of the concrete–steel system have become the subject of recent studies due to the valuable life guaranteed—for at least 50 years—for structures built with reinforced concrete [2]. These are the challenges facing developers and marketing agencies engaged in large-scale construction [3].

Alternative types of cement are obtained by mixing different minerals, residues, and industrial by-products [4]. These materials, on some occasions, do not meet the specifications to be classified as materials suitable for incorporation into cement and concrete [5]. However, using alkaline activation technology, industrial waste and by-products can be transformed into new materials with low energy consumption, high durability, and excellent mechanical performance [6,7]. A specific type of said cement is obtained from blast furnace steel slag, which uses a relatively simple preparation of materials [8]. The metallurgical industry generates considerable amounts of this industrial "waste", but it is not correctly used [9]. Therefore, studying blast furnace slag as a raw material for manufacturing new cement is of great scientific, environmental, and economic interest [10]. The research results aimed at characterizing these materials can benefit both slag-producing companies and those that manufacture construction materials [11]. The former companies can improve the disposal of their waste with less environmental impact. From the low-cost slag, the latter companies will obtain cement with excellent mechanical, durability, and preparation properties [12].

Reinforced concrete is one of the most versatile construction materials, due both to its properties regarding its service and wide range of applications, as well its low cost [12]. However, the reinforcing steel in this structure is susceptible to corrosion, considerably reducing the useful life of structures built with this type of material [13,14].

In this study, the degradation of a type of non-Portland cement obtained from blast furnace steel slag and activated alkali was evaluated. This type of cement is of great interest because it avoids the emission of CO_2 during its manufacture. Estimating the porosity determined the degree of deterioration suffered by the steel embedded in the concrete as a function of the evaluation time. The hydrated samples are also characterized by ^{29}Si magic angle spinning nuclear magnetic resonance (MAS-NMR) to determine the structure of the formed calcium silicate hydrate (C-S-H) gel. This mixture formed a C-S-H gel, constituted mainly of silicon in the middle groups, in chains in the disilicates. The effect of the slag was remarkable in improving the non-Portland cement, i.e., in the porous matrix, the concrete was found to considerably reduce the current passing through as a function of time, showing a reduction in porosity and an increase in impedance because of the generated pozzolanic reaction.

2. Materials and Methods

2.1. Preparation of Concrete Specimen

The study mixtures were obtained by cementing granulated blast furnace slag activated with sodium silicate (Na_2SiO_3) at a 5% Na_2O, expressed as a percentage by slag weight to be incorporated. The SiO_2/Na_2O ratio used was 2.4. The manufacture of the concrete mixtures in both cases contained a cementitious material dosage of 300 kg/m^3. It was assumed that the water + activator/slag solution ratios were equivalent; the ratio was 0.5 to obtain an adequate slump of 80 mm.

The type, composition, and size of the fine and coarse aggregates corresponded to gravel with a maximum size of 19 mm, a specific gravity of 2.94 g/cm^3, a compact unit mass of 1860 g/cm^3, a loose unit mass of 1700 g/cm^3, and an absorption of 1.3%. The sand employed in the mixture had a specific surface area of 2.47 g/cm^3, a compact unit mass of 1670 g/cm^3, a loose unit mass of 1580 g/cm^3, and an absorption of 2.9%. Curing was carried out with a relative humidity of approximately 85% and a constant temperature of 25 °C.

2.2. Microstructural Characterization

The microstructural characteristics of the materials were also studied via ^{29}Si magic angle spinning nuclear magnetic resonance (MAS-NMR) at the age of 90 days. Solid-state ^{29}Si MAS-NMR spectra were recorded using a Bruker Avance-400 pulse spectrometer. Spectra were recorded after irradiation of the samples with a $\pi/2$ (5-µs) pulse; the resonance frequency used was 79.5 MHz, (9.4 T magnetic field), and the spinning rate was 10 kHz. In order to avoid saturation effects, the recycle delay time was set at 10 s. All measurements were taken at room temperature with TMS (tetramethylsilane) as the external standard. The error in chemical shift values was estimated to be lower than 0.5 ppm. NMR spectra deconvolutions were performed using the DMFIT software [15]. Chemical shift (position of the line), intensity (integrated area), width (width at half-height), and line shape (Lorentzian or Gaussian) of the components were then deduced.

2.3. Immersion Test

The samples for the study were manufactured using the materials specified above and were arranged as shown in Figure 1. For the development of the electrochemical tests, $Cu/CuSO_4$ (sat) electrodes were used as external reference electrodes. The working electrode corresponds to a reference structural steel ASTM A706 commonly used in earthquake-resistant constructions, with a diameter of 1/8″ and without any previous treatment.

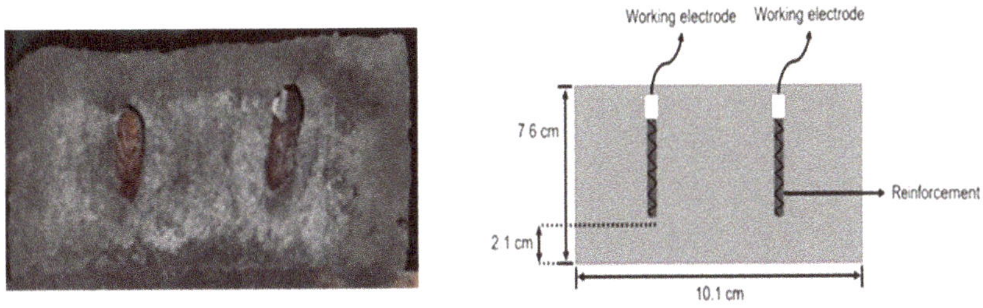

(**a**) (**b**)

Figure 1. Reinforced concrete used in determining the effect of porosity and its correlation with degradation: (**a**) sample; (**b**) schematic.

The samples with the reinforcing steel were immersed in a liquid solution saturated with 3.5% analytical sodium chloride (Panreac PA 131655.1211 NaCl-99.0%), and cycles of agitation of the liquid solution were carried out every 28 days. Due to the evaporation process that occurs, the solution was analyzed and stabilized to the original salt content in order to maintain the initial conditions. This solution then acts as the electrolyte of the electrochemical cell, which was located in a climate controlled environment at 20 °C with a relative humidity of 60%, where it was left to stand for 60 min after carrying out the connections of the assembly in order to avoid distortions during the measurements performed during testing.

2.4. Electrochemical Test

Over the nine-year period, the steel's behavior was monitored using the electrochemical impedance spectroscopy (EIS) technique to determine the characteristics of the concrete–steel interface. The study configuration is shown in Figure 2. The characterization was performed with a Gamry Interface 1010 ETM potentiostat/galvanostat. A 10-mV sinusoidal signal was applied in one sweep at frequencies ranging from 10^5 Hz to 10^{-3} Hz. The modeling of the equivalent electrochemical circuits was carried out with the Echem Analyst software.

2.5. Durability Properties

The properties of water absorption, including total porosity, were evaluated using cylindrical specimens with a diameter of 76.2 mm and a thickness of 50 mm, according to the standard procedure described in ASTM C642-06 [16]. In addition, the resistance to chloride ion penetration was determined through the Rapid Chloride Permeability Test (ASTM C1202-05) by measuring the passage of an electric current through the chloride ion. [17].

2.6. Chloride Migration

The NT Build 492 standard was used to determine the non-steady state chloride migration coefficient in concrete or cement-based materials [18]. The standard requires hardened samples to determine the material's resistance to chloride penetration. The method requires samples with a diameter of 100 mm and a thickness of 50 mm cut from the casting cylinders. The samples are immersed in a saturated solution of calcium hydroxide—$Ca(OH)_2$—and prepared with distilled water, using a desiccator and a vacuum pump, for

25 h. Once the saturation cycle is complete, the anode contains a 0.3 N NaOH solution (12 g NaOH in 1000 mL distilled water), and the cathode contains a 2 N NaCl solution (100 g NaCl in 900 mL water).

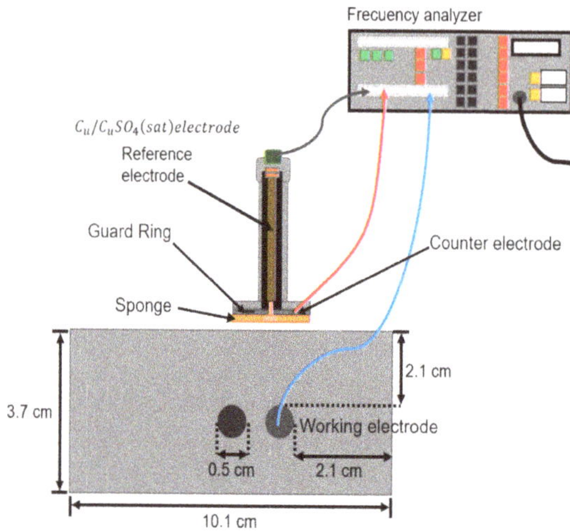

Figure 2. Experimental setup for the characterization of reinforced concrete samples.

3. Results and Discussion

3.1. Microstructural Characterization

The ^{29}Si NMR spectra of the blast-furnace slag are shown in Figure 3, and the deconvolutions of the starting materials are listed in Table 1. The ^{29}Si spectrum of the anhydrous cement showed 5 peaks; the peaks at −57, −71, −80, and −91 are assigned to the glassy component of the slag, and these signals confirmed the presence of natural pozzolans. The last peak in the spectrum is a very wide signal, which is indicative of low crystallinity. The gehlenite −71 ppm signals were still observed. The two signals of the natural pozzolan (−80.3 and −90.8 ppm) were also detected; this implies that the pozzolan is a non-reactive material, as its percentage was maintained after the hydration of the cement, and the peak at −89 ppm corresponded to the crystalline mullite [14]. The signal assigned to the Q4(0Al) units remained very wide, but had shifted slightly downward (greater chemical shift) and was centered at −106 ppm.

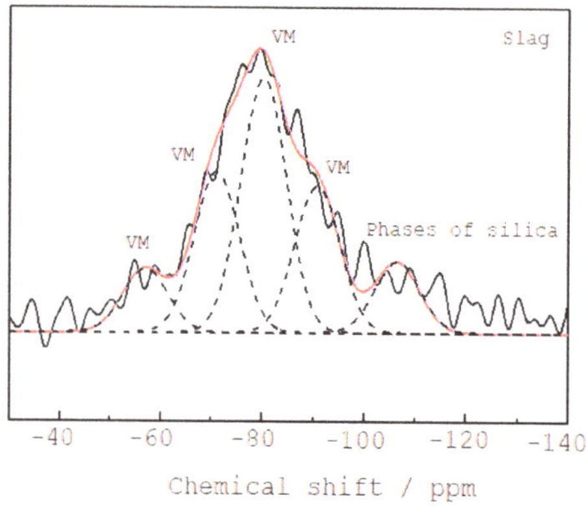

Figure 3. ^{29}Si NMR spectra of the raw materials.

Table 1. Deconvolution of the ^{29}Si MAS-NMR spectra.

		Vitreous Material				Phases of Silica, Q^4(0Al)
Slag	Position (ppm)	−57.1	−71.1	−80.3	−90.8	−106.5
	Width	10.49	10.49	10.49	10.49	10.49
	Integral (%)	9.11	23.13	36.50	21.05	10.22

3.2. Electrochemical Impedance Spectroscopy

Figure 4 shows the equivalent circuit characterized by the resistors and CPEs (constant phase elements) used in the impedance data simulation, indicating the distribution of the ions in the sample-solution system. The first interface corresponds to cementitious material in contact with the saline solution (R_s combined with Y_{p1}-R_1). The second corresponds to the cementitious material containing the reinforcing steel (R_1-Y_{p1}) combined with Y_{p2}-R_2.

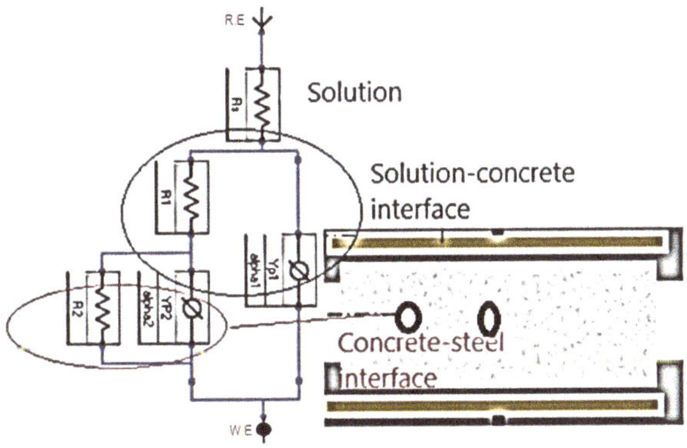

Figure 4. Equivalent circuit corresponding to the different interfaces.

Figure 5 shows the Bode plots as a function of time for the alkaline-activated concrete samples immersed in a 3.5% sodium solution. Table 2 includes the parameters used in the simulation. The values of these parameters have been obtained using a complex least-square nonlinear program (CNLS) [19]. In addition, the EIS test allows the diffusion coefficient to be calculated and the degree of porosity of the material to be related [20]. Finally, the Bode plots have been simulated using the equivalent electrical circuit in Figure 3.

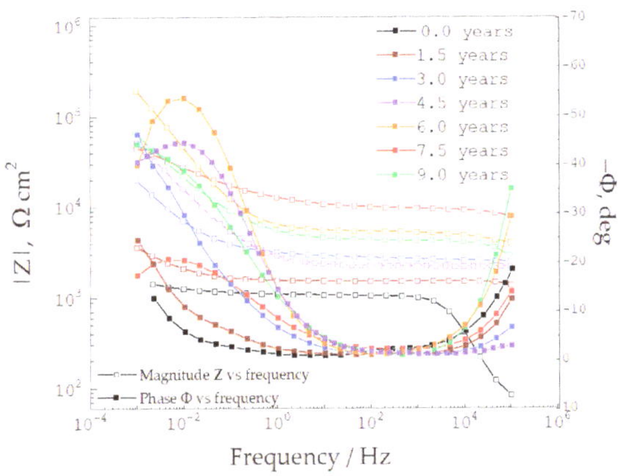

Figure 5. Bode plots for steel samples embedded in concrete and evaluated over a nine-year period.

Table 2. Values of the characteristic parameters used in evaluating the behavior of steel embedded in alkaline-activated concrete, along with its response over time.

Time Years	R_S Ω cm^2	R_1 kΩ cm^2	Y_{P1} μF/cm^2 s$^{1-\alpha1}$	α_1	Y_{P2} μF/cm^2 s$^{1-\alpha2}$	α_2	R_2 kΩ cm^2
0	7.27	0.82	385	0.8421	12	0.8214	1.425
1.5	8.31	1.357	563	0.8448	45	0.7953	3.602
3	12.21	2.158	474	0.8718	58	0.6399	20.316
4.5	25.14	2.479	465	0.8932	69	0.7563	56.521
6	28.65	8.869	1274	0.8510	235	0.6954	189.350
7.5	39.21	3.953	622	0.8714	157	0.7369	63.872
9	48.97	5.77	653	0.8845	155	0.7152	69.233

In the Bode plots, as a function of the evaluation time (Figure 4), a capacitive behavior is observed by two time-constants determined by two flattened mid- and low-frequency inflections in which the center is located below the real axis [21]. The inflection defined at high frequencies is associated with the porosity of the samples [22]. This phenomenon of lower inflection attenuation is associated with a dispersion process in the frequency, since the surface of the steel rod is corrugated. The phase vs. frequency Bode plots define an inflection at high frequencies after 3 years of evaluation. This type of behavior can be interpreted by the barrier effect generated in the system, being more significant at 6 years and stabilizing at 7.5 years [23,24].

The Bode plots showing magnitude vs. frequency (Figure 5) show the evolution of the concrete's protective effect over time when immersed in a solution of 3.5% by weight of NaCl. For 1.5 years, the value of the magnitude of impedance doubled regarding the material after the curing process and without the presence of a chloride ion. This condition was called zero time. Subsequently, the impedance values increased dramatically up to the 6-years point [25]. After 7.5 years of testing, the magnitude of the impedance decreased, stabilizing, and the values were like those obtained at 9 years [26]. The results showed that the impedance values increased in the first years of the test due to the alkaline activation of the aluminosilicates, which played an essential role in the stability of the steel. After the sixth year of testing, the decomposition of the steel's protective layer was accelerated, similar to the behavior noted without protection; the loss of stability was minimal after 7.5 years, and there was no evidence of a decrease in impedance values [27].

The evolution of the impedance occurs in the first 28 days of concrete curing, during which time the formation of the hydration products of the pozzolanic reaction favors the initial values of protection against the corrosion phenomena [28]. Likewise, the samples in contact with the NaCl solution evaluated between 18 months and 9 years showed that the impedance values were related to the decrease in the degree of porosity of the mixtures. This process is attributed to a higher generation of hydration products as a result of the protection of steel. Using materials such as iron and steel slag in concrete mixtures contributes to the optimization of the mechanisms that control the entry of chloride ions into the samples [29]. Using materials such as iron and steel slag in concrete mixes helps optimize the mechanisms that control the entry of chloride ions into the samples [29]. Blast furnace slag, with its pozzolanic, chemical, and physical properties, can reduce the permeability of mixtures during their hydration processes. These characteristics prevent the migration of chloride ions into the samples, improving their durability [30]. Alkaline activation generates dense and uniform zones at the steel–concrete interface owing to the slag activated with Na_2SiO_3. This behavior is attributed to factors such as the water content that reduces the function of Na_2SiO_3. The low porosity as a function of time is explained by the high initial concentrations of SiO_4 in the pore solution and the increasing dissolution of quartz [31]. Another factor is the production of calcium hydroxide in the cementitious hydration processes, resulting in the formation of hydrated calcium silicates that fill the existing pores in the concrete mixtures, reducing porosity [32,33].

The impedances shown in Figure 5 illustrate the different assessment times, indicating that when compared to the average migration of the chloride ion in the concrete, the first years of evaluation show that the reaction was slow, and the migration values were high. However, after 7.5 years of evaluation, it was established that the pozzolanic activity generated a reduction in the migration of the chlorides. Therefore, time was determined to be a factor in reducing the mobility of the chloride ions across the porous matrix.

3.3. Resistance to Chloride Ion Penetration

Figure 6 shows the results obtained by applying the ASTM C1202 standard to determine the permeability of concrete to chloride ions. This methodology measures the passage of an electrical current through concrete samples [31,32]. The behavior of the analyzed concretes as a function of the transferred load is observed (Figure 6). The results show that the alkali-activated concrete evaluated for up to 9 years presented a lower permeability to chlorides than did the samples at 28 days, which was called zero value. Slag concrete activated with sodium silicate is classified as having a low permeability to chlorides, and the behavior is observed as a function of time, up to nine years [33]. This behavior is associated with the plugging of the pores. The permeability of concrete is explained by the relationship or factor that most influences the fixation of chloride ions to the sample. The high concentration of chlorides in the studied systems is due to the exposure time (nine years), which can increase the capacity of the sample to fix chlorides [34]. It was also determined that the concretes evaluated at nine years could still fix chlorides; therefore, the existence of free chloride ions (within the concrete samples) looking for a way to fix themselves to the specimens was noted. These free chloride ions can affect the durability of concrete because they are responsible for initiating the corrosive process in the steel reinforcement within the structures [35]. By correlating this phenomenon with the EIS technique, it was established that the high impedance values are due to the presence of alumina in activated concrete, which is associated with the ability of concrete mixtures to fix and immobilize free chloride ions within the pore solution [36].

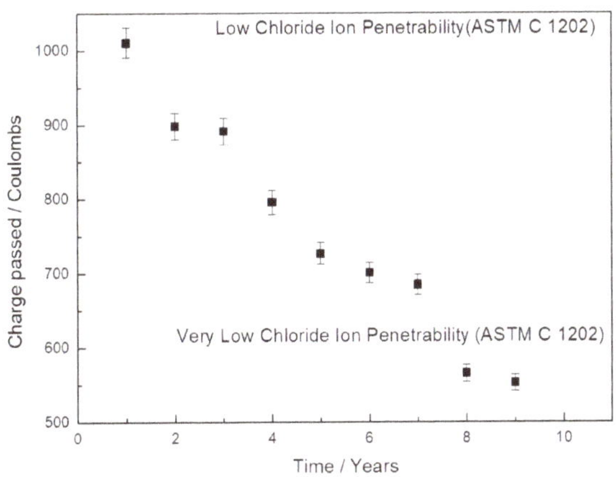

Figure 6. Determination of resistance to chloride ion penetration.

3.4. Volume of Permeable Pores

Figure 7 shows the results obtained for the concrete used in this study, based on the volume of permeable pores. In the graph, the concretes with more exposure time to the saline solution had fewer permeable pores [37]. That is, they are good quality concretes. In the case of alkali-activated iron and steel slag concrete, this property is attributed to the more significant densification of the paste—due to the presence of the hydration products—along the evaluation times, generating a lower percentage of pores [38]. This behavior is compared with the results obtained by chloride ion permeability and electrochemical impedance spectroscopy, since chloride ions can be absorbed or dissolved in the water concentrated in the pores. As previously mentioned, free chlorides are responsible for the corrosion of the steel used to reinforce concrete structures. However, the low volume is associated with the hydration of the cement, so there is no corrosion promoter [39]. This behavior is attributed to the reactivity of the granulated blast furnace slag in the presence of the alkaline activator, causing hydration products that block the pores, resulting in slower chloride ion ingress rates when evaluating the systems as a function of time [40].

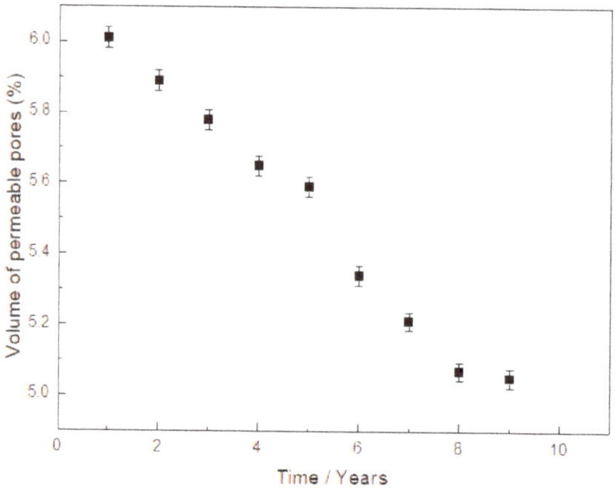

Figure 7. Variation of permeability in concrete with alkaline activation.

3.5. Chloride Ion Migration

Figure 8 shows the average evaluation times of chloride ion migration in concrete. The graph shows high values in the first years of evaluation. However, after the third year of evaluation, it was determined that the pozzolanic activity reduces the migration of chlorides. From the analysis, it can be concluded that the mobility of the chloride ions through the porous matrix is reduced over time.

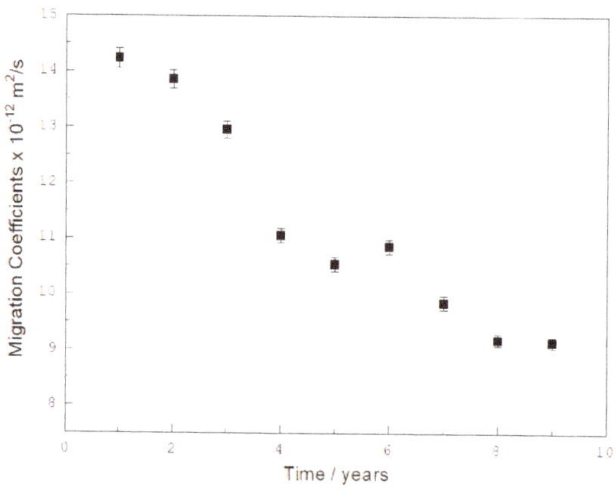

Figure 8. Non-steady-state migration coefficient as a function of time.

3.6. X-ray Diffraction

Figure 9 shows the X-ray diffraction results of the corrosion product characterization once the electrochemical studies have been completed. The procedure consisted of longitudinally cutting the specimens at immersion times of 0, 1.5, 3, 6, and 9 years to study the concrete–steel interface. The spectra in Figure 7 show crystalline quartz (SiO_2), C-S-H formation (CS), calcium carbonate (C), and beta larnite corresponding to Ca_2SiO_4. The corrosion products identified on the steel surface correspond to compounds such as brownmillerite related to CaAlFeO, goethite, magnetite, fayalite, iron oxyhydroxides, and hematite. These compounds are related to iron oxides and hydroxides in all the samples analyzed.

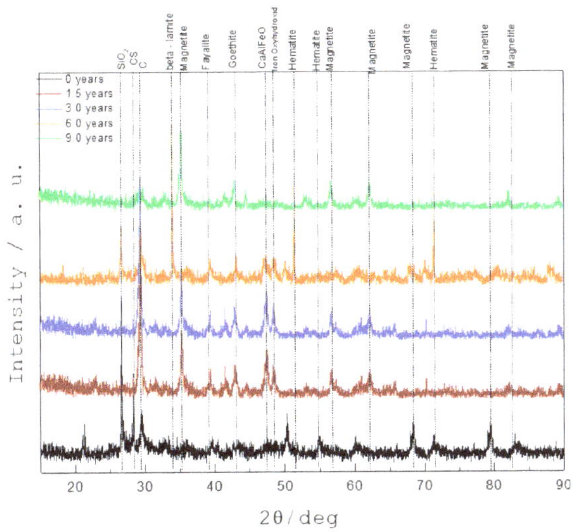

Figure 9. The X-ray diffraction patterns of the concrete–steel interface.

Additionally, the ray diffraction results for the alkali-activated slag concrete samples experiencing at least 1.5 years or more of saline solution immersion time showed overlapping characteristic diffraction peaks. The same intensity values were observed for the calcium silicate-hydrate phase at 28.52° and calcium carbonate at 29.46°. It is also evident that the samples evaluated after up to three years of immersion in the saline solution showed a higher intensity peak due to the stability of the layer of cementitious material present in the steel–concrete interface. This layer was not removed by the pH present in the concrete. When the concrete samples are subjected to salt immersion, the influx of chloride ions accelerates the removal of the oxide layer. The removal degree of the oxides is directly related to the exposure time of the samples to the solution [41]. However, the passivating layer was determined with EIS measurements (Figure 4). The impedance values indirectly indicate the formation of a protective passivating layer over the steel.

The study determined that the corrosion products found on the steel surface are the phases corresponding to iron oxides, especially the ones corresponding to the fayalite $Fe_2(SiO_4)$ at 39.17° formed in the electrochemical reactions. Magnetite compounds ($FeO{\cdot}Fe_2O_3$, $Fe^{2+}Fe^{3+}{}_2O_4$) with peaks located at 35.19°; 56.90°; 62.20°; 68.37°; 79.50°, and 82.58°, and hematite (α-Fe_2O_3), with peaks at 51.40°, 54.91°, and 71.46°, were also noted. These products arise from the reaction process, as well as during the manufacture of steel; they are known as the "steel shell", but they offer low anticorrosive protection. The protective layer is mainly formed by iron oxyhydroxides (48.51°), transforming into a mixture of hydroxides and goethite (42.97°). Although this iron oxide has excellent chemical stability, it has an irregular morphology and protects the steel against electrochemical attacks. In addition, compounds such as magnetite and hematite increase the thickness of the oxide layer, forming a mechanical protective barrier (small scale) [42,43]. Finally, the thick passive layer formed with low protection compounds (iron oxyhydroxides and goethite), the electrochemical stability of the steel, and the low porosity of the concrete generate more excellent anticorrosive protection of the steels embedded in concretes based on activated slag.

4. Conclusions

This work used alkali-activated slag as a raw material to obtain concrete. Monosilicates (Si-OH) and Al-OH groups then formed, which facilitated the polymerization of the silicate chains in the C-S-H gel. These species connected the C-S-H dimers, producing a C-S-H gel richer in Al. This compound contributes to optimizing the mechanisms that control the entry of chlorides into the mixture, reducing the permeability of the concrete during the hydration processes and favoring protection against electrochemical attacks.

It was also established that the protection mechanism in slag concrete improved when it was immersed in saline solution, which evaluated as a function of time. Furthermore, the protection of metal immersed in the concrete is related to the formation of oxide (and hydroxide) layers on the steel due to the low degree of porosity of the samples.

The impedance data increased up to 6 years, and its stabilization at 7.5 years is explained by the simultaneous combination of the uniform formation of protective layers on the steel surface and the restriction of the passage of harmful elements into the system through the ceramic matrix.

Author Contributions: Conceptualization, W.A. and J.B.-R.; methodology, J.S.-M.; software, W.A.; validation, J.B.-R.; formal analysis, J.S.-M.; investigation, W.A. and J.B.-R. All authors have read and agreed to the published version of the manuscript.

Funding: This research work was funded by Vicerrectoria de Investigaciones Universidad Militar Nueva Granada through the INV_ING_3123 project.

Institutional Review Board Statement: Not applicable.

Data Availability Statement: Not applicable.

Acknowledgments: W. Aperador acknowledges support from the Universidad Militar Nueva Granada.

Conflicts of Interest: The authors declare no conflict of interest.

References

1. Schneider, N.; Stephan, D. Reactivation of a Retarded Suspension of Ground Granulated Blast-Furnace Slag. *Materials* **2016**, *9*, 174. [CrossRef] [PubMed]
2. Mujahid-Amran, H.S.; Abdelgader, A.M.; Onaizi, R.F.; Today-Ozbakkaloglu, R.S.M.; Rashid, G. 3D-printable alkali-activated concretes for building applications: A critical review. *Constr. Build. Mater.* **2022**, *319*, 126126. [CrossRef]
3. Mengasini, L.; Mavroulidou, M.; Gunn, M.J. Alkali-activated concrete mixes with ground granulated blast furnace slag and paper sludge ash in seawater environments. *Sustain. Chem. Pharm.* **2021**, *20*, 100380. [CrossRef]
4. Criado, M.; Aperador, W.; Sobrados, I. Microstructural and Mechanical Properties of Alkali Activated Colombian Raw Materials. *Materials* **2016**, *9*, 158. [CrossRef] [PubMed]
5. Jin, Z.; Zhao, X.; Du, Y.; Yang, S.; Wang, D.; Zhao, T.; Bai, Y. Comprehensive properties of passive film formed in simulated pore solution of alkali-activated concrete. *Constr. Build. Mater.* **2022**, *319*, 126142. [CrossRef]
6. Tahri, W.; Hu, X.; Shi, C.; Zhang, Z. Review on corrosion of steel reinforcement in alkali-activated concretes in chloride-containing environments. *Constr. Build. Mater.* **2021**, *293*, 123484. [CrossRef]
7. Ibrahim, M.; Abiodun, B.; Algaifi, H.; Rahman, M.; Nasir, M.; Ewebajo, A. Assessment of acid resistance of natural pozzolan-based alkali-activated concrete: Experimental and optimization modelling. *Constr. Build. Mater.* **2021**, *304*, 124657. [CrossRef]
8. Fořt, J.; Mildner, M.; Koppert, M.; Černý, R. Waste solidified alkalis as activators of aluminosilicate precursors: Functional and environmental evaluation. *J. Build. Eng.* **2022**, *54*, 104598. [CrossRef]
9. Kumar-Das, S.; Adediran, A.; Rodrigue-Kaze, C.; Mohammed-Mustakim, S.; Leklou, N. Production, characteristics, and utilization of rice husk ash in alkali-activated materials: An overview of fresh and hardened state properties. *Constr. Build. Mater.* **2022**, *345*, 128341. [CrossRef]
10. Amran, M.; Al-Fakih, A.; Chu, S.H.; Fediuk, R.; Haruna, S.; Azevedo, A.; Vatin, N. Long-term durability properties of geopolymer concrete: An in-depth review. *Case Stud. Constr. Mater.* **2021**, *15*, e00661. [CrossRef]
11. Ibrahim, M.; Maslehuddin, M. An overview of factors influencing the properties of alkali-activated binders. *J. Clean. Prod.* **2021**, *286*, 124972. [CrossRef]
12. Amran, M.; Fediuk, R.; Abdelgader, H.; Murali, G.; Yong, Y. Fiber-reinforced alkali-activated concrete: A review. *J. Build. Eng.* **2022**, *45*, 103638. [CrossRef]
13. Kathirvel, P.; Mohan-Kaliyaperumal, S.R. Influence of recycled concrete aggregates on the flexural properties of reinforced alkali-activated slag concrete. *Constr. Build. Mater.* **2016**, *102*, 51–58. [CrossRef]

14. Aperador, W.; Mejía de Gutiérrez, R.; Bastidas, D.M. Steel corrosion behaviour in carbonated alkali-activated slag concrete. *Corros. Sci.* **2009**, *51*, 2027–2033. [CrossRef]
15. Massiot, D.; Fayon, F.; Capron, M.; King, I.; Le Calvé, S.; Alonso, B.; Durand, J.O.; Bujoli, B.; Gan, Z.; Hoatson, G. Modelling one- and two-dimensional solid-state NMR spectra. *Magn. Reson. Chem.* **2002**, *40*, 70–76. [CrossRef]
16. *ASTM C642-21(2022)*; Standard Test Method for Density, Absorption, and Voids in Hardened Concrete. ASTM International: West Conshohocken, PA, USA, 2021.
17. *ASTM C1202-22(2022)*; Standard Test Method for Electrical Indication of Concrete's Ability to Resist Chloride Ion Penetration. ASTM International: West Conshohocken, PA, USA, 2022.
18. *NT BUILD 492 (1999)*; Concrete, Mortar and Cement-Based Repair Materials: Chloride Migration Coefficient from Non-steady-state Migration Experiments. NORDTEST: Finland, Norway, 1999.
19. Zic, M.; Pereverzyev, S. Optimizing noisy CNLS problems by using Nelder-Mead algorithm: A new method to compute simplex step efficiency. *J. Electroanal. Chem.* **2019**, *851*, 113439. [CrossRef]
20. Shen, L.; Zhang, H. Corrosion inhibition and adsorption behavior of (3-aminopropyl)-triethoxysilane on steel surface in the simulated concrete pore solution contaminated with chloride. *J. Mol. Liq.* **2022**, *363*, 119896. [CrossRef]
21. Yang, H.; Xiong, C.; Liu, X.; Liu, A.; Li, T.; Ding, R.; Shah, S.P.; Li, W. Application of layered double hydroxides (LDHs) in corrosion resistance of reinforced concrete-state of the art. *Constr. Build. Mater.* **2021**, *307*, 124991. [CrossRef]
22. Biondi, L.; Perry, M.; McAlorum, J.; Vlachakis, C.; Hamilton, A. Geopolymer-based moisture sensors for reinforced concrete health monitoring. *Sens. Actuators B Chem.* **2020**, *309*, 127775. [CrossRef]
23. Rodrigues, R.; Gaboreau, S.; Gance, J.; Ignatiadis, I.; Betelu, S. Reinforced concrete structures: A review of corrosion mechanisms and advances in electrical methods for corrosion monitoring. *Constr. Build. Mater.* **2021**, *269*, 121240. [CrossRef]
24. Criado, M.; Sobrados, I.; Bastidas, J.M.; Sanz, J. Corrosion behaviour of coated steel rebars in carbonated and chloride-contaminated alkali-activated fly ash mortar. *Prog. Org. Coat.* **2016**, *99*, 11–22. [CrossRef]
25. Zhou, X.; Yang, H.; Wang, F. [BMIM]BF4 ionic liquids as effective inhibitor for carbon steel in alkaline chloride solution. *Electrochim. Acta* **2011**, *56*, 4268–4275. [CrossRef]
26. Huyen-Vu, T.; Chi-Dang, L.; Kang, G.; Sirivivatnanon, V. Chloride induced corrosion of steel reinforcement in alkali-activated slag concretes: A critical review. *Case Stud. Constr. Mater.* **2022**, *16*, e01112. [CrossRef]
27. Kaur, H.; Singla, S. Non-Destructive testing to detect multiple cracks in reinforced concrete beam using electromechanical impedance technique. *Mater. Today Proc.* 2022; *in press*.
28. Liu, P.; Wang, W.; Chen, Y.; Feng, X.; Miao, L. Concrete damage diagnosis using electromechanical impedance technique. *Constr. Build. Mater.* **2017**, *136*, 450–455. [CrossRef]
29. Talakokula, V.; Bhalla, S.; Gupta, A. Monitoring early hydration of reinforced concrete structures using structural parameters identified by piezo sensors via electromechanical impedance technique. *Mech Syst. Signal Process* **2018**, *99*, 129–141. [CrossRef]
30. Da Silva, G.F.; Martini, S.; Moraes, J.C.B.; Teles, L.K. AC impedance spectroscopy (AC-IS) analysis to characterize the effect of nanomaterials in cement-based mortars. *Constr. Build. Mater.* **2021**, *269*, 121260. [CrossRef]
31. Bai, R.; Zhang, J.; Yan, C.; Liu, S.; Wang, X.; Yang, Z. Calcium hydroxide content and hydration degree of cement in cementitious composites containing calcium silicate slag. *Chemosphere* **2021**, *280*, 130918. [CrossRef]
32. Ibrahim, M.; Kalimur-Rahman, M.; Megat-Johari, M.A.; Nasir, M.; Adeoluwa-Oladapo, E. Chloride diffusion and chloride-induced corrosion of steel embedded in natural pozzolan-based alkali-activated concrete. *Constr. Build. Mater.* **2020**, *262*, 120669. [CrossRef]
33. Aguirre-Guerrero, A.M.; Mejía de Gutiérrez, R. Alkali-activated protective coatings for reinforced concrete exposed to chlorides. *Constr. Build. Mater.* **2021**, *268*, 121098. [CrossRef]
34. Thomas, R.J.; Ariyachandra, E.; Lezama, D.; Peethamparan, S. Comparison of chloride permeability methods for Alkali-Activated concrete. *Constr. Build. Mater.* **2018**, *165*, 104–111. [CrossRef]
35. Mehta, A.; Siddique, R.; Ozbakkaloglu, T.; Ahmed-Shaikh, F.U.; Belarbi, R. Fly ash and ground granulated blast furnace slag-based alkali-activated concrete: Mechanical, transport and microstructural properties. *Constr. Build. Mater.* **2020**, *257*, 119548. [CrossRef]
36. Zhang, J.X.; Ma, Y.; Zheng, J.; Hu, J.; Fu, J.; Zhang, Z.; Wang, H. Chloride diffusion in alkali-activated fly ash/slag concretes: Role of slag content, water/binder ratio, alkali content and sand-aggregate ratio. *Constr. Build. Mater.* **2020**, *261*, 119940. [CrossRef]
37. Ma, Y.; Hu, J.; Ye, G. The pore structure and permeability of alkali-activated fly ash. *Fuel* **2013**, *104*, 771–780. [CrossRef]
38. Hu, X.; Shi, C.; Shi, Z.; Zhang, L. Compressive strength, pore structure and chloride transport properties of alkali-activated slag/fly ash mortars. *Cem. Concr. Compos.* **2019**, *104*, 103392. [CrossRef]
39. Gao, X.; Yao, X.; Wang, C.; Geng, C.; Yang, T. Properties and microstructure of eco-friendly alkali-activated slag cements under hydrothermal conditions relevant to well cementing applications. *Constr. Build. Mater.* **2022**, *318*, 125973. [CrossRef]
40. Runci, A.; Serdar, M. Effect of curing time on the chloride diffusion of alkali-activated slag. *Case Stud. Constr. Mater.* **2022**, *16*, e00927. [CrossRef]
41. Sun, X.; Liu, J.; Qiu, J.; Wu, P.; Zhao, Y. Alkali activation of blast furnace slag using a carbonate-calcium carbide residue alkaline mixture to prepare cemented paste backfill. *Constr. Build. Mater.* **2022**, *320*, 126234. [CrossRef]

42. Moussadik, A.; Saadi, M.; Diouri, A. Chemical, mineralogical and thermal characterization of a composite alkali-activated binder based on coal gangue and fly ash. *Mater. Today Proc.* **2022**, *58*, 1452–1458. [CrossRef]
43. Jae-Kim, M.; Ik-Hwang, W.; Jung-Cho, W. The influence of alkali activators on the properties of ternary blended cement incorporated with ferronickel slag. *Constr. Build. Mater.* **2022**, *318*, 126174. [CrossRef]

metals

MDPI

Article

Biofilm Development on Carbon Steel by Iron Reducing Bacterium *Shewanella putrefaciens* and Their Role in Corrosion

Sachie Welikala [1], Saad Al-Saadi [1,2], Will P. Gates [3], Christopher Panter [4] and R. K. Singh Raman [1,2,*]

1 Department of Chemical and Biological Engineering, Monash University, Clayton, VIC 3800, Australia; swelikala@hotmail.com (S.W.); saad.al-saadi@monash.edu (S.A.-S.)
2 Department of Mechanical and Aerospace Engineering, Monash University, Clayton, VIC 3800, Australia
3 Melbourne-Burwood Campus, Institute for Frontier Materials, Deakin University, Burwood, VIC 3125, Australia; will.gates@deakin.edu.au
4 CP Microbiology and Analytical Laboratories, 677 Springvale Rd., Mulgrave, VIC 3170, Australia; services@cpmicro.com.au
* Correspondence: raman.singh@monash.edu

Abstract: Microscopic, electrochemical and surface characterization techniques were used to investigate the effects of iron reducing bacteria (IRB) biofilm on carbon steel corrosion for 72 and 168 h under batch conditions. The organic nutrient availability for the bacteria was varied to evaluate biofilms formed under nutritionally rich, as compared to nutritionally deficient, conditions. Focused ion beam-scanning electron microscopy (FIB-SEM) was used to investigate the effect of subsurface biofilm structures on the corrosion characteristics of carbon steel. Hydrated biofilms produced by IRB were observed under environmental scanning electron microscope (ESEM) with minimal surface preparation, and the elemental composition of the biofilms was investigated using energy dispersive spectroscopy (EDX). Attenuated total reflectance-Fourier transform infrared spectroscopy (ATR-FTIR) was used to provide information on the organic and inorganic chemical makeup of the biofilms. Electrochemical techniques employed for assessing corrosion, by open circuit potential, linear polarization and potentiodynamic polarization tests indicated no significant difference in the corrosion resistance for carbon steel in IRB-inoculated, compared to the abiotic solutions of common Postgate C after 72 and 168 h. However, the steel was found to be more susceptible to corrosion when the yeast extract was removed from the biotic environment for the 168 h test. In the absence of yeast nutrient, it is postulated that IRB received energy by transforming the protective film of Fe^{3+} into more soluble Fe^{2+} products.

Keywords: Iron reducing bacteria (IRB); focused ion beam-scanning electron microscopy (FIB-SEM); environmental scanning electron microscopy (ESEM); microbiologically influenced corrosion

Citation: Welikala, S.; Al-Saadi, S.; Gates, W.P.; Panter, C.; Raman, R.K.S. Biofilm Development on Carbon Steel by Iron Reducing Bacterium *Shewanella putrefaciens* and Their Role in Corrosion. *Metals* **2022**, *12*, 1005. https://doi.org/10.3390/met12061005

Academic Editor: Sebastian Feliú, Jr.

Received: 8 May 2022
Accepted: 8 June 2022
Published: 12 June 2022

1. Introduction

Microbiologically influenced corrosion (MIC) is the term designated to the acceleration of a corrosion process due to the activities of microorganisms where the electrochemical reactions are enhanced because of bacterial metabolic products [1–4]. Microorganisms contribute to corrosion processes by either direct influence on the rate of anodic or cathodic reactions or by changing surface film characteristics due to biofilm formation [4]. The structural and chemical heterogeneities of the biofilm most commonly causes localized corrosion [5,6]. Moreover, due to the influence of microorganisms and their metabolic processes, corrosion may be observed in unlikely conditions such as anoxic [7] or low chloride environments where corrosion would otherwise not occur [8].

Iron reducing bacteria (IRB) derive energy from the reduction of ferric ions (Fe^{3+}) using either the electron transport chain in anaerobic respiration or via fermentation [9]. Influence of IRB on corrosion is a less explored topic, and the findings in the reported studies are conflicting, with some observing corrosion protection and others showing corrosion acceleration in presence of IRB [10–12]. Such inconsistencies are due to difference in species of IRB or the composition of electrolyte used in laboratory experiments [10]. In the limited studies on the topic, the corrosion acceleration due to IRB is attributed to the bacteria's ability to produce soluble ferrous irons through reductive dissolution of the protective corrosion film of insoluble ferric iron oxides on the steel surface, thereby exposing the alloy surface to aggressive ions in the environment [10,12,13]. However, a few laboratory studies have reported the protective effect of IRB [14,15]. Lee and Newman reported that even if biofilm coverage of the metal is incomplete, cumulative O_2 respiration from both planktonic and sessile cells rapidly depletes O_2 on the metal surface, thus inhibiting the galvanic process and corrosion [16]. In fact, Dubiel et al., reported that mutant S. oneidensis sp. could be used as a potential method to control corrosion in pipeline systems as the bacteria may colonise the metal surface and aerobically consume oxygen molecules adjacent to the metal surface [15]. As oxygen is depleted, the facultative anaerobic IRB turn to Fe^{3+} reduction and produce Fe^{2+} ions which diffuse into the bulk fluid. This process has been reported to create a chemical shield that further reduces oxygen diffusion, which in turn may inhibit the cathodic reaction due to lower oxygen availability. By electrochemical reaction, Fe^{2+} is oxidized to Fe^{3+} and is available again for reduction by bacterial respiration [17,18].

IRB Shewanella genera are relevant to corrosion as they are among the most common bacteria in oilfield injection water [19]. The bacteria has the ability to reductively dissolve iron oxyhydroxide minerals (that are generally insoluble at neutral pH [20–23]) into soluble ferrous ion or secondary mineralised Fe(II) oxides such as magnetite (Fe_3O_4), siderite ($FeCO_3$), vivianite ($Fe(PO_4)_2$) and green rusts [21,24–26].

The objective of this work is to characterise the IRB biofilm on carbon steel, and its effect on corrosion characteristics. In particular, the biofilms were manipulated by varying the organic carbon nutrients available to the bacteria so that the biofilms developed under: (i) high organic nutritional conditions such as the commonly used Postgate C medium; (ii) where organic nutrients were moderately decreased by the removal of chemically undefined nutrient yeast extract (modified Postgate C medium) and (iii) where the biofilm formed under starvation conditions, where all organic nutrients were removed (i.e., the inorganic medium). Biofilms formed by a representative IRB (*Shewanella putrefaciens*) were investigated using microscopic observation, chemical characterisation and electrochemical techniques. Whether biofilms accelerate or inhibit corrosion processes on carbon steel is highly dependent on environmental conditions available to the bacteria or by the ability of the bacteria to alter the environment. The biofilms formed on carbon steel were investigated for 72 h and 168 h in pure culture of IRB. Open circuit potential (OCP), linear polarization (LP) and potentiodynamic polarization (PDP) tests were used for characterization of corrosion. Environmental scanning electron microscopy (ESEM) was used to investigate the morphologies of the biofilm on the carbon steel specimen. To the best of authors' knowledge this is among the first comprehensive characterisation of cross-sections of biofilms and MIC using focused ion beam-scanning electron microscopy (FIB-SEM) that enabled investigation of the sub-surface features of the biofilm, particularly at the locations of specific surface features recognised by SEM. Attenuated Fourier transformation infrared spectroscopy (ATR-FTIR) was used, alongside microscopic and electrochemical techniques, to observe the extracellular polymeric substances (EPS) produced by IRB.

2. Materials and Methods

2.1. Maintenance of Bacterial Culture

Pure bacterial culture of IRB (*S. putrefaciens*) was grown in test tubes under aerobic conditions in an amended B7 medium at a mesophilic temperature of 30 °C. The composition of the amended B7 medium (henceforth referred to as B7) was (/L): 0.8 g K_2HPO_4, 0.2 g KH_2PO_4, 0.2 g $MgSO_4$, 5 g yeast extract, 5 g peptone, 10 g ferric ammonium citrate, 10 mL $CaSO_4$ (saturated solution). The pH was adjusted to 7.5–7.8 using NaOH solution. All lab glassware pipettes and solutions were autoclaved at 121 °C for 15 min to avoid contamination of the enrichment cultures. The cultures were periodically renewed to keep them active and healthy. A 1 mL sample of a three-day old inoculum was taken from the culture for biofilm development and electrochemical experiments.

2.2. Test Solutions Preparation

Three recipes of Postgate C medium were prepared for biofilm development and corrosion characterization (i.e., commonly used Postgate C, modified Postgate C and inorganic Postgate C solutions). Table 1 shows the composition of the three Postgate C used in the present study. The modified Postgate C medium was prepared by removing the yeast extract from the medium to moderately decrease the amount of organic nutrients available to the bacteria. The inorganic medium was prepared by completely removing all organic ingredients in the medium in order to study the behaviour of the bacteria under starvation conditions. The pH of the culture medium was adjusted to 7.5–7.8 using sterilised NaOH solution and all solutions were autoclaved to avoid contamination. No iron component was included in the test media for electrochemical testing or for biofilm study. The iron in solution that originated from the dissolution of the carbon steel specimen was measured from the 1, 10 phenanthroline method.

Table 1. Chemical composition of the commonly used Postgate C [26], modified Postgate C and inorganic Postgate C solutions.

Chemicals	Postgate C	Modified Postgate C	Inorganic Postgate C
Na_2SO_4 [a]	4.5	4.5	4.5
$CaCl_2{\cdot}2H_2O$ [a]	0.06	0.06	0.06
Lactic acid [b]	4.8	4.8	-
Sodium citrate [a]	0.3	0.3	-
NH_4Cl [a]	1.0	1.0	1.0
K_2HPO_4 [a]	0.5	0.5	0.5
$MgSO_4{\cdot}7H_2O$ [a]	2.0	2.0	2.0
Yeast extract [a]	1.0	-	-

[a] Concentration in g/L. [b] Concentration in mL/L.

2.3. Carbon Steel Sample Preparation

For electrochemical test, the preparation of steel specimens is described in detail in [27–29]. A copper wire was soldered to a square specimen of the carbon steel. The wire was encased in a glass tube to prevent any contact of the wire with the electrolyte. Each specimen was then mounted in epoxy resin so that only one face of the sample was exposed to the test solution. The surface was wet ground with emery paper up to 1200 grit and then polished with diamond paste up to 3 μm. The edges of the resin and the metal were painted with enamel coating to avoid crevice corrosion [27–30]. For samples used for microscopic investigation, a fishing line was used to hang the sample in the test solution. The samples were ultrasonically cleaned for 15 min in ethanol prior to being used as test electrodes in electrochemical experiments or in immersion tests for biofilm characterization.

2.4. Iron Quantification

Due to the importance of ferrous and ferric ion concentrations on the corrosion characteristics of carbon steel, the total iron (Fe^{2+} and Fe^{3+}) concentration in this study was determined using a colorimetric technique using 1, 10-phenanthroline according to the procedure outlined in reference [31].

2.5. Microscopic and Biofilm Cross-Sectional Analysis

2.5.1. Development of Biofilms on Carbon Steel for Microscopic Imaging and FIB-SEM Analysis

The carbon steel specimens, mounted in epoxy resin (as described in Section 2.3) were attached to a polypropylene fishing line, and immersed in 250 mL of sterilized test solutions in a 250 mL Duran bottle. Using aseptic techniques, 1 mL of bacterial culture was taken from a 3-day old stock culture of IRB and inoculated into the Duran bottle as pure culture. The solution was not de-aerated. An abiotic control was also set up under the same conditions. The carbon steel specimen was exposed to the bacterial culture (or abiotic conditions) for a period of 72 h or 168 h at 30 °C. At least two samples were observed for each condition in order to examine reproducibility.

2.5.2. Biofilm Fixation and Environmental Scanning Electron Microscopy (ESEM) Imaging

The morphology of the biofilm on the carbon steel specimen was observed using the 3D Quanta FEI in ESEM mode. Biofilms were fixed using 2% glutaraldehyde solution for 1 h and then subject to two washes in deionised water, each wash for 5 min. The use of relatively high pressures (~650 Pa, or 1 mbar) and water vapor in the specimen chamber in the ESEM mode allowed the samples to be viewed in a hydrated state, and without the need for metallic coating, as is often required in SEM sample preparation which introduces artifacts. Minimum of two samples were observed under ESEM for each condition.

2.5.3. Focused Ion Beam—Scanning Electron Microscopy (FIB-SEM)

FIB-SEM technique was used to investigate sub-surface features of the biofilm. FIB milling was carried out using the FEI Quanta 3D FEG instrument. As this instrument also has SEM and EDS capabilities, the subsurface structure could be viewed, and elemental composition analysed following the milling. Due to the sample being placed in a high vacuum chamber (~10^{-6} Pa), in addition to the sample preparation steps detailed in Section 2.5.2 for ESEM, the sample was also dehydrated and sputter coated with 3 nm of platinum coating in order to avoid charging. The surface of the specimen was first viewed under SEM to locate a site of interest in the biofilm for subsequent cross-sectional analysis. The site chosen for FIB milling was typical and representative for the condition under study. A platinum strip, 1 µm in width and 1 µm in thickness, was applied across the length of the area of interest in order to protect the biofilm from ion beam degradation during the milling process [32]. The sample was then tilted to 52° so that the milling could be performed using current gallium (Ga^{+}) beam (in the range of 5 nA). Beam currents were lowered (in the pA range) for subsequent cleaning of cross-sections to remove material re-deposited on the area during the initial rough milling.

2.5.4. Attenuated Total Reflectance—Fourier Transform Infrared (ATR-FTIR) Spectroscopy

The specimens exposed to biotic and abiotic conditions were removed after 72 h and any loosely attached bacterial cells were rinsed off with phosphate buffer solution. The specimens were left in a desiccator for at least 24 h to dehydrate. FTIR spectra were obtained using a Perkin Elmer Spectrum 100 series spectrometer. The spectral acquisition (128 coadded scans at 8 cm^{-1} of spectral resolution in the 4000 cm^{-1} to 600 cm^{-1} range).

2.6. Electrochemical Measurements

Linear polarization (LP) and potentiodynamic polarization (PDP) were carried out using a Princeton Applied Research Potentiostat (model: 2273) and a three-electrode electrochemical cell. The test cell was a 250 mL Duran bottle filled with 250 mL of the test solution. The design of the electrochemical cell is described in [27–29]. The steel specimens were used as the working electrodes, saturated calomel electrode (SCE) was used as the reference electrode and a platinum mesh was used as the counter electrode. For the biotic electrochemical experiments, 1 mL of inoculum from the maintenance culture was introduced into the test solution. Abiotic control experiments were also conducted alongside the biotic experiments. All the measurements were repeated at least thrice to examine reproducibility. Open circuit potential (OCP) was monitored to confirm its stability with time. For LP test, the working electrode was polarized ±10 mV from E_{corr} at a scan rate of 0.166 mV/s. Potentiodynamic polarization was carried out at a scan rate of 1 mV/s, starting at a potential of 400 mV more negative to the OCP.

3. Results and Discussion

3.1. Iron and Sulfide Measurements in Solution

The B7 medium was used as the enrichment medium for maintaining the IRB culture. This medium is designed especially for the IRB [33], and therefore, contains high contents of ferric ammonium citrate (10 g/L) and yeast extract (5.0 g/L). Figure 1 follows the growth of the IRB in the B7 medium over a 48-day period. Initial colour, at the time of IRB inoculation (at day 0) is brown. This colour gradually changes to dark green, which indicates the ability of IRB to reduce Fe^{3+} ions from ferric ammonium citrate to the ferrous ions (Fe^{2+}) [33]. From day 3 onwards, a gelatinous substance was observed to have deposited at the bottom of the test tube (though not clearly visible in Figure 1). This precipitate was sticky and difficult to remove from the test tubes, at the end of experiments.

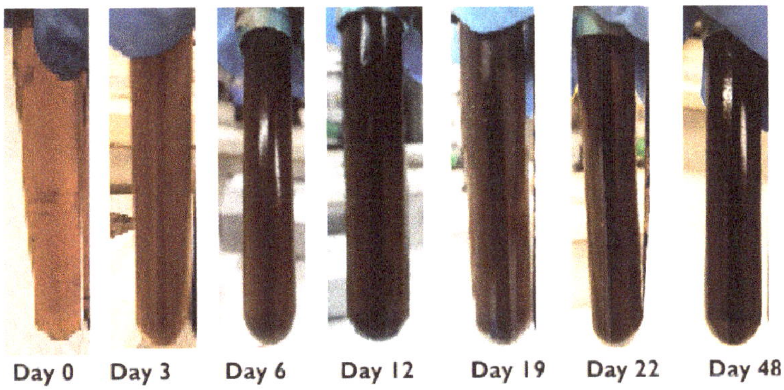

Figure 1. Growth of the IRB, *Shewanella putrefaciens* in B7 medium.

Dissolved sulfide ions were not detected in the solution in the B7 or Postgate C medium containing IRB when tested spectrometrically, using the $CuSO_4$ technique. This confirmed that the bacterium used in this particular study was unable to reduce sulfate, as also reported in the literature [34].

Figure 2 shows quantification of total iron (Fe^{2+} and Fe^{3+}) in the modified Postgate C medium in the presence of the IRB and carbon steel specimen. As suggested, the concentration of iron increased over time in the presence of the IRB, form 2.2 mg/L after 72 h to 21 mg/L after 168 h. The IRB is known to reduce insoluble Fe(III) oxides such as goethite, hematite, ferrihydrite, akaganéite, and lepidocrocite into soluble ferrous complexes [12,21,35] and this is the most likely reason for the much higher dissolved iron concentration in biotic solution containing IRB.

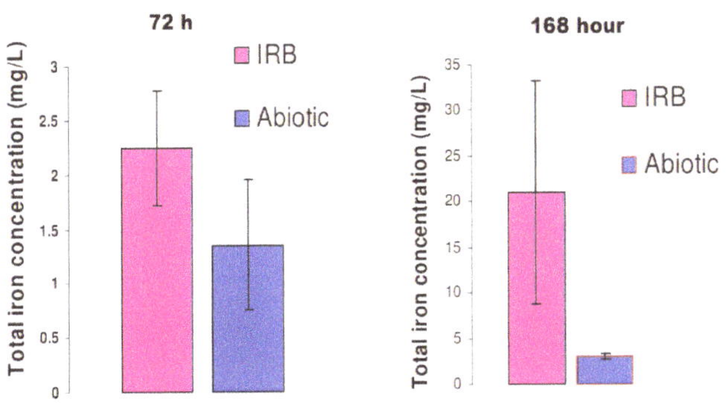

Figure 2. Quantification of total iron (Fe^{2+} and Fe^{3+}) in solution after 72 and 168 h in modified Postgate C medium in the presence of the IRB and carbon steel specimen.

3.2. Biofilm Development and Electrochemical Characterization of Carbon Steel in Biotic Postgate C Solution

3.2.1. Biofilm Development in Commonly Used Postgate C Solution with IRB

Following 72 h exposure of carbon steel to the IRB in the Postgate C medium, the specimen was covered by EPS-like material (Figure 3). Figure 3a shows a globular deposit of ~2 μm diameter. Much smaller globular precipitates can be observed in Figure 3b which are packed close together. Pits covered by loose corrosion product cap, as shown in Figure 3c,d, were also a common occurrence throughout the sample surface.

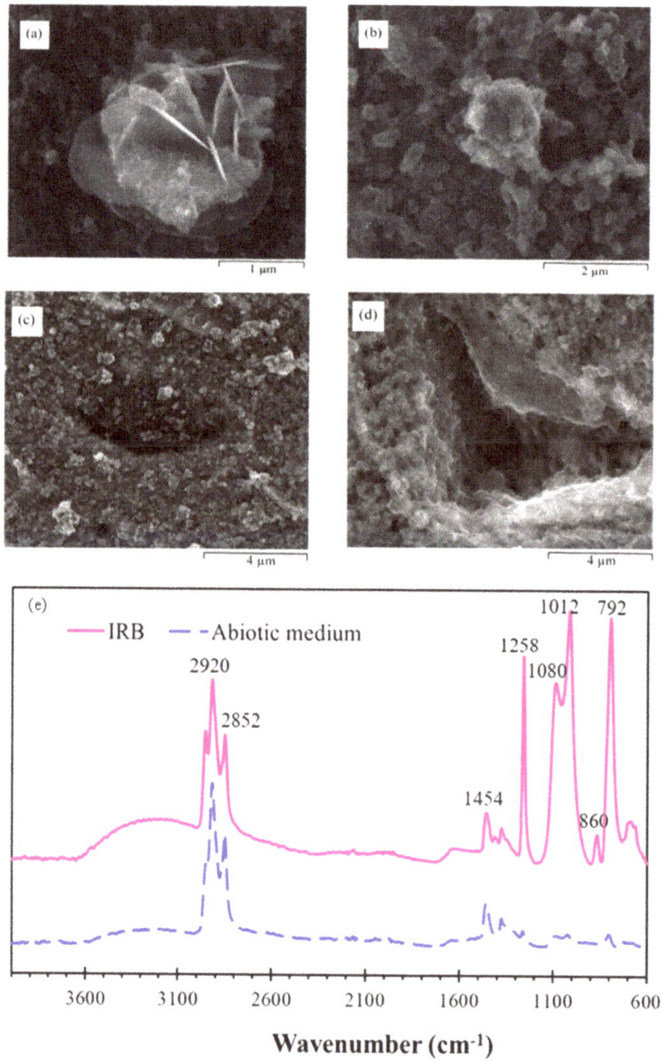

Figure 3. ESEM images of hydrated biofilm formed on carbon steel by a pure IRB culture following 72 h exposure to Postgate C medium, (**a**,**b**) reduced Fe(II) oxide species, (**c**) sites of pitting attack under loose cap of corrosion product and (**d**) pit initiated at a MnS inclusion site. (**e**) IR spectra of IRB biofilm formed on carbon steel upon to exposure to pure IRB culture in Postgate C medium and abiotic Postgate C for a period of 72 h.

The IRB is capable of reducing minerals such as amorphous hydrous ferric oxide (HFO), goethite, lepidocrocite or ferrihydrite into mixed Fe(III)/Fe(II) species [25]. Little et al. [12] also observed similar globular precipitates when *S. putrefaciens* was grown on top of synthetic goethite or ferrihydrite minerals. EDS carried out on these globular particles following 190 h exposure showed that the composition contained iron, phosphorous and magnesium, suggesting that these particles were residual goethite particles [12]. In the current study, EDS could not be used to distinguish between the chemical compositions of different particles in the 72 h IRB biofilm due to the thickness of the film being less than the interaction volume of the EDS beam. Framboidal structures of Fe minerals are very much associated with the biogenic formation. It has to do with the external surface to total volume ratio of a sphere being the most stable condition thermodynamically.

Uncovered, irregular shaped pits (Figure 3d) were also observed and are likely to be initiated due to dissolution of inclusions. The site of the inclusion at the centre of the pit appears to have completely dissolved and pitting has progressed to a greater extent than the micro-pits observed in the abiotic control. ESEM analysis could not confirm the presence or absence of the IRB bacterial cells on the biofilm. Individual bacterial cells are difficult to image using ESEM as bacteria are mostly composed of water and low atomic number elements which are not very electron dense. It is also likely that the IRB cells are unrecognizable due to being encrusted in the iron oxides and/or biofilm. Direct contact between the IRB and the Fe(III) oxide, however, may not always be necessary, as numerous geomicrobiological studies have reported on the ability of *S. putrefaciens* to reduce Fe(III) from a distance without direct contact with the Fe(III) under anaerobic conditions [36,37]. Lies et al. [37] observed that S. oneidensis reduced at least 86.5% of iron from an "iron-bead" system in the absence of direct contact when under nutritionally rich conditions. This iron reduction from a distance is thought to be carried out through ferric iron chelators or extracellular electron shuttles [36–38]. This is not, however, in agreement with all studies, as some authors have reported that iron reduction did not occur when direct contact between the bacterial cells and the iron oxyhydroxides was impeded [10,12].

Figure 3e depicts IR spectrum of IRB biofilm developed on carbon steel exposed to pure culture of IRB in Postgate C medium for a period of 72 h, with the spectrum of abiotic Poctgate C for comparison. Hydrogen and fatty acids are primary products of IRB metabolism [39] and the peaks corresponding to these fatty acids can be clearly identified for the IRB biofilm. Characteristic peaks corresponding to -CH_3 asymmetric stretching (2956 cm^{-1}), asymmetric -CH_2 stretching (2920 cm^{-1}) and symmetric -CH_2 stretching (2852 cm^{-1}) were observed confirming the presence of organic hydrocarbon chains in the IRB biofilm. Complementary bands for these hydrocarbons were found at 1454 cm^{-1} (asymmetric -CH_3 bending/CH_2 scissoring) as well as -CH_3 symmetric bending at 1372 cm^{-1}.

The broad peak at 3200 cm^{-1} is considered to be due to -OH stretch of Fe(III) oxyhydroxides, which are the primary inorganic corrosion product of steel. IRB dissolves Fe(III) oxyhydroxide minerals into Fe(II) oxides as part of anaerobic respiration. These OH groups detected in the IR spectra are attributed to either stoichiometric -OH in iron oxyhydroxide species such as goethite, lepidocrocite, ferrihydrite, but could also be surface bound water present in the iron hydroxide minerals [40]. The broad peak detected at 3200 cm^{-1} in this study could also be due to the formation of the mixed ferric/ferrous species, green rust, due to the bacteria's incomplete reduction of ferric oxyhydroxide and oxide minerals as also observed in other studies in the presence of *S. putrefaciens* [21,41]. Magnetite may also form in the solution if deficient in Fe^{2+} ions as this mineral has smaller Fe(II)/Fe(III) ratio [25,41]. The green rusts precipitated due to IRB bio-mineralization may abiotically re-mineralize into black magnetite precipitates with time [24].

The most intense peaks detected in the IR spectra for the IRB biofilm correspond to the stretching of the phosphate ion. High intensity peaks for asymmetric PO^{2-} stretch at 1258 cm^{-1} and symmetric PO^{2-} stretching at 1080 cm^{-1} were observed. The P-O asymmetric stretch at 860 cm^{-1} was also detected as well as a high intensity peak for P-O symmetric stretch at 792 cm^{-1}. A high intensity vibration at 1012 cm^{-1} was observed in the saccharide region which may correspond to C-O-C, C-O-P, P-O-P ring vibrations of carbohydrates [42–44]. Specific identification is difficult in the saccharide region due to the complex superpositions of the characteristic absorptions of various polysaccharides with the various organophosphates present. The cell wall of gram-negative bacteria, such as *S. putrefaciens*, consists of an outer membrane containing lipopolysaccharides (LPS) [25,45] as well as a periplasm with a thin peptidoglycan layer [45]. Therefore, the C-O-P and P-O-P vibrations may be due to polysaccharides of cell wall. LPS are known to be highly anionic and would therefore contribute to binding of cations such as Fe^{2+} ions from the carbon steel due to electrostatic interactions [24]. Omoike and Chorover [46] have reported that phosphodiester groups of nucleic acids in the EPS may also facilitate the binding of the EPS onto iron oxyhydroxide minerals.

The FIB-SEM cross-sections (Figure 4a,b) show two separate pits covered by a loose corrosion product crust such as that shown in Figure 3c. The depth of these pits was approximately 1.5 µm. The biofilm covering the pit was loosely attached to the corroded carbon steel. The innermost layer was composed of initial layer of colloidal corrosion products. The loose attachment between the metal and biofilm material covering the pit allows chemical species from electrolyte to come in contact with the metal to facilitate pit propagation. High organic material in the IRB biofilm caused charging and sample movement when FIB milling was being carried out. This difficulty in obtaining clean cross-sections from FIB milling only occurred in the IRB pure culture biofilm, due to the higher organic EPS material in the biofilm which were not electron dense.

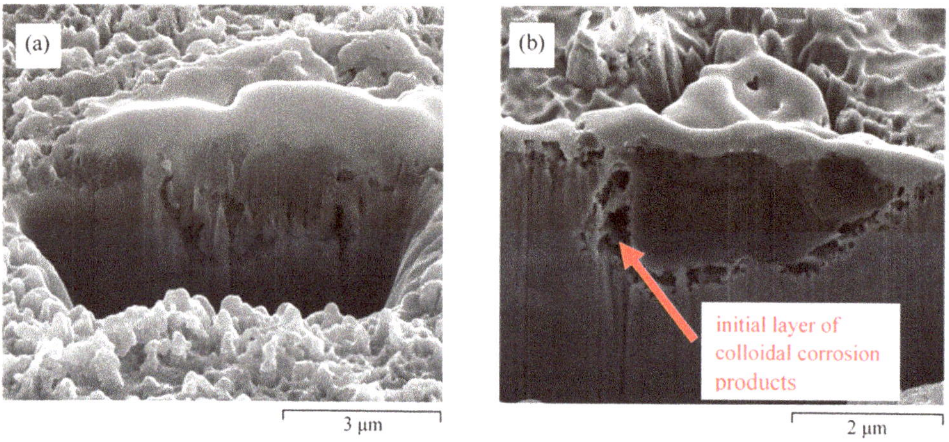

Figure 4. (**a**,**b**) FIB-SEM cross-sections of the biofilm formed on carbon steel and corrosion pitting due to pure culture of IRB in Postgate C medium following a 72 h exposure.

Figure 5a,b show ESEM and SEM images of biofilm formed on carbon steel exposed for 168 h to Postgate C medium inoculated with pure IRB culture. Polishing marks were visible on some areas of the carbon steel surface, indicating a low corrosion rate in considerable parts of the specimen (Figure 5a). The surface was covered with a thin and porous biofilm, with corrosion products scattered throughout the sample when observed under ESEM (Figure 5b). EDS analysis of the biofilm corresponding to the ESEM image showed iron and carbon to be the main elements detected in the biofilm as shown in Figure 5c. The detection of carbon is most likely to be due to the EPS material, which is high in hydrocarbons from fatty acids. Alongside the flat areas observed in ESEM analysis, thicker corrosion deposits also occurred throughout the sample. One such deposit was chosen for the FIB-SEM cross-sectional characterization (Figure 5d,e). At high magnification, the thin biofilm beside the deposit showed initial signs of undercutting attack as the metal was dissolved from underneath.

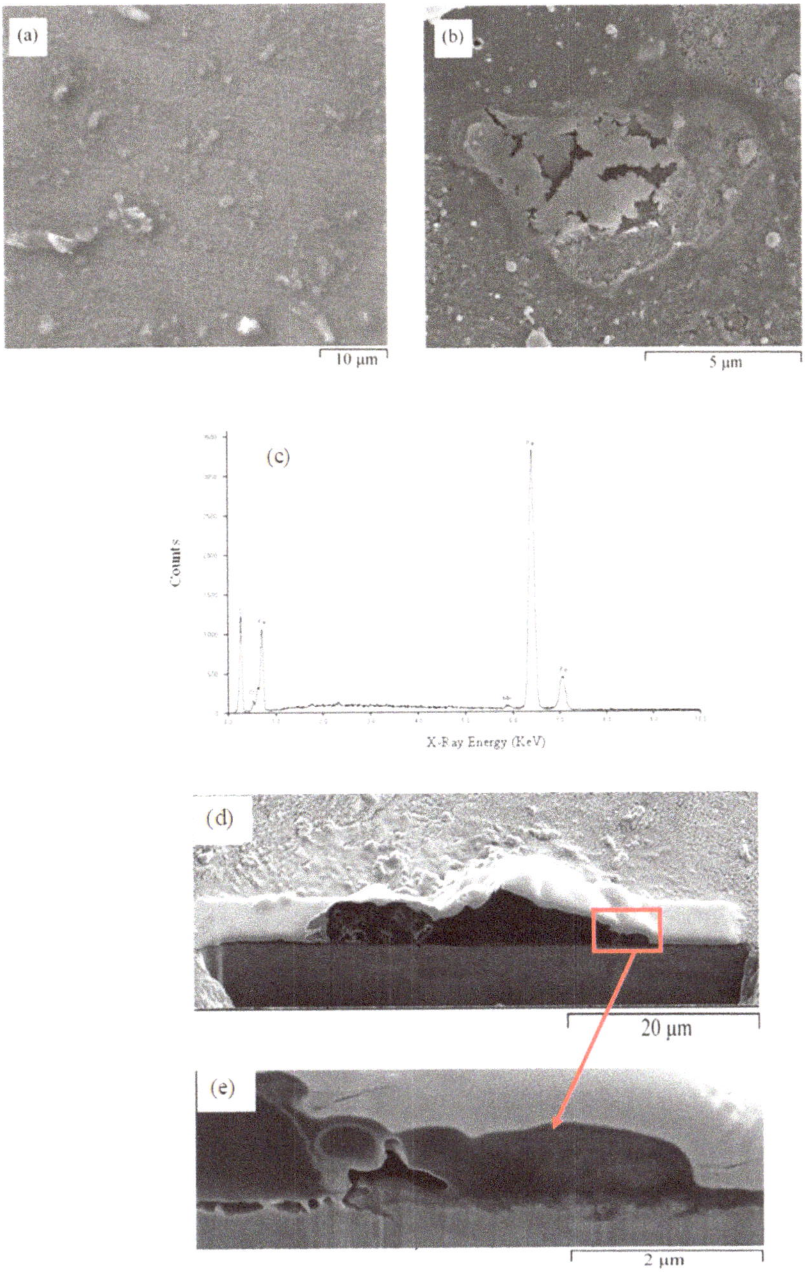

Figure 5. (**a**) ESEM and (**b**) SEM images of biofilm formed on carbon steel by a pure IRB culture (168 h exposure time to Postgate C medium). (**c**) EDS analysis of the biofilm corresponding to the ESEM image. (**d**) FIB-SEM images of the cross-section and (**e**) initial layer of fibrous EPS material produced by IRB.

3.2.2. Electrochemical Behaviour of Carbon STEEL in Postgate C Medium Inoculated with IRB

Figure 6a shows the polarization resistance (R_p) of mild steel exposed during the duration of exposure of 72 h to the Postgate C medium inoculated with IRB culture in comparison to that exposed to the abiotic Postgate C solution. R_p increased from 890 Ω/cm^2 at 0 h to be 7676 Ω/cm^2 at 24 h, presumably due to the removal of oxygen by the IRB as IRB respiration removes oxygen, i.e., an expected cathodic reactant for steel corrosion. From 24 h to 72 h, the R_p remained steady. The R_p of the carbon steel exposed to the IRB had a slightly lower R_p value from around 24 h to 72 h reading than the R_p for abiotic control. This indicated that the IRB biofilm may have somewhat decreased the corrosion protection offered by the abiotic corrosion products, by dissolving iron oxyhydroxides and/or forming porous EPS and reduced Fe(II) oxide species.

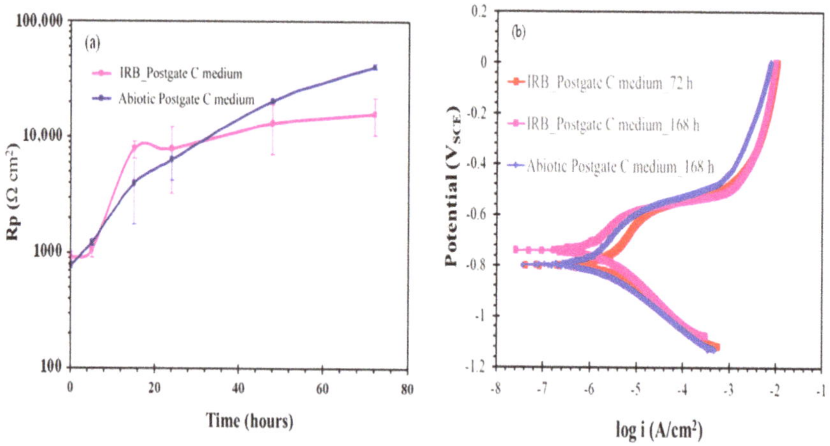

Figure 6. (**a**) Polarization resistance (**a**) of carbon steel exposed to a pure IRB culture and abiotic conditions in Postgate C medium, and (**b**) a potentiodynamic polarization and (**b**) scan of carbon steel after 72 h exposure with and without IRB.

Figure 6b shows potentiodynamic polarization curves of the steel samples pre-exposed to the Postgate C solution with IRB for 72 and 168 h. The polarization curve of the steel sample pre-exposed to the abiotic solution for 168 h is shown for comparison. There was no significant difference in the anodic or cathodic regime of the carbon steel exposed to IRB for 72 h or 168 h or to abiotic condition for 168 h. The E_{corr} of the steel samples exposed to the biotic solution is shifted about 50 mV towards positive direction. In all cases, E_{pit} appeared to be around −550 mV, where a sudden and rapid increase in anodic current was observed. The IRB biofilm consisting of Fe(II)/Fe(III) oxides and EPS did not offer any resistance to anodic dissolution.

3.3. *Effect of Removal of Yeast Extract from Postgate C (Modified Postgate C) on IRB Activity and Biofilm Development*

3.3.1. Biofilm Development in Modified Postgate C Solution with IRB

Exclusion of yeast extract from the culture medium in the modified Postgate C medium resulted in less coverage of the steel specimen with EPS and biominerals. However, as will be evident from subsequent description, the carbon steel surface exposed to the modified Postgate C medium for 72 h confirmed that IRB metabolized and reduced Fe^{3+} ions even in the absence of the yeast extract. Lovely et al. [39] have also reported that yeast extract was not essential for the growth and anaerobic metabolism of *S. putrefaciens*.

Figure 7 presents ESEM analysis of hydrated biofilm formed on carbon steel by a pure IRB culture, following 72 h exposure in modified Postgate C medium. Most parts of the carbon steel surface had a very flat, uniformly corroded appearance (Figure 7a). However, as seen in Figure 7b, a higher magnification reveals the hollow iron oxyhydroxide shell. Lepidocrocite crystals in Figure 7c appear to be a mixture of crystalline and amorphous or poorly crystalline material. O'Loughlin has observed through time-resolved XRD that lepidocrocite was reduced to green rust by *S. putrefaciens* [21]. Green rust is a mixed ferrous/ferric hydroxide which has a structure consisting of alternating layers of positively charged hydroxide and hydrated anion often containing carbonate [21]. The morphology of the green rust crystal is reported to vary with environmental conditions. The green rust crystals are hexagonal in the presence of SO_4^{2-} alone [47,48] but rhombohedral in solutions containing Cl^-, SO_4^{2-} or Br^- [47]. Globular aggregates, similar to that shown in Figure 7d, have been assigned to be siderite in study by Fredrickson et al. [25]. Due to the thin nature of the film, similar EDS spectra were obtained for all minerals across the biofilm. Carbon, oxygen, iron, phosphorous were the common elements detected through EDS (Figure 7e), which suggested that most minerals observed were different morphologies of the same phases (mainly iron oxyhydroxides). Elemental sulphur was also detected in some areas of the film.

The FTIR spectra for the biofilm in the modified Postgate C at 72 h (Figure 7f) was generally similar to that observed in the Postgate C medium (Figure 3e). Due to the low level of organic nutrients available for the bacteria, lower amounts of EPS were produced by the IRB, as suggested by the weaker absorbance signal in the modified Postgate C medium. The presence of hydrocarbon chains of lipids was confirmed by the characteristic peaks corresponding to -CH_3 asymmetric stretching (2958 cm^{-1}), asymmetric -CH_2 stretching (2916 cm^{-1}) and symmetric -CH_2 stretching (2848 cm^{-1}). Complementary bands for these hydrocarbons were found at 1452 cm^{-1} (asymmetric -CH_3 bending/CH_2 scissoring), -CH_2 of rocking vibration of lipids (716 cm^{-1}) as well as -CH_3 symmetric bending at 1372 cm^{-1}. Some contribution to a broad -OH stretch, probably due to iron oxyhydroxides, was observed. However, this peak was not as intense due to the presence of less oxyhydroxides or structurally bound water. Additional peaks detected in the IRB biofilm formed in the modified Postgate C media and not in the Postgate C media included the band at 1600 cm^{-1} corresponding to the COO- symmetric stretch of lipids [44], the band corresponding to amino acid side chains at 1508 cm^{-1}, and a band observed at 1288 cm^{-1} corresponding to α helix of amide III (30% C-N str, 30% N-H bending vibration, 10% O=C-N bending, 20% other) in proteins [44,49]. The most intense peak of the IR spectra, at 1180 cm^{-1} and 1002 cm^{-1}, corresponded to the presence of carbohydrates and possibly LPS bacterial cell wall CO-P, P-O-P components in the IRB biofilm [42–44]. The stretching of the phosphate ions was also present, as indicated by peaks at 1246 cm^{-1} (asymmetric PO_2 stretching) and 1076 cm^{-1} (PO_2 symmetric stretching) and asymmetric P-O stretch at 826 cm^{-1} [44]. The dwarf peaks at 696 cm^{-1} and 674 cm^{-1} assigned to C=O bending and CH_2 rocking vibrations, respectively [50,51].

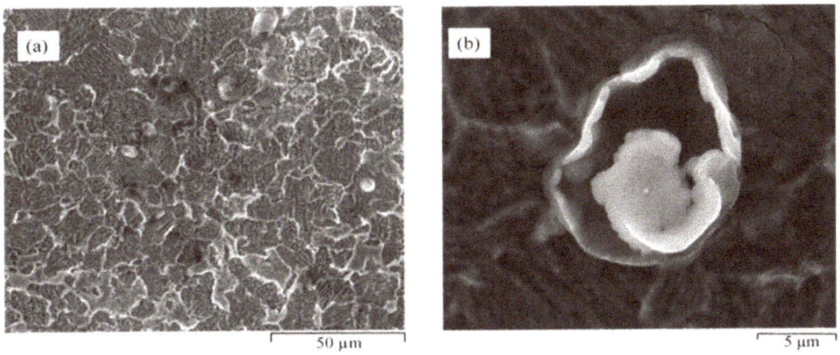

Figure 7. *Cont.*

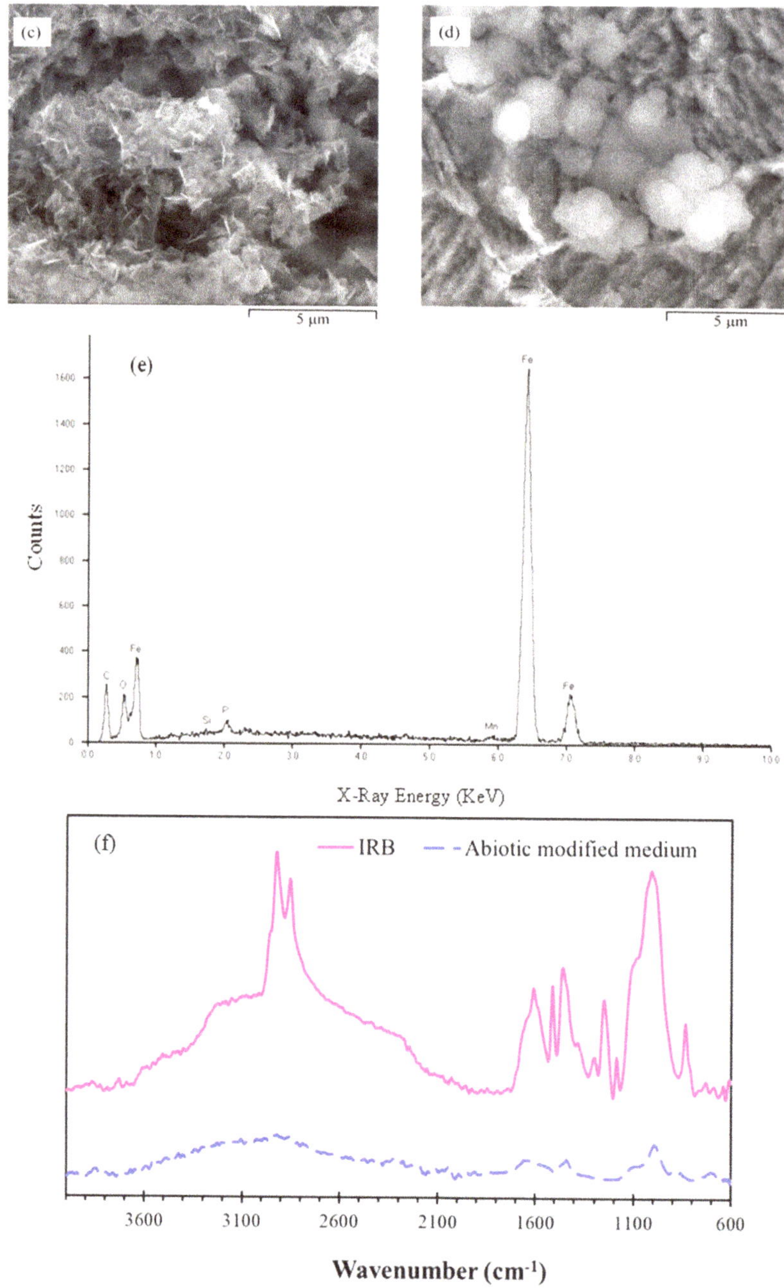

Figure 7. ESEM analysis of hydrated biofilm formed on carbon steel by a pure IRB culture following 72 h exposure to modified Postgate C medium: appearance of uniform corrosion (**a**) across the surface of carbon steel, a hollow iron oxide shell (**b**) at the surface of a corrosion pit, and needle (**c**) and globular (**d**) forms of reduced iron oxyhydroxide compound. (**e**) EDS spectra of the biofilm. (**f**) IR spectra of IRB biofilm formed on carbon steel exposed to pure IRB culture in modified Postgate C medium for a period of 72 h.

IR spectroscopy (Figures 3e and 7f) as well as microscopic observations (Figure 4a) of the pure IRB biofilm showed that organic material in the form of EPS was more prevalent in the presence of *S. putrefaciens*. The adhesive nature of the EPS secreted by the IRB may have a role in facilitating attachment of other bacterial species and/or nutrients to the biofilm as well as in consolidating the biofilm structure. The binding of metal ions by negatively charged EPS carbohydrates and proteins has also been previously reported [52].

FIB-SEM analysis showed the IRB biofilm to be extremely porous (Figure 8a,b). As a result, this film did not provide much protection and was expected to facilitate transport of chemical species between the electrolyte and the carbon steel surface. Therefore, the surface of the carbon steel under the biofilm was rough and corroded. At higher magnification (Figure 8c), an initial colloidal iron oxyhydroxide layer can be observed that appeared to be adhering to the corroding carbon steel with the aid from a sticky EPS-like material.

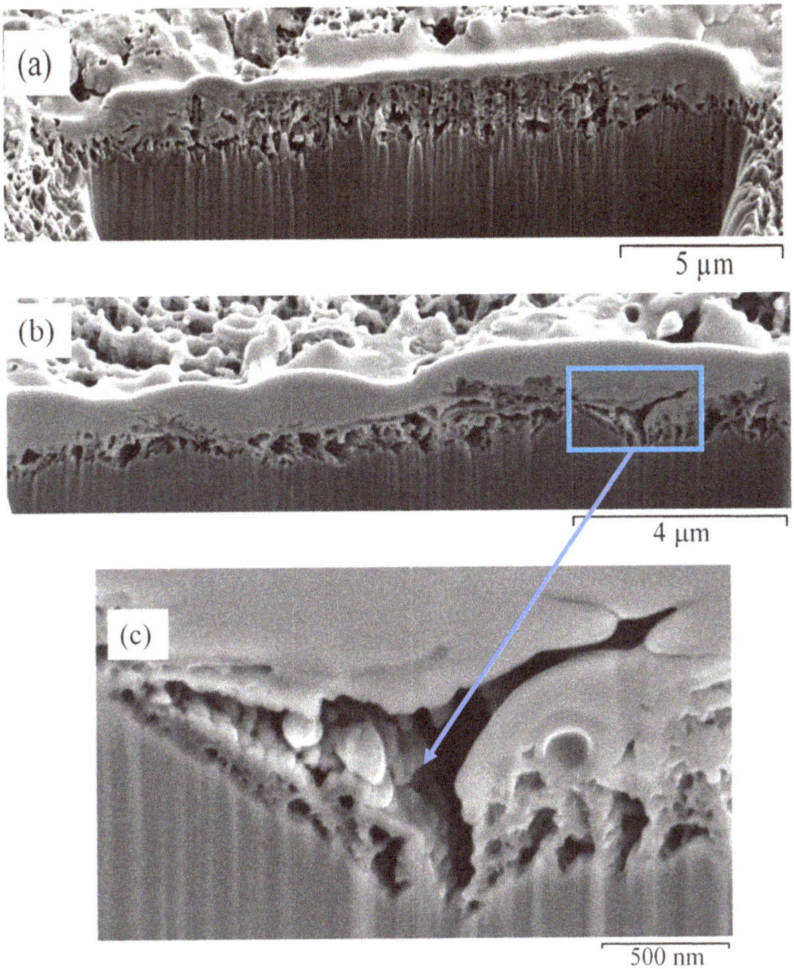

Figure 8. FIB-SEM images of the cross-sections (**a**,**b**) of biofilm formed two areas of a carbon steel surface by pure culture of IRB in modified Postgate C medium following a 72 h exposure a detailed view (**c**) of the corroded area of the cross-section identified in Figure 8b at higher magnification showing channels that could allow electrolyte transport.

The continuity of the IRB biofilm was observed to be of stochastic nature. In certain areas of the same sample, colloidal inorganic/organic precipitates of C, O, Na, P, S, Cl, Mn and Fe were observed (Figure 9a). Iron oxide shells covered in sticky EPS material, such as that shown in Figure 9b, were found scattered throughout the sample.

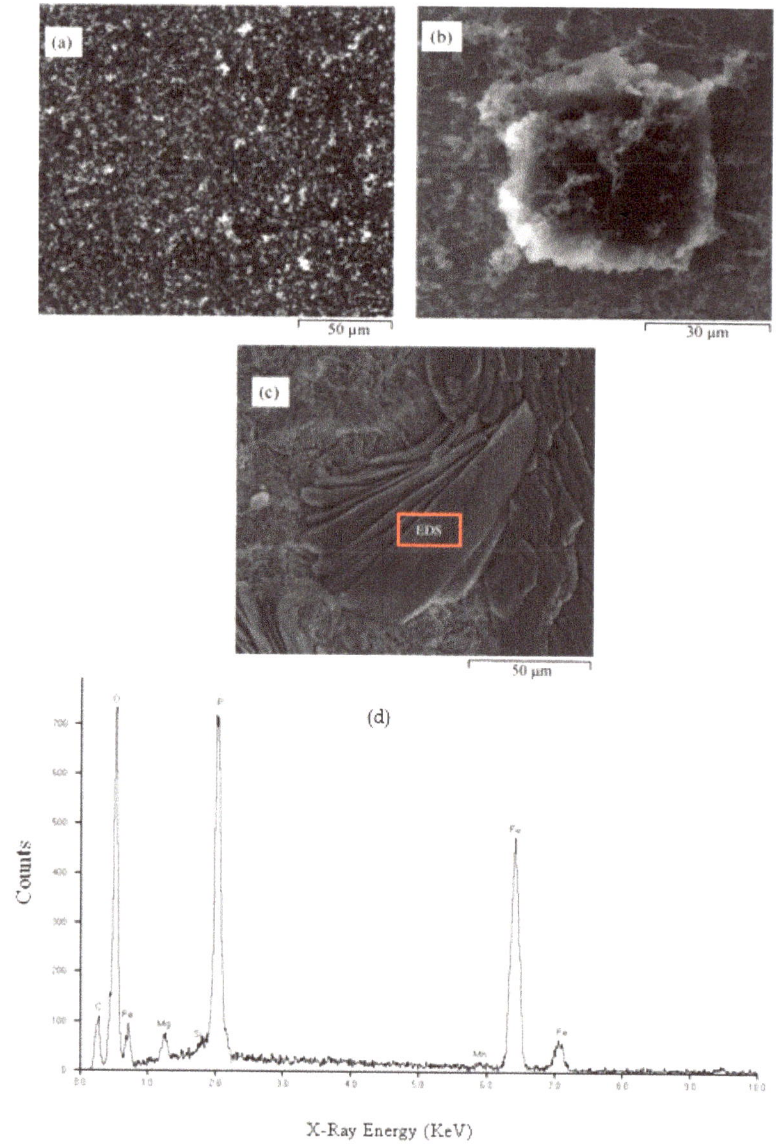

Figure 9. ESEM images of hydrated biofilm formed on carbon steel due to a pure IRB culture following 168 h exposure in the modified Postgate C medium: (**a**) overall covering (inorganic and organic components), (**b**) an iron oxyhydroxide shell covered in adhesive EPS material and (**c**) SEM image of precipitates of an iron phosphate species. (**d**) EDS analysis of the layered, plate-like deposit in Figure 9c.

The layered, plate-like deposit observed in Figure 9c was confirmed to be an iron phosphate species by EDS analysis as shown in Figure 9d. Glasauer et al., have reported that vivianite can form when sufficient Fe^{2+} and phosphates are available [22]. As discussed earlier in Section 3.2.2, the morphology of these minerals was more easily observed under SEM than ESEM. It is likely that dehydration, which is most likely under SEM set up, changed the morphology of the minerals. FIB-SEM cross-section carried out under these iron phosphate deposits showed the total thickness of the layered mineral in Figure 9c to be approximately 10 μm. Pitting corrosion was not observed under iron phosphate deposits.

3.3.2. Electrochemical Behaviour of Steel in Modified Postgate C Medium with IRB

The build-up of biofilm was monitored using the OCP technique for a period of 168 h (Figure 10a). The negative shift in E_{corr} that occurred in the presence of IRB biofilm due to either a decrease of cathodic reactant O_2 caused by IRB aerobic respiration and/or an increase in the anodic dissolution of iron caused by a loss of protective nature of the passive iron oxyhydroxide film. During the OCP measurement, E_{corr} stabilized quickly and did not oscillate a great deal, suggesting low oxygen conditions in solution. O_2 consumption appeared to occur instantaneously as shown by a drop in E_{corr} from initial reading of −744 mV at 0 h down to −842 mV at 72 h. The E_{corr} then became stable during 72–168 h period where anaerobic or microaerobic conditions were set up by the bacteria. Consistent with this results, Sherar et al., reported an $E_{corr} < -800$ mV with anaerobic corrosion of carbon steel [53]. Once anaerobic conditions ensued, the corrosion of carbon steel was controlled by a combination of H_2O, HCO_3^- and Fe^{3+} reduction reactions.

Figure 10b depicts potentiodynamic polarization curves of carbon steel after 72 h exposure to the biotic and abiotic modified Postgate C solution. Active corrosion behaviour of carbon steel occurred in both the IRB and abiotic control. Corrosion resistance was similar in the IRB and abiotic control following 72 h pre-exposure (i.e., there was insignificant change in E_{corr} and I_{corr}). However, following 168 h exposure, the E_{corr} in the presence of the IRB was lower than that observed under abiotic conditions (Figure 10c), indicating the carbon steel to be more susceptible to corrosion in biotic solution without yeast after 168 h exposure. The corrosion current density of carbon steel in biotic solution was found to be about 9 times higher than that in the abiotic solution. These findings are consistent with the FIB-SEM observation, where pits are observed under the biofilm during the 168 h exposure to biotic Postgate C without yeast. Pits appeared to have propagated sideways and vertically. The slower increase in anodic current density in the potential range of −600 to −800 mV can possibly be attributed to the imbedding of corrosion products into the bacterial biofilm, thereby impeding the diffusion across the biofilm, which is consistent with the literature [9]. E_{pit} after 168 h exposure to biotic medium is observed around −575 mV. At applied potentials more positive to E_{pit}, active corrosion similar to the abiotic control occurred (Figure 10c).

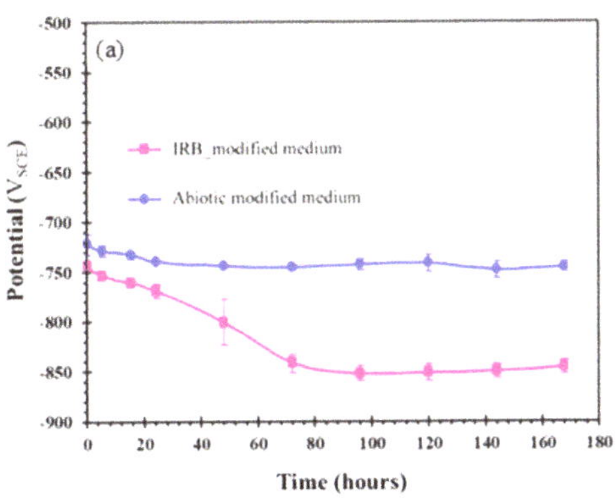

Figure 10. *Cont.*

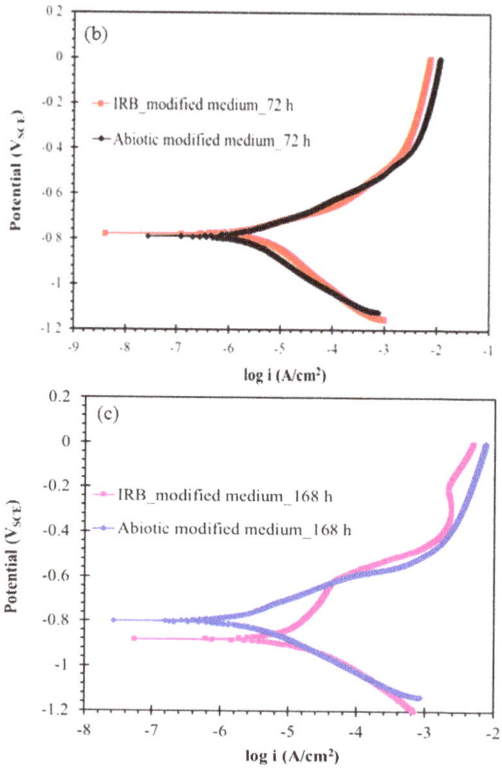

Figure 10. (**a**) Change in OCP of carbon steel with time of exposure to a modified Postgate C medium with pure IRB culture and without the culture (abiotic). (**b**) Potentiodynamic polarization scans of carbon steel after 72 h exposure in modified Postgate C medium with and without IRB. (**c**) Potentiodynamic polarization scans of carbon steel after 168 h exposure in modified Postgate C medium with and without IRB.

3.4. Biofilm Development and Electrochemical Characterization of Carbon Steel in Biotic Postgate C Solution without Organic Nutrients (i.e., Inorganic Medium)

3.4.1. Biofilm Development on Carbon Steel Exposed to IRB in Inorganic Postgate C Solution

In the absence of organic nutrients such as yeast extract, lactic acid and sodium citrate, the most likely electron donor for the IRB is hydrogen [54,55]. Not all IRB species can couple H_2 oxidation with growth [39,56]. However, Lovely et al. [39] have reported on the ability of *S. putrefaciens* to metabolize hydrogen at partial pressures 25-fold lower than that by the pure cultures of SRB. The hydrogen for bacterial metabolism would be generated by the reduction of water Equation (1). The reduction of water would also be the most likely cathodic reaction at neutral pH and limited O_2 environments for the corrosion of the carbon steel.

$$2H_2O + 2e^- \rightarrow H_2 + 2OH^- \tag{1}$$

The IRB may then utilise the H_2 generated through Equation (2) and couple with the reduction of Fe^{3+} ions from ferric oxyhydroxides with the oxidation of H_2 Equation (2) [39];

$$H_2 + 2Fe^{3+} \rightarrow H^+ + 2Fe^{2+} \tag{2}$$

Previous studies have shown that IRB require direct contact with the Fe(III) minerals for bioreduction under nutritionally limiting conditions [21,37]. In contrast, other studies have observed that microbial siderophores (extracellular chelating agents) were involved in reductive dissolution of ferric oxyhydroxides and oxides [57]. Glasauer et al. [22] observed that *S. putrefaciens* grown under nutrient-limited conditions only reduced poorly crystalline hydrous ferric oxides. Due to the absence of lactate in the inorganic medium, the formation of siderite, according to Equation (3), is not expected as the availability of bicarbonate in solution occurs due to oxidation of lactate coupled with iron oxyhydroxide reduction [25]. However, the precipitation of magnetite Equation (4), vivianite Equation (5) or green rust (GR) Equation (6) is still possible under different conditions of redox potential, available Fe(III) and/or Fe (II) and appropriate ligands.

$$Fe^{2+} + HCO_3^- \rightarrow FeCO_3 + H^+ \quad (3)$$

$$2Fe(OH)_3 + Fe^{2+} \rightarrow Fe_3O_4 + 2H_2O + 2H^+ \quad (4)$$

$$3Fe^{2+} + 2HPO_4^{2-} + 8H_2O \rightarrow Fe_3(PO_4)_2.8H_2O + 2H^+ \quad (5)$$

$$2Fe(OH)_3 + 4Fe^{2+} + A^{2-}\left(Cl^-, CO_3^{2-}, SO_4^{2-}\right) + 9H_2O \rightarrow GR + 6H^+ \quad (6)$$

In contrast to the IRB biofilms formed in complete Postgate C and the modified Postgate C mediums, there was lesser incidence of crystalline iron oxyhydroxides observed on the carbon steel surface when exposed to IRB in the inorganic medium. This may be due to the rate of crystalline iron oxyhydroxide formation being slower than the biotic iron reduction reaction [25]. Some colloidal particles observed in the IRB biofilm (Figure 11a) are speculated to be magnetite. EDX analysis shows the presence of Fe, C and O corresponding to the elemental composition of typical area, such as that shown in Figure 11a, indicating that iron phosphates may also be present as was also observed earlier under inorganic abiotic condition. The presence of iron oxyhydroxide shells in the biofilm became more recognisable in SEM analysis (Figure 11b) than from hydrated biofilms observed through ESEM. The shells in this case were observed to be broken down, either by the environmental conditions in the IRB solution or due to sample dehydration under SEM chamber. FIB cross-sectional analysis could not be performed with the intact hollow shells due to the fragile nature of the shells. However, as shown in Figure 12, FIB-SEM cross-sectional analysis could be performed on the broken iron oxyhydroxide shell (Figure 11b).

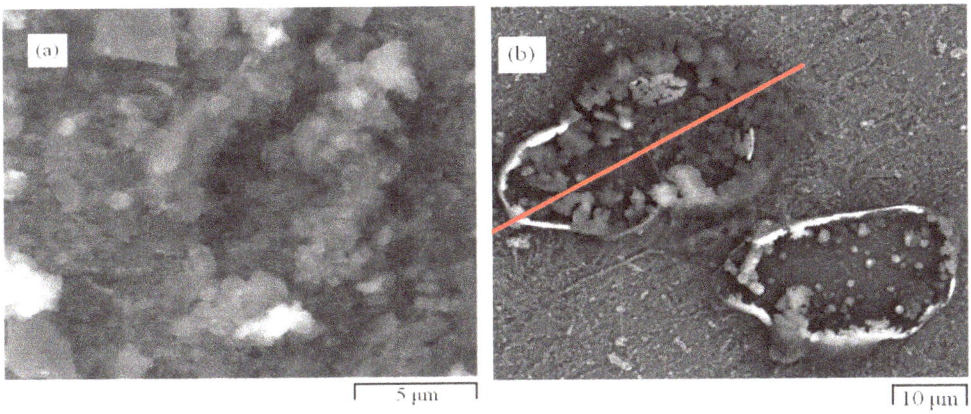

Figure 11. ESEM images of hydrated biofilm formed on carbon steel by a pure IRB culture following 72 h exposure to IRB and Postgate C without organic constituents: (**a**) amorphous corrosion product deposits containing carbon, oxygen phosphorous and iron. (**b**) SEM image of iron oxyhyrdroxide shell. FIB cross-section was carried out across the area indicated by the red line.

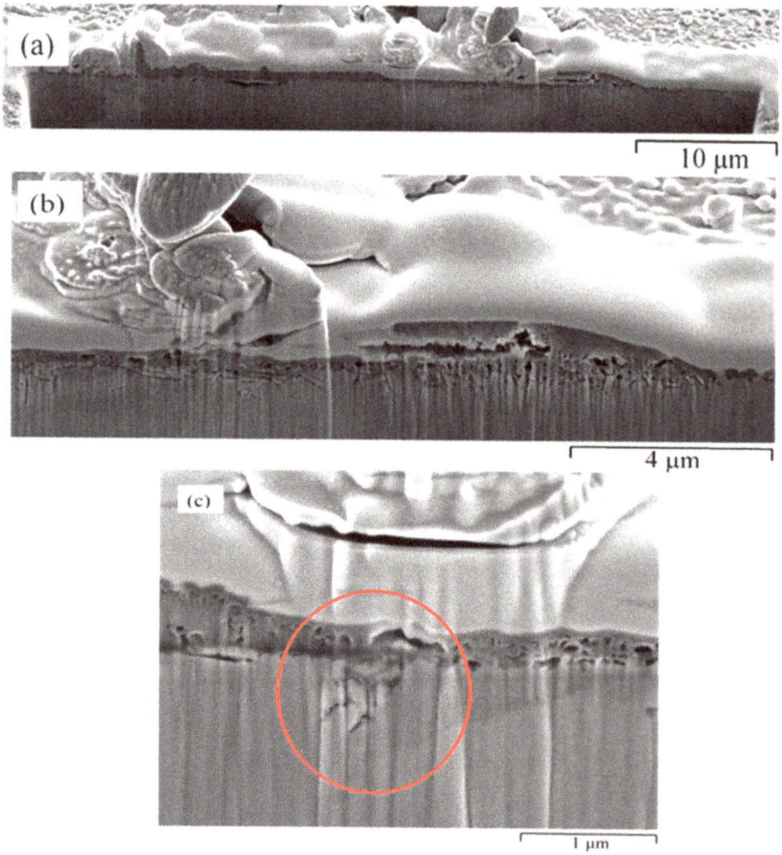

Figure 12. FIB-SEM cross-sectional analysis of biofilm formed on carbon steel by pure culture of IRB in inorganic medium following a 72 h exposure: (**a**) overview of the biofilm cross-section, and (**b**,**c**) the extent of cracking attack observed under the IRB biofilm at higher magnification.

The IRB biofilm formed under the inorganic environment was very thin. FIB-SEM cross-sectional analysis (Figure 12a) showed that the thickness was in the nm scale. When the cross-section was observed under higher magnification, extensive grooving into the steel surface was observed (Figure 12b). Figure 12c highlights an area of localized attack that progressed along susceptible points in the steel microstructure such as voids, inclusions or grain boundaries. Biofilm material was observed to have filled the grooves produced by the localized attack. There have been limited studies on the influence of IRB on corrosion, and this type of localized attack has not been previously reported in literature.

The localised attack shown in Figure 12c appears like a crack, which might occur because of H^+ cation build-up. H^+ is generated due to H_2 oxidizing ability of the IRB, coupled with ferric hydroxide reduction at localized areas. H^+ cations can transform into H atoms that may lead to the build-up of adsorbed hydrogen atom in the carbon steel subsurface. The H atoms may then combine to form H_2 molecules, particularly at the high energy locations such as voids and grain boundaries. The H_2 molecules are too big to diffuse out of the voids and the resulting build-up of gas pressure would cause cracking. Such cracking does not require externally applied tensile stress to progress [58]. Although initial signs of undercutting pitting attack were observed under the IRB biofilm in nutritionally rich conditions (Figure 5), such cracking was only observed for the carbon steel exposed to IRB in the inorganic medium. This may be due to the IRB oxidizing organic lactate in these two nutritionally rich media removes 7 moles of H^+ for every lactate molecule oxidized, as per Equation (7), and controls the build-up of H^+ ions. While previous studies have not investigated the MIC behaviour under the IRB biofilm, the acidic H_2S formation in the presence of the SRB has been reported to lead to MIC. H_2S enhances the adsorption of hydrogen atoms into the metal by poisoning the recombination of H atoms on the metal surface [58]. A previous study has used hydrogen permeation experiments and has observed enhanced hydrogen permeation due to cultures of SRB isolated from oil field waters [59].

$$4Fe(OH)_3 + CH_3CHOHCOO^- + 7H^+ \rightarrow 4Fe^{2+} + CH_3COO^- + HCO_3{}^- + 10H_2O \quad (7)$$

After 168 h exposure of the carbon steel to the IRB in the inorganic medium, micro-pitting could be observed (Figure 13a). The colloidal precipitates, such as these shown in Figure 13b, were scatted throughout the sample surface, and were identified as an iron phosphates species based on EDS. The polishing marks visible alongside these micro-pits suggest a non-equal distribution of the corrosion rate. Low corrosion rate has occurred in certain areas of the sample (Figure 13c), whereas deep localized pitting attack was also observed, as discussed below.

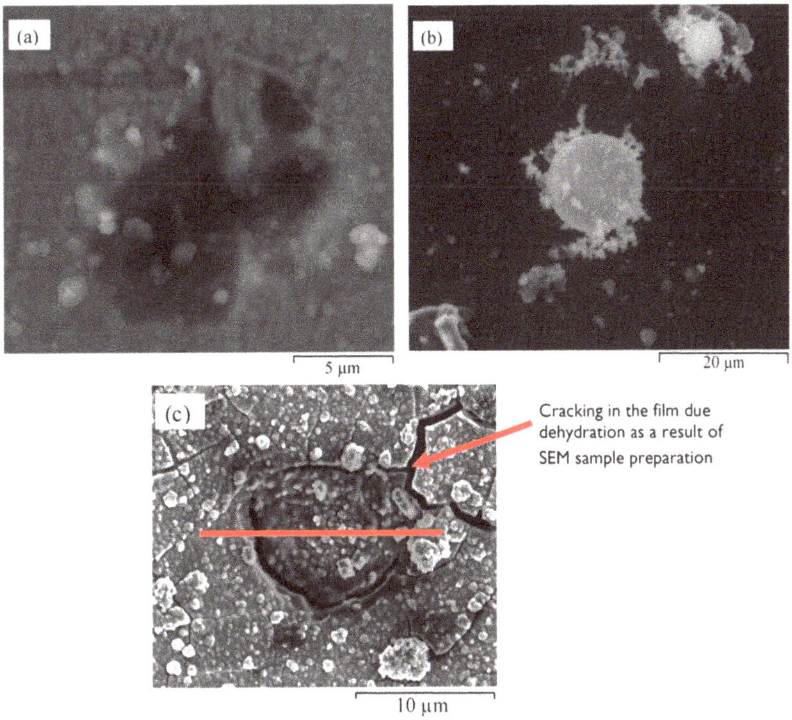

Figure 13. ESEM images of hydrated biofilm formed on carbon steel by a pure IRB culture following 168 h exposure in inorganic medium: (**a**) pits covered under corrosion deposit, (**b**) micro-pits in an area of low corrosion rate, and (**c**) SEM image showing the location where the cross-sectional FIB milling was performed.

EDS carried out at the centre of one of the dark patches observed in ESEM (Figure 13a) detected iron, oxygen and phosphorous to be the main elements. FIB-SEM cross-sectional analysis was carried out on such an area to investigate if these were sites of localized pitting attack. Figure 13c is a SEM image of the area and the red line is the location of FIB cross-sectioning. Cracking of the film observed in the SEM images (Figure 13c) were not observed under ESEM for the same sample and hence, the conclusion that the cracks were a result of dehydration of IRB biofilm in the SEM chamber.

FIB-SEM in Figure 14a confirmed the existence of a deep pit under the area shown in Figure 13c. A pit (~10 μm deep) developed underneath the biofilm. In previous studies on stainless steels, growth of the pit may occur back towards the outer surface to where the pit originated [60,61]. The mechanism for such behaviour has been attributed to the passivation of the surface near the pit mouth due to concentration of metal ions being below the critical concentration for pit propagation (C_{crit}). However, in the pit adjacent to the passivated area, Fe^{2+} ions are generated and the dissolution front progresses through the sidewall until the passive film is punctured [61]. EDS analysis from the area highlighted in Figure 14a showed a high intensity peak for carbon that can only be attributed to the high content of organic material in the biofilm, since the test medium of Postgate C only consists of inorganic constituents. Inorganic sulphate and chloride were also present. It is assumed that SO_4^- (as indicated by S and O in EDS) and Cl^- ions moved to the corroding site to maintain electrical neutrality by balancing out the positive Fe^{2+} generated from the corroding carbon steel. They may also be incorporated into green rust minerals according to Equation (6). Another notable mineral present on the pit covering material was the hollow iron oxyhydroxides shells. These hollow shells were observed near the side walls (Figure 14b) as well as being incorporated between other corrosion product layers as outlined in Figure 14c.

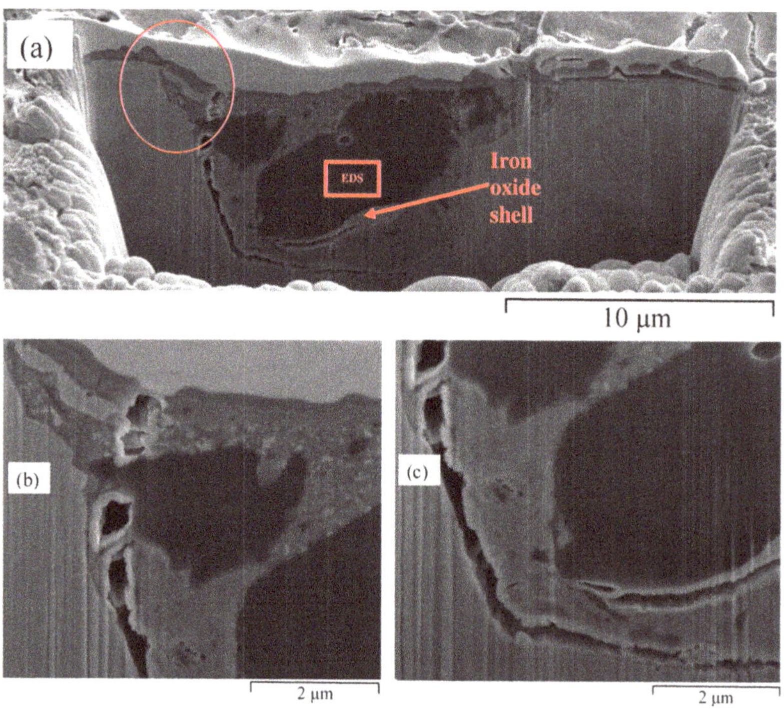

Figure 14. FIB-SEM cross-section of specific features of carbon steel exposed to the pure culture of IRB in the inorganic medium (i.e., Postgate C without organic components) for 168 h: (**a**) an overview of the cross-section (the circled area marks undercutting pitting attack making its way back towards the outer surface, EDS was performed in the area in the rectangle), (**b**) a close up view of the undercutting pit, and (**c**) a close up view of the side walls of the pit where biofilm is loosely attached in this area allowing electrolyte transport.

FIB-SEM cross-section characterization was carried out also for another location in the sample exposed for 168 h (Figure 15). In this case, the sample suffered some charging during SEM imaging, due probably to the high organic content in the IRB biofilm. Pit was covered under the corrosion product deposit (Figure 15b). The morphology showing an undercutting attack in Figure 15b is similar to pitting in acidic environments [62]. The cracking in the biofilm that was observed in the 72 h exposure to the inorganic medium occurred also in the 168 h exposure (Figure 15c). Biofilm extended into the cracks, as seen in the higher magnification images (Figure 15c). Similar to the 72 h exposure sample, EDS of the corrosion deposit covering the pit shown in Figure 15b detected a high intensity peak for carbon, indicating the presence of organic molecules. Other elements detected were sulphur, chloride and iron.

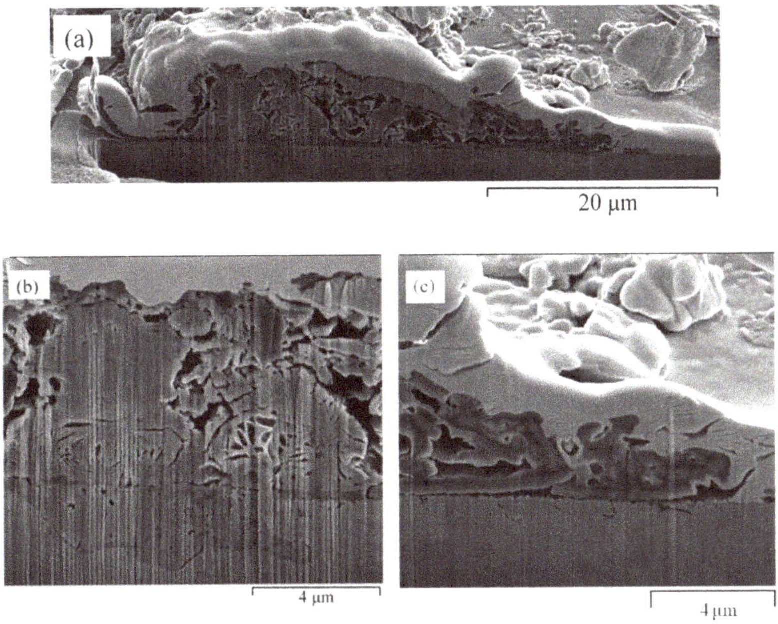

Figure 15. FIB-SEM cross-section of more common features of carbon steel exposed to the pure culture of IRB in the inorganic medium (i.e., Postgate C without organic components) for 168 h: (**a**) overview of a cross-section under an area of a corrosion deposit, (**b**) undercutting pitting has occurred under the biofilm, and (**c**) localised corrosion penetration and biofilm.

3.4.2. Electrochemical Behaviour Carbon Steel Exposed to Inorganic Postgate C Solution with IRB

Potentiodynamic polarization scans of carbon steel following 72 h exposure in the inorganic medium with and without the IRB are shown in Figure 16a. The corrosion potential of steel in inorganic biotic medium shifted by about 50 mV in a cathodic direction compared to that in the abiotic solution. The corrosion current density of the steel in biotic solution was at least an order of magnitude lower than that in the abiotic solution. As mentioned earlier, IRB are able to reduce the oxygen level, and limit the oxygen available for corrosion. It is also reported that in case of nutrient deficiency, IRB fail to survive; hence, there is no need of IRB forcing Fe(III) reduction to soluble Fe(II). Thus, the insoluble Fe(III) oxides (such as goethite) [63] persist and provide protection.

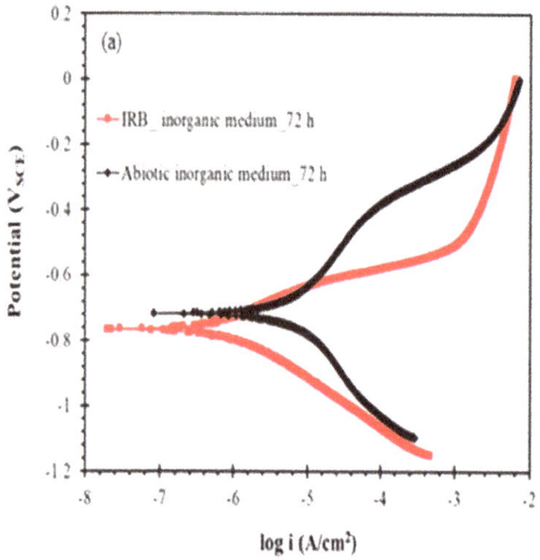

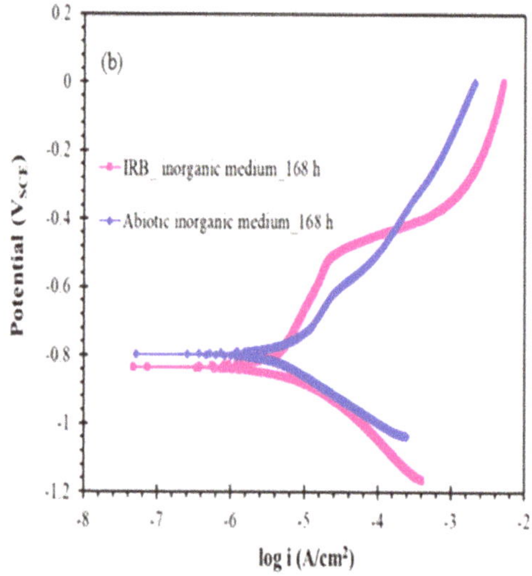

Figure 16. (**a**,**b**) Potentiodynamic polarization scans of carbon steel following 72 and 168 h exposure, with and without the IRB, to inorganic Postgate C solution.

Figure 16b shows the potentiodynamic polarization scan following 168 h exposure to IRB and the abiotic control inorganic Postgate C solutions. Though the corrosion resistance of steel decreased when the duration of exposure to inorganic IRB solution increased to 168 h, the corrosion current density of steel specimen in biotic solution still greater by >1.5 times than that of steel exposed to the abiotic solution. The sluggish increase in anodic current density at the anodic potential range of 0.74–0.58 V_{SCE} could be attributed to the formation of the ferrous species and rust during the anodic polarization. E_{pit} was observed at −550 mV, and as the applied potential was increased to more positive values from E_{pit}, active corrosion was observed.

No previous studies have investigated the influence of nutrients on the IRB biofilm formation or study the cross-section of the IRB biofilm. Results of this study show that the accelerated corrosion observed in IRB environments, when cultured under starvation conditions, requires further investigation.

4. Conclusions

S. putrefaciens was observed to be biofilm forming bacteria. The highest coverage to EPS material as well as reduced iron/oxyhydroxides were observed in the IRB biofilm in the presence of yeast extract in the Postgate C medium. The IRB contained a higher carbohydrate portion in the biofilm than the abiotic corrosion product film.

The build-up of organic material and reduced minerals did not provide any significant protection to the metal and active corrosion occurred when external potential was applied. Dissolution of Fe(III) oxyhydroxides occurred in the presence of the IRB which led to a greater dissolution of Fe^{2+} ions into the solution than under abiotic conditions. The IRB were also able to set up anaerobic conditions in solution by the removal of residual O_2 present in solution.

Undercutting pitting corrosion morphology was observed in the presence of the IRB in the nutrient-rich media, and to a greater extent, in nutrient-limited conditions. Deep pits on carbon steel were observed when IRB were cultured in the inorganic medium. From the results of this study, it can be concluded that the IRB biofilm does not protect carbon steel from corrosion and was observed to accelerate abiotic corrosion processes in a localised manner.

Author Contributions: S.W.; Conceptualization, Data curation, Formal analysis, Investigation, Methodology, Software, Validation, Visualization, Writing—review & editing. S.A.-S.; Investigation, Methodology, Formal analysis, Validation, Writing—review & editing. W.P.G.; Conceptualization, Investigation, Methodology, Formal analysis, Validation, Writing—review & editing. R.K.S.R.; Conceptualization, Investigation, Methodology, Formal analysis, Validation, Writing—review & editing. C.P.; Investigation, Methodology, Writing—review & editing. All authors have read and agreed to the published version of the manuscript.

Funding: This research received no external funding.

Informed Consent Statement: Not applicable.

Data Availability Statement: Data is contained within the article.

Acknowledgments: Authors acknowledge the use of microscopic facilities at Monash Centre for Electron Microscopy (MCEM).

Conflicts of Interest: The authors declare no conflict of interest.

References

1. Dawuda, A.-W.; Taleb-Berrouane, M.; Khan, F. A probabilistic model to estimate microbiologically influenced corrosion rate. *Process Saf. Environ. Prot.* **2021**, *148*, 908–926. [CrossRef]
2. Javaherdashti, R. Microbiologically influenced corrosion (MIC). In *Microbiologically Influenced Corrosion*; Springer: London, UK, 2017; pp. 29–71.
3. Little, B.J.; Lee, J.S. *Microbiologically Influenced Corrosion*; John Wiley & Sons: Hoboken, NJ, USA, 2007; Volume 3.
4. Sequeira, C.A.; Tiller, A. *Microbial Corrosion-1*; Elsevier Applied Science: London, UK, 1988.
5. Padilla-Viveros, A.; Garcia-Ochoa, E.; Alazard, D. Comparative electrochemical noise study of the corrosion process of carbon steel by the sulfate-reducing bacterium Desulfovibrio alaskensis under nutritionally rich and oligotrophic culture conditions. *Electrochim. Acta* **2006**, *51*, 3841–3847. [CrossRef]
6. Sun, Z.; Moradi, M.; Chen, Y.; Bagheri, R.; Guo, P.; Yang, L.; Song, Z.; Xu, C. Simulation of the marine environment using bioreactor for investigation of 2507 duplex stainless steel corrosion in the presence of marine isolated Bacillus Vietnamensis bacterium. *Mater. Chem. Phys.* **2018**, *208*, 149–156. [CrossRef]
7. Enning, D.; Garrelfs, J. Corrosion of Iron by Sulfate-Reducing Bacteria: New Views of an Old Problem. *Appl. Environ. Microbiol.* **2014**, *80*, 1226–1236. [CrossRef]
8. Little, B.J.; Lee, J.S. Microbiologically influenced corrosion. In *Kirk-Othmer Encyclopedia of Chemical Technology*; Wiley: Hoboken, NJ, USA, 2000; pp. 1–42.
9. Usher, K.M.; Kaksonen, A.H.; Cole, I.; Marney, D. Critical review: Microbially influenced corrosion of buried carbon steel pipes. *Int. Biodeterior. Biodegrad.* **2014**, *93*, 84–106. [CrossRef]
10. Herrera, L.K.; Videla, H.A. Role of iron-reducing bacteria in corrosion and protection of carbon steel. *Int. Biodeterior. Biodegrad.* **2009**, *63*, 891–895. [CrossRef]
11. Gaylarde, C.C.; Videla, H.A. Localised corrosion induced by a marine vibrio. *Int. Biodeterior.* **1987**, *23*, 91–104. [CrossRef]
12. Little, B.; Wagner, P.; Hart, K.; Ray, R.; Lavoie, D.; Nealson, K.; Aguilar, C. The role of biomineralization in microbiologically influenced corrosion. *Biogeochemistry* **1998**, *9*, 1–10. [CrossRef] [PubMed]
13. Kalajahi, S.T.; Rasekh, B.; Yazdian, F.; Neshati, J.; Taghavi, L. Corrosion behaviour of X60 steel in the presence of sulphate-reducing bacteria (SRB) and iron-reducing bacteria (IRB) in seawater. *Corros. Eng. Sci. Technol.* **2021**, *56*, 543–552. [CrossRef]
14. Perego, P.; Fabiano, B. Microbiological Induced Corrosion and Inhibition. In *Encyclopedia of Industrial Biotechnology: Bioprocess, Bioseparation, and Cell Technology*; John Wiley & Sons: Hoboken, NJ, USA, 2009; pp. 1–21.
15. Dubiel, M.; Hsu, C.H.; Chien, C.C.; Mansfeld, F.; Newman, D.K. Microbial Iron Respiration Can Protect Steel from Corrosion. *Appl. Environ. Microbiol.* **2002**, *68*, 1440–1445. [CrossRef] [PubMed]
16. Lee, A.K.; Newman, D.K. Microbial iron respiration: Impacts on corrosion processes. *Appl. Microbiol. Biotechnol.* **2003**, *62*, 134–139. [CrossRef] [PubMed]
17. AlAbbas, F.M.; Bhola, S.M.; Spear, J.R.; Olson, D.L.; Mishra, B. The shielding effect of wild type iron reducing bacterial flora on the corrosion of linepipe steel. *Eng. Fail. Anal.* **2013**, *33*, 222–235. [CrossRef]
18. Sharma, M.; Voordouw, G. MIC detection and assessment: A holistic approach. In *Microbiologically Influenced Corrosion in the Upstream Oil and Gas Industry*; CRC Press: Boca Raton, FL, USA, 2017; pp. 177–212.
19. Videla, H.A.; Herrera, L. Biocorrosion in oil recovery systems: Prevention and protection-An update. *Edición Espec.* **2007**, *30*, 272–279.
20. Kim, H.J.; Park, H.S.; Hyun, M.S.; Chang, I.S.; Kim, M.; Kim, B.H. A mediator-less microbial fuel cell using a metal reducing bacterium, *Shewanella putrefaciens*. *Enzym. Microb. Technol.* **2002**, *30*, 145–152. [CrossRef]
21. O'Loughlin, E.J. Effects of electron transfer mediators on the bioreduction of lepidocrocite (γ-FeOOH) by *Shewanella putrefaciens* CN32. *Environ. Sci. Technol.* **2008**, *42*, 6876–6882. [CrossRef]

22. Glasauer, S.; Weidler, P.G.; Langley, S.; Beveridge, T.J. Controls on Fe reduction and mineral formation by a subsurface bacterium. *Geochim. Cosmochim. Acta* **2003**, *67*, 1277–1288. [CrossRef]
23. Igarashi, K.; Kato, S. Reductive Transformation of Fe (III) (oxyhydr)Oxides by Mesophilic Homoacetogens in the Genus Sporomusa. *Front. Microbiol.* **2021**, *12*. [CrossRef] [PubMed]
24. Frankel, R.B.; Bazylinski, D.A. Biologically Induced Mineralization by Bacteria. *Rev. Mineral. Geochem.* **2003**, *54*, 95–114. [CrossRef]
25. Fredrickson, J.K.; Zachara, J.M.; Kennedy, D.W.; Dong, H.; Onstott, T.C.; Hinman, N.W.; Li, S.-M. Biogenic iron mineralization accompanying the dissimilatory reduction of hydrous ferric oxide by a groundwater bacterium. *Geochim. Cosmochim. Acta* **1998**, *62*, 3239–3257. [CrossRef]
26. Postgate, J.R. *The Sulphate-Reducing Bacteria*; CUP Archive: Cambridge, UK, 1979.
27. Al-Saadi, S.; Banerjee, P.C.; Raman, R.S. Corrosion of bare and silane-coated mild steel in chloride medium with and without sulphate reducing bacteria. *Prog. Org. Coat.* **2017**, *111*, 231–239. [CrossRef]
28. Al-Saadi, S.; Raman, R.S.; Panter, C. A Two-Step Silane Coating Incorporated with Quaternary Ammonium Silane for Mitigation of Microbial Corrosion of Mild Steel. *ACS Omega* **2021**, *6*, 16913–16923. [CrossRef] [PubMed]
29. Al-Saadi, S.; Raman, R.S. A long aliphatic chain functional silane for corrosion and microbial corrosion resistance of steel. *Prog. Org. Coat.* **2018**, *127*, 27–36. [CrossRef]
30. Cheng, Y.F.; Rairdan, B.R.; Luo, J.L. Features of electrochemical noise generated during pitting of inhibited A516-70 carbon steel in chloride solutions. *J. Appl. Electrochem.* **1998**, *28*, 1371–1375. [CrossRef]
31. American Public Health Association. *Standard Methods for the Examination of Water and Wastewater*; American Public Health Association: Washington, DC, USA, 2005.
32. Blankemeier, A.R. Characterization of Pseudomonas Fluorescens Biofilm. Ph.D. Thesis, The Ohio State University, Columbus, OH, USA, 2011.
33. Simpson, W.J. Isolation and Characterisation of Thermophilic Anaerobes from Bass Strait oil Production Waters. Master's Thesis, Monash University, Clayton, Australia, 1999. [CrossRef]
34. Hamilton, W.A. Microbially Influenced Corrosion as a Model System for the Study of Metal Microbe Interactions: A Unifying Electron Transfer Hypothesis. *Biofouling* **2003**, *19*, 65–76. [CrossRef] [PubMed]
35. O'Loughlin, E.J.; Boyanov, M.I.; Gorski, C.A.; Scherer, M.M.; Kemner, K.M. Effects of Fe(III) Oxide Mineralogy and Phosphate on Fe(II) Secondary Mineral Formation during Microbial Iron Reduction. *Minerals* **2021**, *11*, 149. [CrossRef]
36. Bonneville, S.; Behrends, T.; Van Cappellen, P.; Hyacinthe, C.; Röling, W.F. Reduction of Fe(III) colloids by *Shewanella putrefaciens*: A kinetic model. *Geochim. Cosmochim. Acta* **2006**, *70*, 5842–5854. [CrossRef]
37. Lies, D.P.; Hernandez, M.E.; Kappler, A.; Mielke, R.E.; Gralnick, J.A.; Newman, D.K. *Shewanella oneidensis* MR-1 Uses Overlapping Pathways for Iron Reduction at a Distance and by Direct Contact under Conditions Relevant for Biofilms. *Appl. Environ. Microbiol.* **2005**, *71*, 4414–4426. [CrossRef]
38. Fuller, S.J.; McMillan, D.; Renz, M.B.; Schmidt, M.; Burke, I.T.; Stewart, D.I. Extracellular Electron Transport-Mediated Fe (III) Reduction by a Community of Alkaliphilic Bacteria That Use Flavins as Electron Shuttles. *Appl. Environ. Microbiol.* **2014**, *80*, 128–137. [CrossRef]
39. Lovley, D.R.; Phillips, E.J.P.; Lonergan, D.J. Hydrogen and Formate Oxidation Coupled to Dissimilatory Reduction of Iron or Manganese by *Alteromonas putrefaciens*. *Appl. Environ. Microbiol.* **1989**, *55*, 700–706. [CrossRef]
40. Pinney, N.; Morgan, D. Ab initio study of structurally bound water at cation vacancy sites in Fe- and Al-oxyhydroxide materials. *Geochim. Cosmochim. Acta* **2013**, *114*, 94–111. [CrossRef]
41. Ona Nguema, G.; Abdelmoula, M.; Jorand, F.; Benali, O.; Géhin, A.; Block, J.C.; Génin, J.M. Microbial reduction of lepidocrocite gamma-FeOOH by *Shewanella putrefaciens*: The formation of green rust. In *Industrial Applications of the Mössbauer Effect*; Springer: Dordrecht, The Netherlands, 2002.
42. Bosch, A.; Serra, D.; Prieto, C.; Schmitt, J.; Naumann, D.; Yantorno, O. Characterization of Bordetella pertussis growing as biofilm by chemical analysis and FT-IR spectroscopy. *Appl. Microbiol. Biotechnol.* **2005**, *71*, 736–747. [CrossRef] [PubMed]
43. Schmitt, J.; Flemming, H.-C. FTIR-spectroscopy in microbial and material analysis. *Int. Biodeterior. Biodegrad.* **1998**, *41*, 1–11. [CrossRef]
44. Socrates, G. *Infrared and Raman Characteristic Group Frequencies Tables and Charts*; John Wiley & Sons Ltd.: Chichester, UK, 2001.
45. Furukawa, Y.; Dale, J.R. The surface properties of *Shewanella putrefaciens* 200 and S. oneidensis MR-1: The effect of pH and terminal electron acceptors. *Geochem. Trans.* **2013**, *14*, 3. [CrossRef] [PubMed]
46. Omoike, A.; Chorover, J.; Kwon, K.D.; Kubicki, J. DAdhesion of Bacterial Exopolymers to α-FeOOH: Inner-Sphere Complexation of Phosphodiester Groups. *Langmuir* **2004**, *20*, 11108–11114. [CrossRef]
47. McGill, I.R.; McEnaney, B.; Smith, D.C. Crystal structure of green rust formed by corrosion of cast iron. *Nature* **1976**, *259*, 200–201. [CrossRef]
48. Refait, P.; Memet, J.-B.; Bon, C.; Sabot, R.; Génin, J.-M. Formation of the Fe (II)–Fe (III) hydroxysulphate green rust during marine corrosion of steel. *Corros. Sci.* **2003**, *45*, 833–845. [CrossRef]
49. Meyers, R.A. *Encyclopedia of Molecular Cell Biology and Molecular Medicine*; Wiley Online Library: Weinheim, Germany, 2004.
50. Abdel-Khalek, N.A.; Selim, K.A.; Amir, A.; Farahat, M.M.; El-Hendawy, H.H. Enhancement of Selective Bioflocculation of (Quartz-PyrolusiteHematite) System Using Paenibacillus polymyxa Bacteria. *Bio Tech.* **2015**, *89*, 36712–36717.
51. Natarajan, K.; Deo, N. Role of bacterial interaction and bioreagents in iron ore flotation. *Int. J. Miner. Process.* **2001**, *62*, 143–157. [CrossRef]
52. Beech, I.B.; Sunner, J. Biocorrosion: Towards understanding interactions between biofilms and metals. *Curr. Opin. Biotechnol.* **2004**, *15*, 181–186. [CrossRef] [PubMed]

53. Sherar, B.W.A.; Keech, P.G.; Qin, Z.; King, F.; Shoesmith, D.W. Nominally Anaerobic Corrosion of Carbon Steel in Near-Neutral pH Saline Environments. *Corrosion* **2010**, *66*, 045001–045001-11. [CrossRef]
54. Nealson, K.H.; Myers, C.R. Iron reduction by bacteria: A potential role in the genesis of banded iron formations. *Am. J. Sci.* **1990**, *290*, 35–45.
55. Berthelin, J.; Ona-Nguema, G.; Stemmler, S.; Quantin, C.; Abdelmoula, M.; Jorand, F. Bioreduction of ferric species and biogenesis of green rusts in soils. *Comptes Rendus Geosci.* **2006**, *338*, 447–455. [CrossRef]
56. Hedrich, S.; Johnson, D.B. Aerobic and anaerobic oxidation of hydrogen by acidophilic bacteria. *FEMS Microbiol. Lett.* **2013**, *349*, 40–45. [CrossRef] [PubMed]
57. Gram, L. Siderophore-Mediated Iron Sequestering by *Shewanella putrefaciens*. *Appl. Environ. Microbiol.* **1994**, *60*, 2132–2136. [CrossRef] [PubMed]
58. Kane, R.D.; Cayard, M.S. Roles of H_2S in the Behavior of Engineering Alloys: A Review of Literature and Experience. In Proceedings of the Corrosion-National Association OF Corrosion Engineers Annual Conference, San Diego, CA, USA, 22–27 March 1998.
59. Jack, T.R. *Biological Corrosion Failures*; ASM International: Russell, OH, USA, 2002.
60. Pickering, H.W. Important Early Developments and Current Understanding of the IR Mechanism of Localized Corrosion. *J. Electrochem. Soc.* **2003**, *150*, K1. [CrossRef]
61. Ghahari, S.M. In Situ Synchrotron X-ray Characterisation and Modelling of Pitting Corrosion of Stainless Steel. Ph.D. Thesis, University of Birmingham, Birmingham, UK, 2012.
62. Esaklul, K.A. *Handbook of Case Histories in Failure Analysis*; ASM International: Almere, The Netherlands, 1992.
63. Hansel, C.M.; Benner, S.G.; Neiss, J.; Dohnalkova, A.; Kukkadapu, R.K.; Fendorf, S. Secondary mineralization pathways induced by dissimilatory iron reduction of ferrihydrite under advective flow. *Geochim. Cosmochim. Acta* **2003**, *67*, 2977–2992. [CrossRef]

metals

Article

Oxidation Behavior of Intermetallic Phase and Its Contribution to the Oxidation Resistance in Fe-Cr-Zr Ferritic Alloy

Shenghu Chen * and Lijian Rong

Key Laboratory of Nuclear Materials and Safety Assessment, Institute of Metal Research, Chinese Academy of Sciences, 72 Wenhua Road, Shenyang 110016, China; ljrong@imr.ac.cn
* Correspondence: chensh@imr.ac.cn; Tel.: +86-24-23971981

Abstract: Oxidation resistance of the ferritic Fe-Cr-W-Zr alloy with dispersed Fe_2Zr phase were investigated in stagnant air and in static oxygen-saturated liquid Pb-Bi eutectic separately considering the service environment in the advanced generation IV nuclear reactors. A duplex structure including an outer Fe_2O_3 layer and an inner $(Fe,Cr,Zr)_2O_3$ layer is developed after oxidation in air, while a three-layered structure consisting of outer magnetite layer, inner Fe-Cr spinel layer, and internal oxidation zone is formed after oxidation in liquid Pb-Bi eutectic. The dispersed Fe_2Zr phase shows delayed oxidation with respect to the α-Fe in air and in liquid Pb-Bi eutectic, which significantly affects the oxidation behaviors of the alloy. After oxidation in air at 923 K, the incorporated Fe_2Zr phase in the scale would obstruct the diffusion of metal/oxygen across the scale, resulting in the nonuniform oxidation behavior. After oxidation in static oxygen-saturated liquid Pb-Bi eutectic at 823 K, a reduction in the Fe supply to the magnetite/Fe-Cr spinel interface is present adjacent to the Fe_2Zr phase, which might lead to the creation of cavities in the outer magnetite layer with prolonged oxidation time.

Keywords: Fe-Cr-W-Zr alloy; Fe_2Zr phase; oxidation; air; liquid Pb-Bi eutectic

Citation: Chen, S.; Rong, L. Oxidation Behavior of Intermetallic Phase and Its Contribution to the Oxidation Resistance in Fe-Cr-Zr Ferritic Alloy. *Metals* **2022**, *12*, 827. https://doi.org/10.3390/met12050827

Academic Editor: Renato Altobelli Antunes

Received: 22 April 2022
Accepted: 9 May 2022
Published: 11 May 2022

1. Introduction

Ferritic/martensitic steels with a chromium content of 9~12 wt.% have been considered as candidate structural materials in future advanced nuclear reactors due to their higher resistance to irradiation swelling, lower thermal expansion coefficients, and higher thermal conductivity [1–4]. Ferritic/martensitic steels are usually characterized by the tempered martensitic structure, consisting of a high density of tangled dislocations within laths and dispersion of carbides along their boundaries and within their matrix. During the long-term exposure at temperature above 823 K, reduction in the dislocation density and coarsening of carbide would lead to the recovery of martensitic structure and a significant reduction in creep strength [5–7]. Afterwards, dispersed oxide nano-particles with a high number density are introduced into the ferrite matrix to develop the oxide dispersion strengthened (ODS) steels [8–10]. These highly stabilized oxide nano-particles are responsible for the excellent tensile strength and creep properties at 923~1173 K. The ODS steels are produced by much more complicated and expensive powder-metallurgy techniques.

Moreover, intermetallic precipitates are potentially considered as the strengthening phase during the design of high-temperature ferritic alloys. K. Yamamoto et al. reported the Fe-Cr-Nb ferritic heat-resistant alloy strengthened by the Fe_2Nb Laves phase, and found that the presence of Fe_2Nb phase could significantly improve the high-temperature strength [11]. D.G. Morris et al. reported a Fe-Al-Zr ferritic alloy with coherent Fe_3Zr phase, and Fe_3Zr phase with excellent stability contributed to the great improvement in creep strength at 973 K [12,13]. Thermodynamic modeling of Fe-Cr-Zr system through the Calphad approach found that Fe-Zr Laves phase could form in the ferritic Fe-Cr-Zr system [14]. Our recent investigation found that dispersed Fe_2Zr phase were introduced into equiaxed α-Fe matrix in the Fe-9Cr-2W alloy with the Zr content of 7~10 wt.%, and enhanced creep-rupture properties was achieved up to 973 K in comparison with the typical 9Cr2WVTa ferritic/martensitic steel [15]. L. Tan et al. found that only a small amount of radiation-induced precipitates was observed in Fe-9Cr-1W-11Zr alloy after Fe ion irradiation to ~50 dpa at 673 K, demonstrating promising radiation resistance [16,17]. It was found that the Fe-Cr-Zr alloy presented better high-temperature creep properties and superior radiation resistance in comparison with the ferritic/martensitic steel. In addition, corrosion is a life-limiting property when the alloy is exposed to the service environment. Liquid metal are the primary coolants for the advanced generation IV nuclear reactors. For instance, liquid Pb-Bi eutectic are the coolant for the advanced lead fast reactors. The liquid Pb-Bi eutectic is very corrosive towards the structural material [18–24]. Therefore, the corrosion properties in contact with liquid Pb-Bi eutectic should be considered. Until now, the corrosion properties of ferritic alloy with the dispersed intermetallic phase have not been clearly understood. In this study, the high-temperature oxidation resistance of Fe-Cr-Zr alloy in air and in static liquid Pb-Bi eutectic were investigated in view of the service environment in the advanced generation IV nuclear reactors.

2. Materials and Methods

2.1. Materials

A Fe-Cr-Zr alloy with a weight of 10 kg is produced by vacuum induction melting technique, and the chemical composition is given in Table 1. The ingot is hot forged and then hot rolled to ~12 mm thick plate. The plate is annealed at 1373 K for 15 min followed by air cooling.

Table 1. Chemical compositions of Fe-Cr-Zr alloy.

Element	C	Cr	W	Zr	Fe
wt.%	0.0039	8.96	2.01	9.86	Bal.

2.2. Oxidation Tests in Air and in Liquid Pb-Bi Eutectic

Samples with dimensions of 8 mm × 8 mm × 8 mm are ground with SiC abrasive papers up to 2000 grit, mechanically polished with 0.5 μm diamond powder, and ultrasonically cleaned by ethanol prior to oxidation tests. The oxidation behaviors in air are evaluated by continuous isothermal oxidation tests, which are conducted at 923 K for different durations up to 1000 h in a furnace with stagnant air. The weight measurement for evaluating the oxidation rate is performed by using a Sartorius BP211D electric balance (Sartorius, Goettingen, Germany) with an accuracy of 0.01 mg.

Furthermore, the oxidation behaviors in liquid metal are evaluated by the isothermal oxidation tests in oxygen-saturated stagnant liquid Pb-Bi eutectic at 823 K using a self-developed liquid Pb-Bi eutectic corrosion test apparatus [19,20]. Ar + 5% O_2 serves as the cover gas to ensure the oxygen-saturated condition and avoid excessive oxidation of the liquid Pb-Bi eutectic. The presence of thin PbO on the liquid metal surface indicated that liquid Pb-Bi eutectic is oxygen-saturated. The samples are taken out after immersion in the oxygen-saturated Pb-Bi eutectic for 500 h, 1000 h, and 2000 h, respectively.

2.3. Characterization

The phase structure of the Fe-Cr-Zr alloy is identified using x-ray diffraction (XRD, Rigaku D/max-2400PC, Rigaku, Tokyo, Japan) with Cu Kα radiation source. Microstructure of Fe-Cr-Zr alloy is characterized by FEI Inspect F50 scanning electron microscope (SEM) (FEI company, Oregon, America) and JEOL JEM-2100F transmission electron microscopy (TEM) (JEOL, Tokyo, Japan).

After oxidation tests in stagnant air and in oxygen-saturated stagnant liquid Pb-Bi eutectic, phase structure of oxide scale is identified using XRD. The surface and cross-sectional morphologies of oxide scales are observed by SEM, and chemical compositions are analyzed by energy-dispersive X-ray spectrometer (EDS) (Oxford Instruments, Oxford, UK). The element distribution analysis is conducted by Shimadzu EPMA-1610 electron probe microanalysis (EPMA) (Shimadzu, Kyoto, Japan). For preparing the cross-sectional samples after oxidation tests in stagnant air, the sample surface is protected by electroless Ni plating. The residual Pb-Bi eutectic on the surface of sample is retained after oxidation tests in oxygen-saturated stagnant liquid Pb-Bi eutectic to protect the oxide scale during the sample preparation process.

3. Results

3.1. Microstructure of Fe-Cr-Zr Alloy

Figure 1 shows the XRD pattern of Fe-Cr-Zr alloy. The peak of Fe_2Zr phase with the hexagonal polymorphs (C14/C36) could be observed besides the sharp diffraction peaks of α-Fe. It is confirmed that the microstructure of Fe-Cr-Zr alloy mainly compromises α-Fe and Fe_2Zr phase.

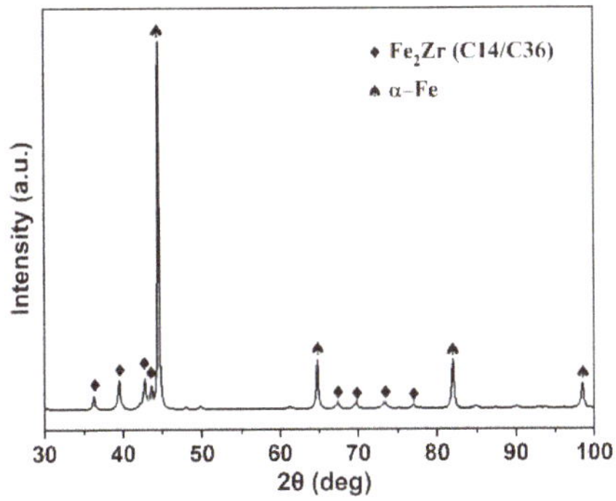

Figure 1. XRD patterns of the Fe-Cr-Zr alloy.

As shown in Figure 2a, the SEM image reveals that the micrometer-sized Fe_2Zr phases are nearly uniformly distributed in the α-Fe matrix, and the area fraction of Fe_2Zr phases is ~9.5%. EDS point analysis shows that Fe_2Zr phases are composed primarily of Fe and Zr, and low contents of Cr and W are present at the phase (Figure 2b). Figure 2c shows the bright-field TEM micrograph and corresponding selected area electron diffraction (SAED) pattern. The TEM micrograph shows the coexistence of light α-Fe and dark Fe_2Zr phases, and the α-Fe and Fe_2Zr phases are both in the micrometer-sized range. The SAED pattern (inset in Figure 2c) of Fe_2Zr phases confirms the Laves structure. The high-resolution electron microscopy (HRTEM) image of α-Fe/Fe_2Zr clearly shows that their interface is incoherent (Figure 2d).

3.2. Oxidation Behavior in Air

3.2.1. Structure and Morphology of Oxide Scale

Figure 3 shows the SEM surface morphologies of Fe-Cr-Zr alloy after oxidation at 923 K for 10 min, 20 h, 500 h, and 1000 h, respectively. It is found that at the initial oxidation stage within 10 min, numerous oxide particles are distributed on the metal surface. However, an uneven oxide surface is formed (Figure 3a and inset), which is indicative of a nonuniform oxidation process. As the oxidation time is prolonged, the surface flatness of oxide scale is aggravated (Figure 3b–d). After oxidation of 1000 h, a number of pits could be found on the surface due to the nonuniform growth of oxide (arrows in Figure 3d and inset). The structure of oxide scale was identified by XRD after oxidation at 923 K for 1000 h, as shown in Figure 4. Single $(Fe,Cr)_2O_3$ phase is detected in the oxide scale, which indicates that Cr_2O_3 could not form in the Fe-Cr-Zr alloy with a Cr content of ~9 wt.%.

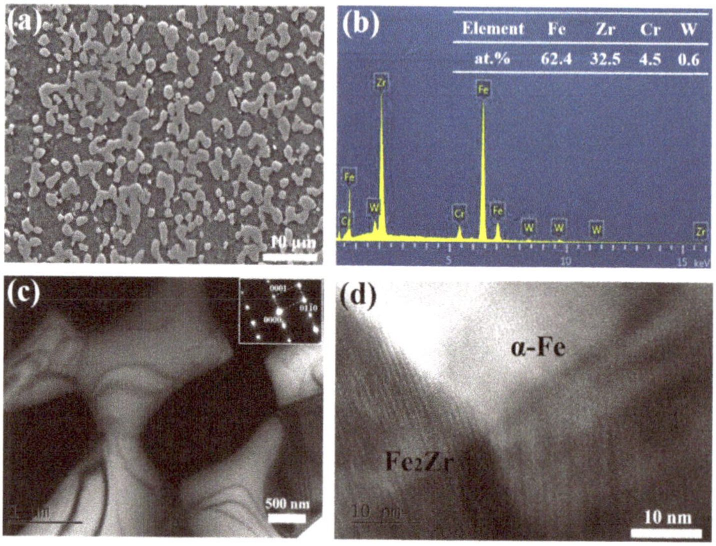

Figure 2. (**a**,**b**) SEM/EDS, (**c**) TEM, and (**d**) HRTEM image of the Fe-Cr-Zr alloy. (**b**) EDS point analysis of Fe_2Zr phase. The inset in (**c**) represents the SAED patterns of Fe_2Zr phase.

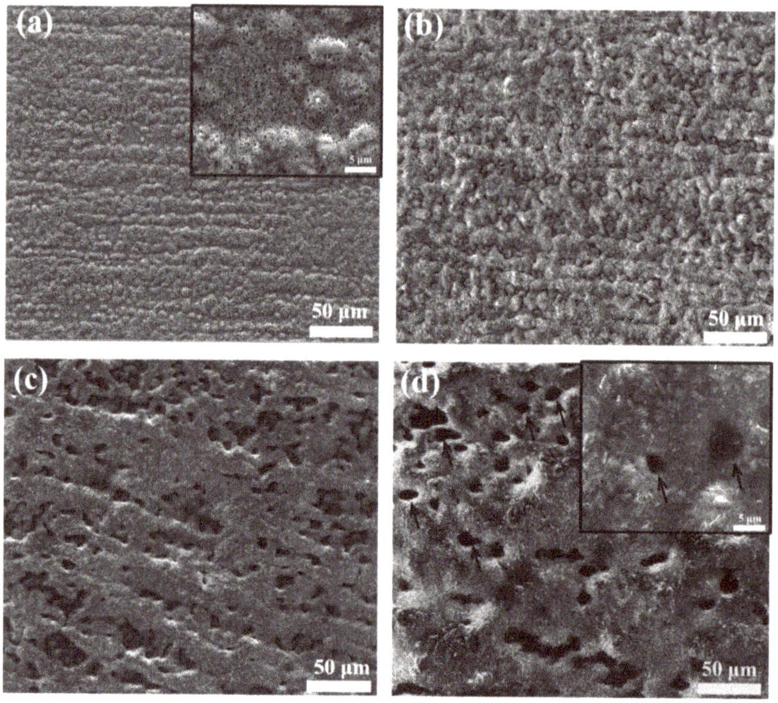

Figure 3. SEM surface morphologies of the Fe-Cr-Zr alloy oxidized in air at 923 K for (**a**) 10 min, (**b**) 20 h, (**c**) 500 h, and (**d**) 1000 h.

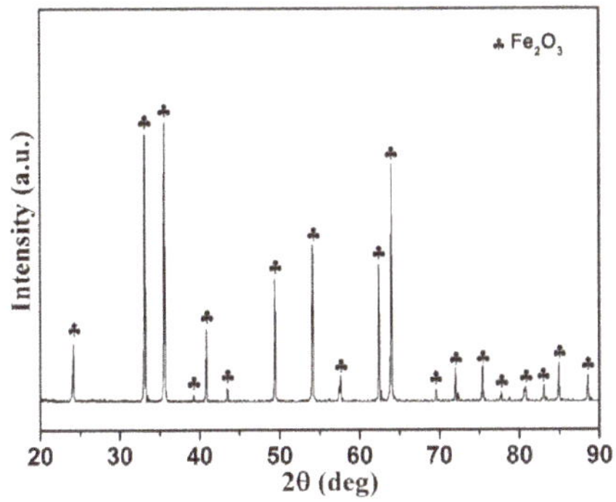

Figure 4. XRD patterns of the Fe-Cr-Zr alloy oxidized in air at 923 K for 1000 h.

3.2.2. Cross-Sectional Morphology

Figure 5a shows a backscattered detector (BSE) image of the cross-sectional morphologies of oxide scale formed on Fe-Cr-Zr alloy after oxidation at 923 K for 10 min. The matrix/scale interface appears to roughen, and incorporated Fe_2Zr phase could be found at the oxidation front due to their white contrast compared to the oxide scale (black arrows in Figure 5a). EDS point analysis in the vicinity of the incorporated Fe_2Zr phase reveal that ~5.2 at.% of O is detected at the incorporated phase (Figure 5b), and concentrations of Fe, Zr, and Cr of the incorporated phase are similar to the composition of Fe_2Zr phase obtained in the matrix (Figure 2b). In contrast, a much higher concentration of O (~59.1 at.%) is detected in the gray scale around the incorporated phase, and the atomic percent is consistent with the stoichiometric ratio of $(Fe,Cr,Zr)_2O_3$ oxide (Figure 5c). Besides, a low concentration of O (~5.9 at.%) is present in the α-Fe adjacent to the gray scale (Figure 5d). Therefore, the incorporated Fe_2Zr phase in the oxide scale close to the matrix/scale interface exhibits delayed oxidation.

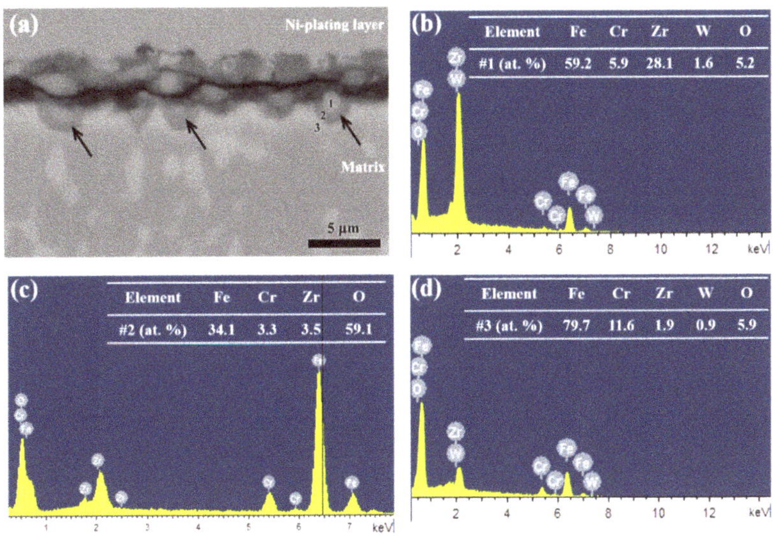

Figure 5. (**a**) BSE images of cross-sectional morphologies of Fe-Cr-Zr alloy after oxidation in air at 923 K for 10 min. (**b**–**d**) EDS spectrums for point 1, point 2, and point 3 marked in (**a**).

After oxidation for 20 h, the average thickness of the oxide scale is ~10 μm, and the incorporation of dark phase in the inner part of white scale could be observed by optical microscope (OM) (Figure 6a). BSE image shows that Fe_2Zr phase are observed in the inner part of oxide scale and some of the phase in the scale are much smaller than those in the matrix (arrows in Figure 6b), which might be resulted from the gradual oxidation of Fe_2Zr phase. EDS point analysis at the Fe_2Zr phase sites in the scale reveals that the ratio of Fe to Zr corresponds to about 2 while the measured oxygen concentration is ~33.5 at.% (Figure 6c), indicating that the Fe_2Zr phase are oxidized. By contrast, the oxide scale surrounding the Fe_2Zr phase having an oxygen concentration of 59.3 at.% (Figure 6d), is found to be fully oxidized into $(Fe,Cr,Zr)_2O_3$. Therefore, it is concluded that the incorporated Fe_2Zr phase in the oxide scale exhibit too low oxygen concentration to be fully oxidized.

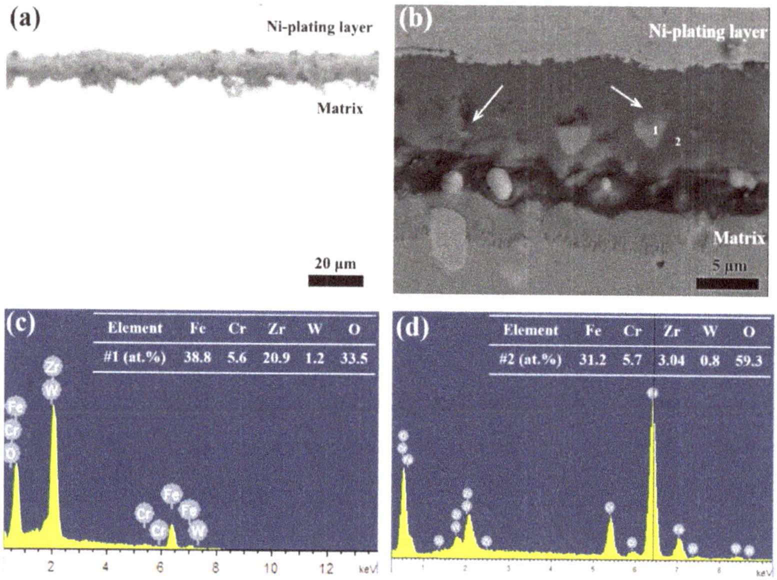

Figure 6. (**a**) OM and (**b**) BSE images of polished cross-sectional morphologies of Fe-Cr-Zr alloy after oxidation in air at 923 K for 20 h. (**c**,**d**) EDS spectrums for point 1 and point 2 marked in (**b**).

OM image of the cross-sectional morphologies of scale after oxidation of 1000 h is shown in Figure 7a. A roughened surface is observed, which is corresponding to the uneven oxide surface morphology (Figure 3d). The incorporated dark phases are also observed in the inner part of the white scale. The BSE image reveals the presence of Fe_2Zr phase in the inner part of oxide scale, and a significant decrease in number density of incorporated phase is observed from the oxide/matrix interface to the outer surface, as shown in Figure 7b. The compositional profiles by EDS point analysis show that the oxide scale has a duplex-layered structure (Figure 7c). The outer layer, with a thickness of ~20 μm, consists mainly of the 40Fe-60O (at.%), which is in line with the stoichiometry of Fe_2O_3. The inner layer, with a thickness of ~25 μm, consists mainly of the 30Fe-6Cr-3Zr-1W-60O (at.%), which is in line with the stoichiometry of $(Fe,Cr,Zr)_2O_3$. Within the inner layer, particles with higher concentration of Zr (20~26 at.%) and lower concentration of O (10~40 at.%) are the delayed oxidized Fe_2Zr phase. In addition, the concave surface (arrows in Figure 7b) is in accordance with the pits on the oxide surface morphologies (arrows in Figure 3d). It is shown that the density of incorporated Fe_2Zr phase is higher in the oxide scale beneath the concave surface than that beneath the convex surface (Figure 7b), which is reminiscent of the fact that the inhomogeneous cross-sectional morphology is related with the number density of the Fe_2Zr phase.

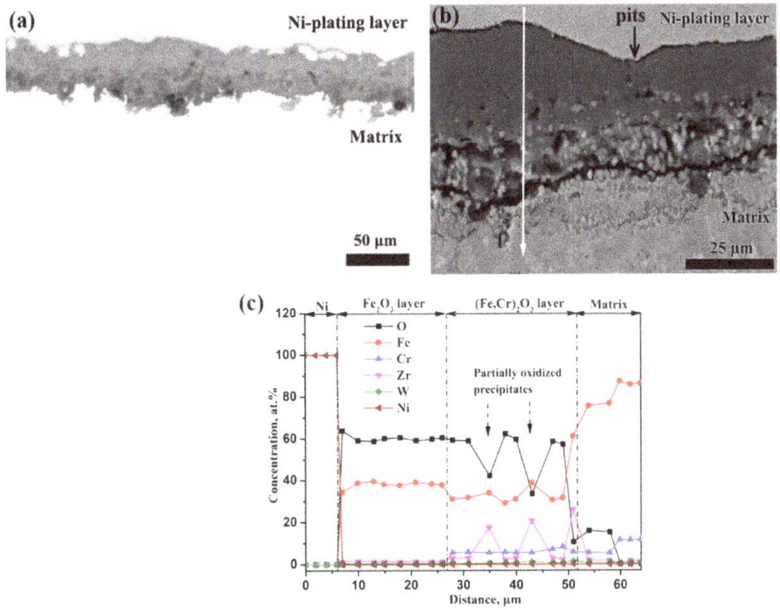

Figure 7. (**a**) OM and (**b**) BSE images of polished cross-sectional morphologies of Fe-Cr-Zr alloy after oxidation in air at 923 K for 1000 h. (**c**) Compositional profiles with the distance across the oxide scale marked in (**b**).

3.3. Oxidation Behavior in Stagnant Liquid Pb-Bi Eutectic

Figure 8 shows the BSE cross-sectional image of oxide scale formed on Fe-Cr-Zr alloy after exposure to oxygen-saturated Pb-Bi eutectic at 823 K for 500 h, 1000 h, and 2000 h, respectively. As the exposure time increases from 500 h to 2000 h, the thickness of oxide scale increases from ~10 μm to ~30 μm. The outer layer of the oxide scale seems porous. In comparison, the inner layer of the oxide scale seems compact, and the incorporation of Fe_2Zr phase in the inner layer is also observed. It could be found that the cavities are mainly present at the inner/outer layer interface of oxide scale and formed adjacent to the incorporated Fe_2Zr phase (white arrows in Figure 8), which implies cavity nucleation around the Fe_2Zr phase.

EPMA analysis of the oxide scale after exposure to oxygen-saturated Pb-Bi eutectic at 823 K for 1000 h is shown in Figure 9. In the matrix, the Fe-rich α-Fe and Zr-rich Fe_2Zr phase could be clearly identified. The oxide scale formed on the Fe-Cr-Zr alloy exhibits a three-layered structure. An enrichment of Fe and O is observed in the outer oxide layer while Cr and Zr could not be detected in this layer, revealing that the outward diffusion of Fe from matrix to the external interface. The outer layer composed of the iron oxide is identified as magnetite (Fe_3O_4) according to the SEM/EDS [19], XRD [21] and glow discharge optical emission spectroscopy (GD-OES) [22]. The penetration of Pb/Bi into the formed magnetite layer could also be observed. In comparison with the outer oxide layer, the inner oxide layer is a Fe-Cr spinel which is enriched by Cr and depleted in Fe. An inhomogeneous distribution of O is present in the inner oxide layer, and the region with higher concentration of O is found exactly to be at the α-Fe site (black arrows in Figure 9). The lower concentration of O at the Fe_2Zr phase site may be resulted from the delayed oxidation of the phase. The ratio of outer magnetite layer thickness on the inner Fe-Cr spinel layer thickness is ~1.2, which is in agreement with the oxidation results of ferritic/martensitic steels in liquid Pb-Bi eutectic [22–24]. An internal oxidation zone (IOZ) with a significant decreased concentration of O develops between the matrix and inner oxide scale, and a relative homogeneous distribution of O exists in the vicinity of the inner oxide layer/IOZ interface.

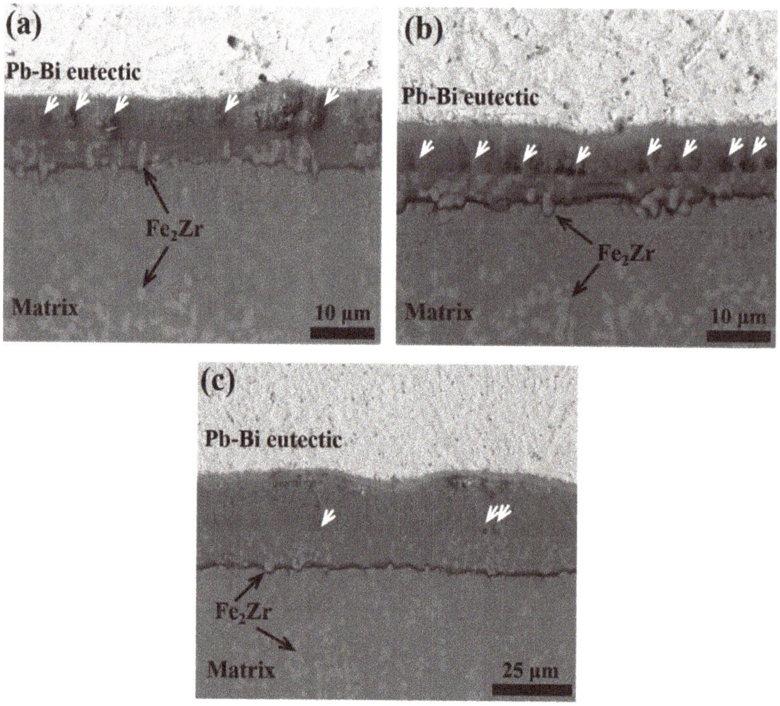

Figure 8. BSE images of cross-sectional morphologies of Fe-Cr-Zr alloy after exposure to oxygen-saturated Pb-Bi eutectic at 823 K for (**a**) 500 h, (**b**) 1000 h, and (**c**) 2000 h.

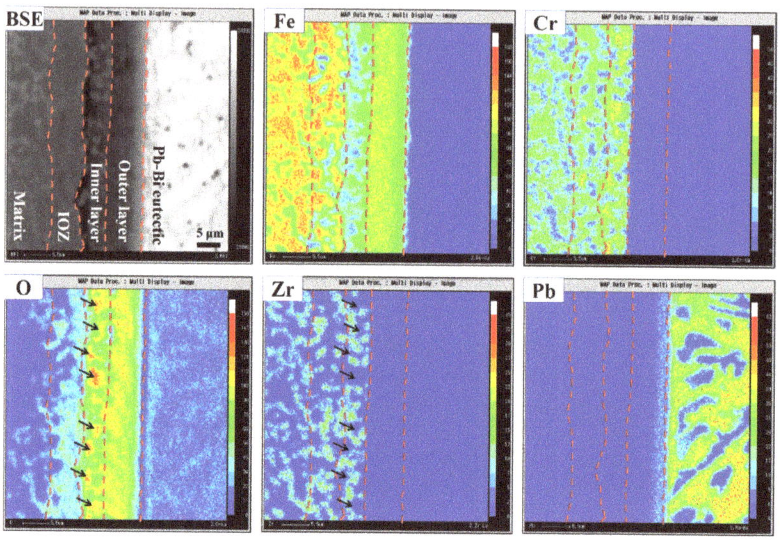

Figure 9. EPMA analysis of the cross-sectional area of Fe-Cr-Zr alloy after exposure to oxygen-saturated Pb-Bi eutectic at 823 K for 1000 h.

4. Discussion

4.1. Oxidation Behavior of Fe_2Zr Phase

The oxidation behaviors of the Fe-Cr-Zr alloy reveal the delayed oxidation of the Fe_2Zr phase with respect to the α-Fe matrix. The incorporated Fe_2Zr phase with low oxygen concentration could be observed in the oxide scale close to the matrix/scale interface (Figure 5). With increased oxidation time, the incorporated Fe_2Zr phase with increased oxygen concentration in the scale indicates the gradual oxidation of Fe_2Zr phase. It is shown that iron oxide in the form of Fe_2O_3 is determined by EDS and XRD during oxidation in air at 923 K, and Zr is reported to be oxidized to ZrO_2 [25]. The oxidation reaction of the Fe_2Zr phase could be written as follows:

$$Fe_2Zr\ (s) + 5/2O_2\ (g) = Fe_2O_3\ (s) + ZrO_2\ (s) \tag{1}$$

It is concluded that the oxygen concentration should be between 60 and 66 at.% at the fully oxidized Fe_2Zr phase site. It is assumed that the lower measured oxygen concentration at the incorporated Fe_2Zr phase site is caused by the partially oxidized Fe_2Zr phase. An approximate fraction of oxidized Fe_2Zr phase can be calculated according to the quantitative EDS results at the incorporated Fe_2Zr phase sites considering that the ratio of Fe to Zr is close to 2 at these sites (Figure 6c). This calculation is made assuming that each analysis point inside the incorporated Fe_2Zr phase consists of oxidized Fe_2Zr and metallic Fe_2Zr. According to the hypotheses, compositions of the incorporated Fe_2Zr phase can be calculated by using the following relations:

$$x_{Fe} = 2x_{oxidized} + 2x_{metallic} \tag{2}$$

$$x_{Zr} = x_{oxidized} + x_{metallic} \tag{3}$$

$$x_O = 5x_{oxidized} \tag{4}$$

The resolution of Equations (2)–(4) leads to:

$$x_{oxidized}/(x_{oxidized} + x_{metallic}) = 0.2(x_O/x_{Zr}) = 0.4(x_O/x_{Fe}) \tag{5}$$

For each analysis point inside the incorporated Fe_2Zr phase, the ratio of O to Fe or O to Zr could be obtained by EDS. As a consequence, the mole fraction of oxidized Fe_2Zr at the incorporated Fe_2Zr phase sites can be determined by the Equation (5). For the incorporated Fe_2Zr phase (marked position #1 in Figure 5a) in the oxide scale close to the matrix/scale interface after oxidation of 10 min, the calculated mole fraction of oxidized Fe_2Zr phase is ~4%. For the incorporated Fe_2Zr phase (marked position #1 in Figure 6b) in the scale after oxidation of 2 h, the calculated mole fraction of oxidized Fe_2Zr phase reaches ~34.5%.

Based on the oxidation reaction of Fe_2Zr (1), standard Gibbs free energy change of the oxidation of metallic Fe_2Zr phase to oxides could be calculated as

$$\Delta G^\circ ox(Fe_2Zr) = -744268 + 175.8T\ J/mol \tag{6}$$

where the standard Gibbs free energy changes of the formation of Fe_2O_3 and ZrO_2 are obtained from the Ellingham/Richardson diagram [26,27] and the standard Gibbs free energy changes of the Fe_2Zr formation from the constituting elements (2Fe + Zr = Fe_2Zr) is given as $\Delta G^\circ = -30400 + 12.7T$ J/mol [28]. At 923 K, $\Delta G^\circ ox(Fe_2Zr) = -582004$ J/mol, and the calculated equilibrium oxygen partial pressure value of p_{O2} is 1.2×10^{-33} atm. For the alloy matrix with the chemical composition of 86Fe-12Cr-2Zr (at.%), the calculated equilibrium oxygen partial pressure value of p_{O2} for the oxidation of Cr solute and Zr solute is 5.2×10^{-34} atm and 2.6×10^{-51} atm, assuming that Cr and Zr solutes would oxidize to their respective oxides. The maximum value of p_{O2} available to a dilute Fe-Cr alloy could be set by Fe-Fe_2O_3 equilibrium because a scale forms on the alloy surface. Accordingly, the maximum value of p_{O2} available is 1.3×10^{-22} atm at 923 K. It is demonstrated that oxidation of metallic Fe_2Zr phase and alloy matrix at 923 K in stagnant air is thermodynamically favorable as evidenced by the initial oxidation behavior (Figure 5). As the oxidation process continues, the oxide scale formed around the Fe_2Zr phase would prevent the inward diffusion of O into the phase, leading to a lower oxygen partial pressure value inside the Fe_2Zr phase. Once the value of p_{O2} is lower than 1.2×10^{-33} atm inside the Fe_2Zr phase, there would be delayed oxidation of the Fe_2Zr phase with respect to the alloy matrix (Figure 6).

Oxidation results of Fe-Cr-Zr alloy in air and liquid Pb-Bi eutectic also show that the presence of incorporated Fe_2Zr phase could significantly affect the oxidation behaviors of the alloy. In order to evaluate the oxidation resistance of the Fe-Cr-Zr alloy, the oxidation behaviors of Fe-Cr-Zr alloy in air and liquid Pb-Bi eutectic are compared with that of ferritic/martensitic steels and ODS steels with the similar Cr content of 9~12 wt.% [29–36]. As shown in Figure 10, the weight gain curves after oxidation at 923 K in air reveal that the weight gain of Fe-Cr-Zr alloy is significantly greater than that of ferritic/martensitic steels, and the obtained parabolic rate constant of Fe-Cr-Zr alloy (~3.8×10^{-2} mg^2 cm^{-4} h^{-1}) is much higher than that of ferritic/martensitic steels (4.4×10^{-3}~9.4×10^{-6} mg^2 cm^{-4} h^{-1}) according to the parabolic kinetic law. However, the thickness of the oxide scale after oxidation in oxygen-saturated liquid Pb-Bi eutectic at 823 K shows that the growth rate of oxide scale thickness of Fe-Cr-Zr alloy is similar to that of ferritic/martensitic steels and ODS steels (Figure 11), demonstrating that the experimental parabolic rate constants of Fe-Cr-Zr alloy, ferritic/martensitic steels and ODS steels are in the range of 1.9×10^{-7} μm^2 h^{-1} to 4.4×10^{-7} μm^2 h^{-1} according to the parabolic kinetic law. Therefore, the oxidation mechanism of the Fe-Cr-Zr alloy in air and liquid Pb-Bi eutectic will be discussed separately.

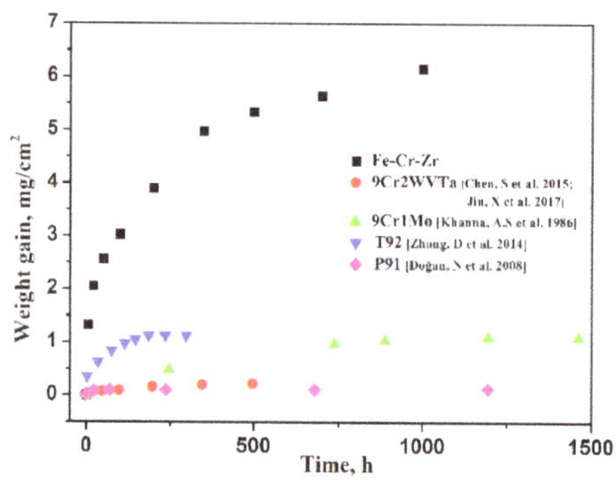

Figure 10. The weight gain curves of Fe-Cr-Zr alloy oxidized in air at 923 K compared with the ferritic/martensitic steels (Data from [29–33]).

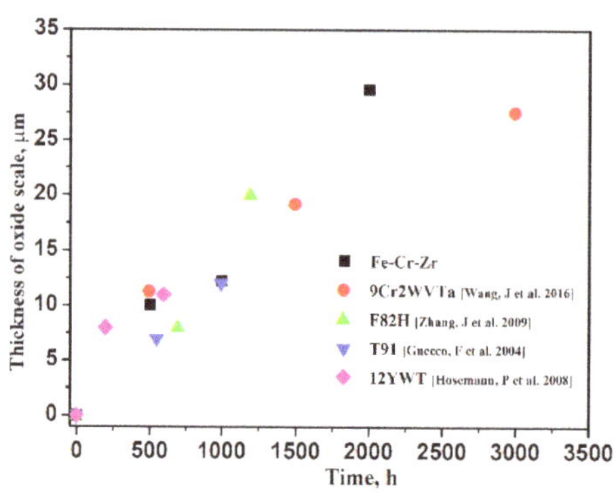

Figure 11. Thickness of oxide scale of Fe-Cr-Zr alloy obtained by oxidation in oxygen-saturated liquid Pb-Bi eutectic at 823 K compared with the ferritic/martensitic steels (Data from [19,34–36]).

4.2. Oxidation Mechanism in Air

During oxidation in air, oxidation process involves the diffusion of reactants through the oxide scale (i.e., solute is transported through the matrix to the scale/air interface and oxygen is transported to the scale/matrix interface). Generally, diffusion of solutes in the matrix is usually correlated with the bulk diffusion and grain boundary diffusion, and the grain boundary diffusion rate is widely considered to be much larger than the bulk diffusion rate [37–39]. For the Fe-Cr-Zr alloy, the presence of incoherent α-Fe/Fe_2Zr interface could also affect the diffusion rate of solutes. Investigations on the interface diffusivities along the metal/ceramic interface, metal/SiC interface, and metal/SiN interface have revealed that interfaces were the high-diffusivity paths for metal atoms, which was much faster than the bulk diffusion [40,41]. Therefore, diffusion along the α-Fe/Fe_2Zr interface might promote the outward diffusion of solutes.

In the early stage of oxidation, Cr is preferentially oxidized to form Cr_2O_3 due to its higher affinity to oxygen according to Ellingham/Richardson diagram. However, a continuous Cr_2O_3 oxide scale could not be formed due to that the low content of Cr (~9 wt.%) in the Fe-Cr-Zr alloy. Fe would also diffuse outward to form Fe_2O_3 oxide, and thus $(Fe,Cr)_2O_3$ oxide scale are generated in the scale. Compositional analysis in the vicinity of the matrix/scale interface reveals that the incorporated Fe_2Zr phase in the oxide scale exhibits delayed oxidation with respect to the α-Fe (Figure 5a). With increased oxidation time, Fe/Cr and O continually penetrate through the less compact $(Fe,Cr)_2O_3$ scale. The inadequate supply of Cr in the matrix could not replenish the Cr consumed by the scale due to the lower concentration of Cr. As a result, Fe_2O_3 is formed in the outer oxide layer and $(Fe,Cr,Zr)_2O_3$ is present in the inner oxide layer (Figure 7c). The Fe_2Zr phases showing delayed oxidation are incorporated in the scale (Figure 7b), and the presence of incorporated phase would obstruct the diffusion of solute/oxygen through the scale. It can be concluded that the oxidation process would be affected by the number density of the incorporated Fe_2Zr phase. A higher density of the incorporated Fe_2Zr phase could effectively retard the outward diffusion of Fe during oxidation process. Thus, a thinner Fe_2O_3 layer is formed in the area with higher density of Fe_2Zr phase due to the lower external oxidation rate, which is consistent with the concave surface on the cross-sectional morphologies (Figure 7b) and pits on the surface morphologies (Figure 3d). The schematic illustration of oxidation behaviors in air is shown in Figure 12a.

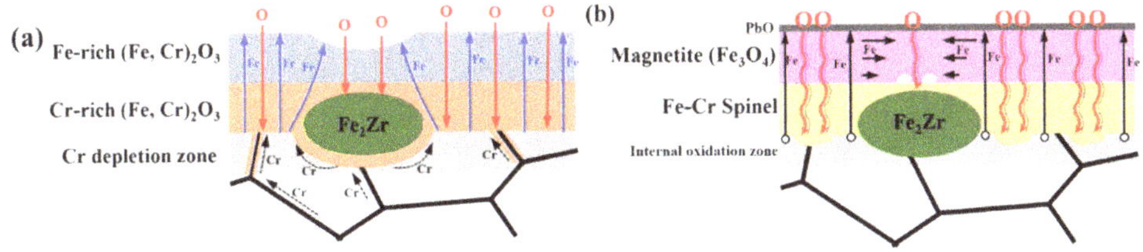

Figure 12. The schematic illustration of oxidation mechanism of Fe-Cr-Zr alloy after oxidation (**a**) in air and (**b**) in oxygen-saturated liquid Pb-Bi eutectic.

It is shown that the growth rate of oxide scale in Fe-Cr-Zr alloy after oxidation in air is mainly controlled by diffusion of solutes/oxygen through the scale, which is similar to that of ferritic/martensitic steels [29,30]. A large disparity in the oxidation rates between the Fe-Cr-Zr alloy and ferritic/martensitic steels is correlated with their different microstructural features. Firstly, the grain size of Fe-Cr-Zr alloy (in the micrometer-sized range) is much larger than the sub-grain size (in the nanometer-sized range) of ferritic/martensitic steels [7], and as a consequence, higher diffusivity of Cr from matrix could promote the formation of continuous Cr-rich $(Fe,Cr)_2O_3$ inner oxide layer via short-circuit diffusion of sub-boundary in ferritic/martensitic steels. Secondly, a higher density of dislocations in the lath martensite of ferritic/martensitic steels could facilitate the bulk diffusion and then accelerate the delivery of solutes into fringes around the grain boundaries, which also promotes the formation of continuous Cr-rich $(Fe,Cr)_2O_3$ inner oxide layer. Finally, α-Fe/Fe_2Zr interfaces serve as the preferential oxidation sites could promote the internal oxidation, and the incorporated Fe_2Zr phase might causes the thermal and growth stress in the oxide scale.

4.3. Oxidation Mechanism in Pb-Bi Eutectic

Oxidation results of Fe-Cr-Zr alloy in oxygen-saturated Pb-Bi eutectic also show that Fe_2Zr phase shows delayed oxidation with respect to α-Fe. It is found that the magnetite/Fe-Cr spinel interface is the original alloy/Pb-Bi interface according to the spatial position of Fe_2Zr phase in the oxide scale (Figure 8), demonstrating that outward diffusion of Fe until oxide/liquid Pb-Bi interface to form the outer magnetite layer while the inner Fe-Cr spinel layer grows at the scale/matrix interface. Unlike the air oxidation in Section 4.2, inward diffusion of O is not achieved by the diffusion inside the oxide lattice, but the presence of nano-channels caused by the Pb-Bi penetrations (Figure 9) are considered as a fast diffusion path for O in the Pb-Bi eutectic [22–24]. The constant ratio (~1.2) of outer magnetite layer thickness on the inner Fe-Cr spinel layer thickness is observed throughout the oxidation process (Figure 8), demonstrating that fast O diffusion paths inside nano-channels do not account for the growth of inner Fe-Cr spinel layer (since it is suggested that the growth rates of outer magnetite layer and inner Fe-Cr spinel layer are directly correlated). It is shown that growth rate of the outer magnetite layer is controlled by the outward diffusion of Fe across the scale by EPMA, and the outward diffusion of Fe would generate vacancies at the scale/matrix interface. The generated vacancies would be accumulated to form nano-cavities at the scale/matrix interface, which is verified by the porous morphology at this interface (Figure 8). The presence of Cr and W with a much slower diffusion rate than Fe could impede the vacancies annihilation through inhibiting Fe vacancies movement to the oxide/metal interface [23]. Meanwhile, O transported by diffusion in the nano-channel could react with the alloy inside the nano-cavities. An equivalent amount of Cr in the matrix and in the Fe-Cr spinel through EPMA analysis (Figure 9) also revealed that the diffusion of Cr in both matrix and Fe-Cr spinel is negligible. It could be deduced that thickness of the newly-formed Fe-Cr spinel is equivalent to the consumed matrix volume, implying that the growth rate of Fe-Cr spinel is also dependent on the outward diffusion of Fe across the scale.

The incorporated Fe_2Zr phase in the inner Fe-Cr spinel scale would impede the diffusion of Fe across the Fe-Cr spinel scale. As a result, a reduction in the Fe supply to the magnetite/Fe-Cr spinel interface is present adjacent to the Fe_2Zr phase, and looser magnetite would be formed around the Fe_2Zr phase, resulting in the creation of cavities in the outer magnetite layer with prolonged oxidation time (arrows in Figure 8). Moreover, grain dissociation of scale above the nano-cavities is responsible for the nano-channels formation in the scale according to dissociative/perforative growth theory [22]. The presence of Fe-Cr spinel/Fe_2Zr interface is the preferential sites for grain dissociation, and would promote the formation of nano-channels at these interfaces. The schematic illustration of oxidation behaviors in oxygen-saturated Pb-Bi eutectic is shown in Figure 12b.

Unlike the large disparity in the oxidation rates between the Fe-Cr-Zr alloy and other ferritic/martensitic steels after air oxidation, the growth rate of oxide scale thickness of Fe-Cr-Zr alloy in oxygen-saturated Pb-Bi eutectic is similar to that of ferritic/martensitic steels and ODS steels. On the one hand, the growth rates of outer magnetite layer and inner Fe-Cr spinel layer in oxygen-saturated Pb-Bi eutectic are both related with the Fe diffusion across the oxide scale, but differences in the microstructural features between Fe-Cr-Zr alloy, ferritic/martensitic steels and ODS steels have little effect on the out-ward diffusion behavior of Fe. On the other hand, diffusion of Cr in both matrix and Fe-Cr spinel is negligible due to the rapid diffusion of O inside the nano-channels, and similar Cr content of 9~12 wt.% in these materials could lead to the same growth rate of inner Fe-Cr spinel layer. Finally, the incorporated Fe_2Zr phase could affect the compactness of the outer magnetite layer, but do not significantly affect the growth rate of scale.

5. Conclusions

A Fe-Cr-Zr alloy with a dispersed Fe_2Zr phase in the α-Fe matrix was produced, and the oxidation behaviors of the Fe-Cr-Zr alloy were investigated in stagnant air and in oxygen-saturated stagnant liquid Pb-Bi eutectic, respectively.

Delayed oxidation of Fe_2Zr phase with respect to the α-Fe matrix is found during oxidation in air and in liquid Pb-Bi eutectic at 823 K and 923 K based on the experimental results and the thermodynamic calculations.

After oxidation in air at 923 K, a duplex structure including an outer Fe_2O_3 layer and an inner $(Fe,Cr,Zr)_2O_3$ layer was developed due to the inadequate supply of Cr. The incorporated Fe_2Zr phase with delayed oxidation in the scale would obstruct the diffusion of solute/oxygen across the scale, resulting in the nonuniform oxidation behavior. The greater oxidation rate of Fe-Cr-Zr alloy than that of ferritic/martensitic steels is directly correlated with their microstructural differences.

After oxidation in oxygen-saturated stagnant liquid Pb-Bi eutectic at 823 K, a three-layered structure consisting of an outer magnetite layer, inner Fe-Cr spinel layer, and internal oxidation zone is formed. The incorporated Fe_2Zr phase in the inner Fe-Cr spinel would obstruct the diffusion of Fe across the Fe-Cr spinel, which is a limiting step for the scale growth. A reduction in the Fe supply to the magnetite/Fe-Cr spinel interface is present adjacent to the Fe_2Zr phase, leading to the creation of cavities in the outer magnetite layer with prolonged oxidation time. Comparable growth rates of oxide scale thickness between Fe-Cr-Zr alloy and ferritic/martensitic steels are due to that the differences in the microstructural features have little effect on the limiting step of scale growth.

Author Contributions: Conceptualization, S.C. and L.R.; methodology, S.C.; validation, S.C. and L.R.; formal analysis, S.C.; investigation, S.C.; resources, L.R.; data curation, S.C.; writing—original draft preparation, S.C.; writing—review and editing, S.C.; visualization, S.C.; supervision, L.R.; project administration, L.R.; funding acquisition, L.R. All authors have read and agreed to the published version of the manuscript.

Funding: This work is financially supported by the National Natural Science Foundation of China (No. 51871218), Youth Innovation Promotion Association, CAS (No. 2018227), and Natural Science Foundation of Liaoning Province (No. 2020-MS-010).

Data Availability Statement: Not applicable.

Acknowledgments: The authors are grateful to Xiaojie Jin for help with oxidation tests in oxygen-saturated stagnant liquid Pb-Bi eutectic.

Conflicts of Interest: The authors declare no conflict of interest.

References

1. Klueh, R.; Nelson, A. Ferritic/martensitic steels for next-generation reactors. *J. Nucl. Mater.* **2007**, *371*, 37–52. [CrossRef]
2. Guérin, Y.; Zinkle, S.J.; Was, G.S. Materials challenges for advanced nuclear energy systems. *MRS Bull.* **2009**, *34*, 10–19.
3. Zinkle, S.J.; Was, G.S. Materials challenges in nuclear energy. *Acta Mater.* **2013**, *61*, 735–758. [CrossRef]
4. Chen, S.; Rong, L. Effect of silicon on the microstructure and mechanical properties of reduced activation ferritic/martensitic steel. *J. Nucl. Mater.* **2015**, *459*, 13–19. [CrossRef]
5. Agamennone, R.; Blum, W.; Gupta, C.; Chakravartty, J.K. Evolution of microstructure and deformation resistance in creep of tempered martensitic 9–12%Cr–2%W–5%Co steels. *Acta Mater.* **2006**, *54*, 3003–3014. [CrossRef]
6. Ghassemi-Armaki, H.; Chen, R.; Maruyama, K.; Yoshizawa, M.; Igarashi, M. Static recovery of tempered lath martensite microstructures during long-term aging in 9–12% Cr heat resistant steels. *Mater. Lett.* **2009**, *63*, 2423–2425. [CrossRef]
7. Chen, S.; Jin, X.; Rong, L. Microstructural Evolution and Mechanical Properties of Ultrafine-Grained Ferritic-Martensitic Steel During Thermal Aging. *Met. Mater. Trans. A* **2020**, *51*, 5154–5168. [CrossRef]
8. Ukai, S.; Fujiwara, M. Perspective of ODS alloys application in nuclear environments. *J. Nucl. Mater.* **2002**, *307*, 749–757. [CrossRef]
9. Hayashi, T.; Sarosi, P.; Schneibel, J.; Mills, M. Creep response and deformation processes in nanocluster-strengthened ferritic steels. *Acta Mater.* **2008**, *56*, 1407–1416. [CrossRef]
10. Alinger, M.; Odette, G.; Hoelzer, D. On the role of alloy composition and processing parameters in nanocluster formation and dispersion strengthening in nanostuctured ferritic alloys. *Acta Mater.* **2009**, *57*, 392–406. [CrossRef]
11. Yamamoto, K.; Kimura, Y.; Wei, F.; Mishima, Y. Design of Laves phase strengthened ferritic heat resisting steels in the Fe–Cr–Nb(–Ni) system. *Mater. Sci. Eng. A* **2002**, *329*, 249–254. [CrossRef]

12. Morris, D.G.; Muñoz-Morris, M.A.; Requejo, L.M. New iron–aluminium alloy with thermally stable coherent intermetallic nanoprecipitates for enhanced high-temperature creep strength. *Acta Mater.* **2006**, *54*, 2335–2341. [CrossRef]
13. Morris, D.G.; Gutierrez-Urrutia, I.; Muñoz-Morris, M.A. The high-temperature creep behaviour of an Fe–Al–Zr alloy strengthened by intermetallic precipitates. *Scr. Mater.* **2007**, *57*, 449–452. [CrossRef]
14. Yang, Y.; Tan, L.; Bei, H.; Busby, J.T. Thermodynamic modeling and experimental study of the Fe–Cr–Zr system. *J. Nucl. Mater.* **2013**, *441*, 190–202. [CrossRef]
15. Jin, X.; Chen, S.; Rong, L. Effect of Fe2Zr phase on the mechanical properties and fracture behavior of Fe-Cr-W-Zr ferritic alloy. *Mater. Sci. Eng. A* **2018**, *722*, 173–181. [CrossRef]
16. Tan, L.; Yang, Y.; Chen, T.; Sridharam, K.; He, L. *Mechanical Properties and Radiation Resistance of Nanoprecipitates-Strengthened Advanced Ferritic Alloys*; Oak Ridge National Lab.(ORNL): Oak Ridge, TN, USA, 2017. [CrossRef]
17. Tan, L.; Yang, Y.; Sridharan, K. *Accelerated Development of Zr-Containing New Generation Ferritic Steels for Advanced Nuclear Reactors*; Oak Ridge National Lab.(ORNL): Oak Ridge, TN, USA, 2015. [CrossRef]
18. Chen, S.; Rong, L. Oxide scale formation on ultrafine-grained ferritic-martensitic steel during pre-oxidation and its effect on the corrosion performance in stagnant liquid Pb-Bi eutectic. *Acta Metall. Sin.* **2021**, *57*, 989–999.
19. Wang, J.; Lu, S.; Rong, L.; Li, D.; Li, Y. Effect of silicon on the oxidation resistance of 9 wt.% Cr heat resistance steels in 550 °C lead-bismuth eutectic. *Corros. Sci.* **2016**, *111*, 13–25. [CrossRef]
20. Lu, Y.; Song, Y.; Chen, S.; Rong, L. Effects of Al and Si on mechanical properties and corrosion resistance in liquid Pb-Bi eutectic of 9Cr2WVTa steel. *Acta Metall. Sin.* **2016**, *52*, 298–306.
21. Gong, X.; Short, M.P.; Auger, T.; Charalampopoulou, E.; Lambrinou, K. Environmental degradation of structural materials in liquid leadand lead-bismuth eutectic-cooled reactors. *Prog. Mater. Sci.* **2022**, *126*, 100920. [CrossRef]
22. Martinelli, L.; Balbaud-Célérier, F.; Terlain, A.; Delpech, S.; Santarini, G.; Favergeon, J.; Moulin, G.; Tabarant, M.; Picard, G. Oxidation mechanism of a Fe–9Cr–1Mo steel by liquid Pb–Bi eutectic alloy (Part I). *Corros. Sci.* **2008**, *50*, 2523–2536. [CrossRef]
23. Martinelli, L.; Balbaud-Célérier, F.; Terlain, A.; Bosonnet, S.; Picard, G.; Santarini, G. Oxidation mechanism of an Fe–9Cr–1Mo steel by liquid Pb–Bi eutectic alloy at 470 °C (Part II). *Corros. Sci.* **2008**, *50*, 2537–2548. [CrossRef]
24. Martinelli, L.; Balbaud-Célérier, F.; Picard, G.; Santarini, G. Oxidation mechanism of a Fe–9Cr–1Mo steel by liquid Pb–Bi eutectic alloy (Part III). *Corros. Sci.* **2008**, *50*, 2549–2559. [CrossRef]
25. Proff, C.; Abolhassani, S.; Lemaignan, C. Oxidation behaviour of zirconium alloys and their precipitates—A mechanistic study. *J. Nucl. Mater.* **2013**, *432*, 222–238. [CrossRef]
26. Young, D.J. *High. Temperature Oxidation and Corrosion of Metals*, 2nd ed.; Elsevier: Sydney, Australia, 2008; ISBN 9780081001011.
27. Robino, C.V. Representation of mixed reactive gases on free energy (Ellinghan-Richardson) diagrams. *Metall. Mater. Trans. B* **1996**, *27*, 65–69. [CrossRef]
28. Matsui, T. Vaporization study on M2Zr (M=Fe and Cr) by a mass-spectrometric method. *J. Nucl. Mater.* **1993**, *201*, 278–283. [CrossRef]
29. Chen, S.; Jin, X.; Rong, L. Improvement in High Temperature Oxidation Resistance of 9 %Cr Ferritic–Martensitic Steel by Enhanced Diffusion of Mn. *Oxid. Met.* **2015**, *85*, 189–203. [CrossRef]
30. Jin, X.; Chen, S.; Rong, L. Effects of Mn on the mechanical properties and high temperature oxidation of 9Cr2WVTa steel. *J. Nucl. Mater.* **2017**, *494*, 103–113. [CrossRef]
31. Khanna, A.S.; Rodriguez, P.; Gnanamoorthy, J.B. Oxidation Kinetics, Breakaway oxidation, and inversion phenomenon in 9Cr—lMo steels. *Oxid. Met.* **1986**, *26*, 171–200. [CrossRef]
32. Zhang, D.; Liu, J.; Xue, Z.; Mao, X. Oxidation behavior of T92 steel with NiCoCrAlY coating by EB-PVD. *Surf. Coat. Technol.* **2014**, *252*, 179–185. [CrossRef]
33. Doğan, N.; Holcomb, G.R.; Alman, D.E.; Jablonski, P.D. *Steamside Oxidation Behavior of Experimental 9% Cr Steels, DOE/NETL-IR-2008-007*; U.S. Department of Energy, National Energy Technology Laboratory: Pittsburgh, PA, USA, 2008.
34. Zhang, J. A review of steel corrosion by liquid lead and lead–bismuth. *Corros. Sci.* **2009**, *51*, 1207–1227. [CrossRef]
35. Gnecco, F.; Ricci, E.; Bottino, C.; Passerone, A. Corrosion behaviour of steels in lead–bismuth at 823 K. *J. Nucl. Mater.* **2004**, *335*, 185–188. [CrossRef]
36. Hosemann, P.; Thau, H.; Johnson, A.; Maloy, S.; Li, N. Corrosion of ODS steels in lead–bismuth eutectic. *J. Nucl. Mater.* **2008**, *373*, 246–253. [CrossRef]
37. Mehrer, H. *Diffusion in Solids*; Springer: Berlin, Germany, 2007.
38. Raman, R.K.S.; Gupta, R.K. Oxidation resistance of nanocrystalline vis-à-vis microcrystalline Fe–Cr alloys. *Corros. Sci.* **2009**, *51*, 316–321. [CrossRef]
39. Chen, S.; Rong, L. Roles of Mn in the High-Temperature Air Oxidation of 9Cr Ferritic–Martensitic Steel After Severe Plastic Deformation. *Oxid. Met.* **2017**, *89*, 415–428. [CrossRef]
40. Kumar, A.; Barda, H.; Klinger, L.; Finnis, M.; Lordi, V.; Rabkin, E.; Srolovitz, D.J. Anomalous diffusion along metal/ceramic interfaces. *Nat. Commun.* **2018**, *9*, 5251. [CrossRef]
41. Amram, D.; Klinger, L.; Gazit, N.; Gluska, H.; Rabkin, E. Grain boundary grooving in thin films revisited: The role of interface diffusion. *Acta Mater.* **2014**, *69*, 386–396. [CrossRef]

metals

Article

Corrosion Mechanism of L360 Pipeline Steel Coated with S8 in CO_2-Cl^- System at Different pH Values

Fan Wang [1], Jinling Li [1,2,*], Chengtun Qu [1,2], Tao Yu [1,2], Yan Li [1], Shidong Zhu [2,3,*], Bo Yang [1,2] and Frank Cheng [4]

1 School of Chemistry and Chemical Engineering, Xi'an Shiyou University, Xi'an 710065, China; 15929739473@163.com (F.W.); xianquct@xsyu.edu.cn (C.Q.); 180708@xsyu.edu.cn (T.Y.); liyan85@xsyu.edu.cn (Y.L.); yangbo@xsyu.edu.cn (B.Y.)
2 Shaanxi Oil and Gas Pollution Control and Reservoir Protection Key Laboratory, Xi'an Shiyou University, Xi'an 710065, China
3 School of Materials Science and Technology, Xi'an Shiyou University, Xi'an 710065, China
4 Department of Mechanical and Manufacturing Engineering, University of Calgary, Calgary, AB T2N 1N4, Canada; fcheng@ucalgary.ca
* Correspondence: lijinling@xsyu.edu.cn (J.L.); zhusd@xsyu.edu.cn (S.Z.)

Abstract: The corrosion behavior of L360 pipeline steel coated with or without elemental sulfur (S8) in CO_2–Cl^- medium at different pH was studied. An autoclave was used to simulate the working conditions for forming the corrosion scale, and an electrochemical workstation with a three-electrode cell was used to analyze the electrochemical characterization of the corrosion scale. A wire beam electrode was used to determine the potential and current distribution, and scanning electron microscopy and X-ray diffraction were used to characterize the morphology and composition of the corrosion scale. The results showed that the deposition of S8 on the surface of the electrodes caused serious localized corrosion, especially under acidic conditions. The morphology and localized corrosion intensity index further proved that the deposition of S8 significantly promoted corrosion, especially pitting corrosion. Finally, a novel corrosion mechanism of L360 pipeline steel coated with S8 in a CO_2-Cl^- environment under acidic conditions was proposed, and we then modeled the theoretical mechanisms that explained the experimental results.

Keywords: L360 pipeline steel; wire beam electrode; electrochemical impedance spectroscopy (EIS); S8 deposition; localized corrosion

Citation: Wang, F.; Li, J.; Qu, C.; Yu, T.; Li, Y.; Zhu, S.; Yang, B.; Cheng, F. Corrosion Mechanism of L360 Pipeline Steel Coated with S8 in CO_2-Cl^- System at Different pH Values. *Metals* **2021**, *11*, 1975. https://doi.org/10.3390/met11121975

Academic Editor: Renato Altobelli Antunes

Received: 1 November 2021
Accepted: 3 December 2021
Published: 8 December 2021

1. Introduction

During the exploitation of oil and gas, some solid particles are deposited on the surfaces of pipelines owing to a decrease in the flow rate and pressure [1]. Under deposit corrosion (UDC) usually occurs and is driven by the differences in the chemistry at the interface between the sediment, the substrate, and the bulk solution. This can cause catastrophic failures, such as a reduction in the equipment's integrity and pipeline perforation, because UDC is difficult to detect [2,3].

Meanwhile, more sour oil and gas fields need to be developed to resolve energy shortages [4]. The dissolution of carbon dioxide produces corrosive carbonic acid, and the hydrolysis of elemental sulfur causes the solution composition to become complex. Elemental sulfur, a yellow powder of S8, is an inorganic sediment generated by the catalytic pyrolysis product of ferrous sulfide under a high temperature and high pressure at the reservoir and deposited due to the reduction in temperature and pressure in the process of fluid production [5]. Zheng et al. [6] proposed that high pressure and temperature were conducive to the formation of polysulfides. Therefore, the chemical equilibrium reaction was changed in a high-temperature and high-pressure environment, and then the decomposition of polysulfides into elemental sulfur and hydrogen sulfide was promoted.

In contrast to the general UDC, the corrosion products of sulfur-containing sediments have semiconductor properties, which can not only promote the formation of the concentration cells but also cause galvanic corrosion with metal substrates [7,8]. Furthermore, Cl^- with electronegativity is present in the pipeline, which easily adsorbs to the positive metal surface and hinders the formation rate of a passivation film on the surface of the metal [9]. Electrochemical tests showed that the increase in Cl^- concentration in the solution could accelerate the anodic dissolution of the metal and the negative shift of the cathode potential [10].

Previous studies have revealed that when the temperature is higher than 60 °C, sulfur reacts with water, resulting in serious acidification of the corrosive solution [11]. The existence of the S8 deposition layer, the distribution of ferrous ions, and the concentration of Cl^- eventually lead to the heterogeneity of the solution [12]. The corrosion of pipeline steel is a very complex process, which is generally affected by the type of sediment and the internal solution of the pipeline. The hydrolysis of the S8 deposition layer and the dissolution of carbon dioxide acidify the solution. The pH distribution in the pipelines is not uniform, and more H^+ is accumulated in the pitting pits. Therefore, the pH distribution at different depths of the pit is heterogeneous, and the pH of the interface between the deposits and the substrate is also different from that of the surface of the bare steel. To maintain electrical neutrality, corrosive Cl^- causes the pitting corrosion pits to continue to be excavated downward, which causes fatal damage to the pipeline steel [13]. To mitigate pipeline failure, it is necessary to understand the corrosion mechanism under S8 deposition by adopting effective measuring technologies. Zagal et al. [14] suggested that acid formation caused by sulfur hydrolysis was the main factor controlling corrosion in the presence of S8. Gong et al. [15] argued that with an increase in immersion time, the uniform corrosion rate in the absence of S8 increased slightly, whereas the corrosion rate in the presence of S8 decreased with an increase in immersion time in a supercritical carbon-dioxide-saturated aqueous environment. Zhang et al. [16] studied the galvanic effect between the covered electrode and the bare electrode of mixed sediments in formation water containing CO_2 through electrochemical measurements and surface characterization. Therefore, the presence of S8 has a great influence on the corrosion behavior of oil country tubular goods (OCTG), and the degree of influence is also different. However, few studies have focused on the influence of the deposition of S8 on the corrosion behavior of steel at different pH values.

A wire beam electrode (WBE) provides a new method for monitoring the processes of localized corrosion and estimating the rate of localized corrosion [17]. A WBE can connect the cathode and anode corrosion processes and the cathode and anode areas of the corrosion surface, respectively [18]. Therefore, it can be used to study the anode and cathode processes of localized corrosion cells [19]. Wu et al. [20] used a WBE to study the galvanic corrosion of mild steel under calcium carbonate deposition and potential and galvanic mappings. They found that the polarity of the electrode covered by calcium carbonate changed over time. Chen et al. [21] used a WBE to track the development process of SRB, inducing localized corrosion of 907 steel.

In this study, the effect of the different pH values on the localized corrosion of L360 pipeline steel coated with or without S8 in a 3.5 wt% NaCl solution containing CO_2 was studied by a potentiodynamic polarization curve, EIS measurement, and a WBE. The surface morphologies of the corrosion samples were observed by scanning electron microscopy (SEM). After removing the corrosion products, the corrosion morphology of the steel substrate was observed using an OLYMPUS DSX500 optical digital microscope. The composition of the corrosion products was analyzed using X-ray diffraction.

2. Materials and Methods

2.1. Solution

The test solution was 3.5 wt% NaCl (analytical-grade reagent) solution; to avoid introducing extra chloride ions, deoxidized dilute sulfuric acid and sodium hydroxide were used to adjust the pH to 3.10, 5.15, and 7.18. The pH meter was PHS-25 (Shanghai INESA & Scientific Instrument Co., Ltd., Shanghai, China). The solution was deoxygenated with N_2 for 4 h and then pumped in CO_2 for 12 h before testing.

2.2. Corrosion Scale Preparation

An autoclave experimental setup (FCZ3-24/320, Dalian Science and Trade Experimental Equipment Co., Ltd., Dalian, China) was used to prepare the corrosion scale. The CO_2 pressure in the autoclave was maintained at 5 MPa. The material was L360 pipeline steel, and the test sample was machined to the dimensions of 50 mm × 10 mm × 3 mm, with a chemical composition (wt%) of C 0.16%, Si 0.45%, Mn 1.60%, P 0.025%, S 0.015%, V 0.06%, Nb 0.05%, Ti 0.04%, and Fe the balance. The samples were sequentially ground with 400, 800, 1200, and 2000 grit SiC papers, rinsed with distilled water, dehydrated in alcohol, and dried in cool air. The fused sublimation sulfur was coated on the electrode surface, and its thickness was controlled within 1.0–1.5 mm by $2000^{\#}$ sandpaper grinding (coating thickness was determined using a vernier caliper) to ensure that there were no cracks in the sulfur layer and the working area of the electrode was covered by the sulfur layer. The average deposition mass of the sample was calculated to be 0.9312 g using the subtraction method.

2.3. Electrode Preparation

The WBE was made of a total of 100 (10 × 10) matrices by epoxy resin array; the electrode wires were made of L360 pipeline steel wires with a diameter of 1.5 mm, which were soldered on a copper wire, and each wire's separation was 0.5 ± 0.05 mm; the total exposed area was 1.766 cm^2. The size of the conventional working electrode was 10 mm × 10 mm × 3 mm, and copper wires were soldered on the working electrodes to ensure conductivity, leaving an area of 10 mm × 10 mm as the experimental surface, and the rest of the surfaces were sealed with epoxy resin. All experimental electrodes were mechanically ground with a series of silicon carbide papers down to $2000^{\#}$ grade, washed in acetone and ethanol, dried with cold air, and prepared for testing. The fused sublimation sulfur was coated on the electrode surface to prepare the sulfur-coated electrode. After solidification, the thickness of the coating was controlled within 1.0–1.5 mm by $2000^{\#}$ sandpaper grinding (coating thickness was determined using a vernier caliper) to ensure that there were no cracks in the sulfur layer, and the working area of the electrode was covered by a sulfur layer. The average coated mass of the sample was calculated to be 0.1050 g using the subtraction method. To better simulate the deposition of S8 on the pipeline, electrochemical measurements were carried out at the end of 2 h of pre-immersion under S8 deposition to stabilize the open circuit potential (OCP).

2.4. Electrochemical Measurement

A WBE scanner (CST520, Wuhan Corrtest Instruments Corp. Ltd., Wuhan, China) was used, and the acquisition time interval for each data point was 3 s. Conventional electrochemical experiments were carried out using a Parstat 4000^+ electrochemical workstation with a three-electrode cell, a saturated calomel reference electrode (SCE), a platinum plate larger than the working electrode area as the counter electrode, and L360 steel as the working electrode. The EIS measurement frequency range spanned from 10 mHz to 100 kHz, the scan rate of the potentiodynamic polarization curve was 0.3 mV/s, and the potential was between −500 mV(vs. OCP) and +500 mV(vs. OCP). Zview software was used to simulate the impedance data using an equivalent circuit. All the tests were carried out at 60 °C.

2.5. Surface Characterization

The autoclave corrosion samples were analyzed by SEM, EDS, and XRD before removing the corrosion products. In the XRD analysis, the 2θ was 20–80° and the scanning rate was 1 °/min. The morphology of the corrosion products was observed using SEM with energy-dispersive spectroscopy (EDS). Using an optical digital microscope (OLYMPUS DSX500, Olympus Optical Industry Co., Ltd., Tokyo, Japan) in the bright field, at 200× magnification, to collect samples after removing corrosion products showed the morphology.

3. Results

3.1. Potentiodynamic Polarization Curves

Figure 1 shows the polarization curves of the bare electrode and S8-coated electrode at different pH levels. In the pH range of 3.10 to 7.18, the polarization curves of the electrode in the corrosive solution were similar. It can be observed that the potential moved toward a more negative direction and the passive range became narrower with the increase in pH, and the Ecorr decreased from −550 to −611 mV (vs. SCE) owing to the deposition of S8 on the electrode. As shown in Figure 1a, the anode was actively dissolved. In contrast to the bare electrode, the Tafel slope (b_a) of the electrode coated with S8 was smaller, and the cathode coated with S8 was greatly affected by diffusion (Figure 1b). The main reason is that when S8 contacts the metal, the electron transfer between them is more direct and rapid, and the corrosion rate is higher. Based on the Marcus theory [22], the deposition of S8 on the surface of steel will lead to the weakening of the metal–metal bonds, thereby reducing the activation energy barrier for the dissolution of the surface metal atoms.

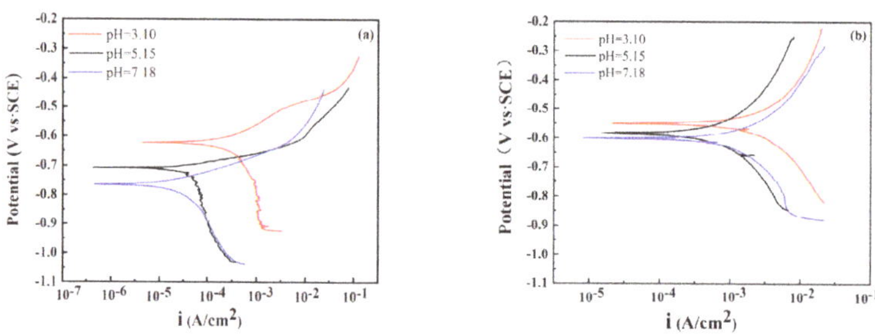

Figure 1. Potentiodynamic polarization curves with 3.5% NaCl at different pH values: (**a**) bare steel, (**b**) S8-coated steel.

The parameters, such as the self-corrosion potential, corrosion current density, and anodic and cathodic Tafel slopes (b_a, b_c), are listed in Table 1. With an increase in pH, both the current density and potential decreased, and the current density of the electrode coated with S8 was significantly higher than that of the bare electrode at the same pH. At pH = 3.10, the corrosion current density of the S8-coated electrode was twice as high as that of the bare electrode and reached 14 times at pH = 7.18. The increase in the corrosion current may be related to the hydrolysis of S8. It was proven that the corrosion potential under acidic conditions was more positive than that under neutral conditions; because of the decrease in pH, the acidity of the solution increased, and more electropositive H^+ accumulated on the surface of the metal, resulting in a positive shift of the potential. The Tafel slope shows that the b_a is greater than the b_c, and the corrosion process is controlled by the anode.

Table 1. Parameters calculated by potentiodynamic polarization curve.

Electrode	pH	E_{corr} (mV vs. SCE)	I_{corr} (μA/cm^2)	b_a (mV vs. SCE/dec)	b_c (mV vs. SCE/dec)
Bare	3.10	−640	316	264	135
	5.15	−710	42.0	432	38.1
	7.18	−761	34.4	266	67.2
S8-coated	3.10	−550	678	73.7	66.3
	5.15	−573	536	218	88.7
	7.18	−611	490	133	126

3.2. Electrochemical Impedance Spectroscopy

To further explore the effect of pH on the electrode coated with S8, EIS measurements were systematically performed, and the impedance spectra of the bare electrode and S8 coated electrode at different pH values are presented in Figure 2. The Nyquist spectrum has a capacitance semi-circular arc in the high-frequency area, and the arc center pressure is below the x-axis, which is attributed to the interface charge transfer reaction [23]. As can be seen from Figure 2a,d, there were obvious differences in capacitance diameter at different pH values, and the radius of the arc increased significantly with the increase in pH. This shows that a high pH has larger resistance, which slows down the corrosion process of the electrode surface.

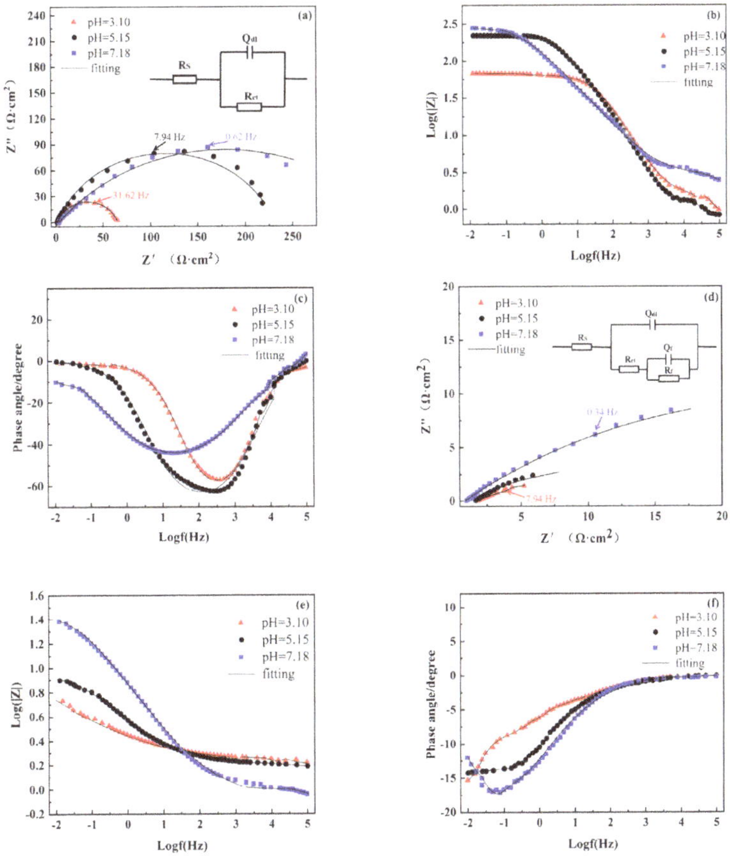

Figure 2. Nyquist plots and Bode plots of the bare electrode and S8-coated electrode at different pH values: (**a–c**) bare electrode, (**d–f**) S8-coated electrode.

From the Bode diagram (Figure 2b,c), the phase angle vs. frequency curves of the bare electrode conformed to a one-time constant model. For the S8-coated electrode (Figure 2e,f), there were two arcs on the electrode surface, which were distributed in low-frequency and high-frequency areas, respectively, and the high frequency was mainly related to the defects caused by S8 coated on the surface, and the low-frequency region was related to the electrochemical corrosion process of the electrode.

In order to fit the EIS data, all the data were fitted using the appropriate equivalent circuit as given in the insert, in which R_s is the solution resistance, R_{ct} is the charge transfer resistance, Q_{dl} is the double-layer capacitance, n is the dispersion index, and R_f and Q_f represent the resistance and capacitance of the elemental sulfur deposition layer, respectively [24]. Table 2 lists the fitting results of electrochemical impedance spectroscopy; the Chi-Squared values were between 10^{-3} and 10^{-4}, which suggests that the fitting results are reliable. With the increase in pH, both R_{ct} and R_f showed a rising tendency. The R_{ct} of the bare electrode far exceeded the sum of R_{ct} and R_f of the S8-coated electrode. Generally, the R_{ct} of the electrode is inversely proportional to the corrosion rate and can be used to characterize the corrosion rate [25]. At pH = 3.10, the R_{ct} value of the bare steel electrode was 16 times higher than that of the S8-coated electrode. Therefore, EIS measurements further showed that the corrosion of the electrode coated with S8 was more severe under acidic conditions, and the EIS results were in good agreement with the results of the polarization curves.

Table 2. EIS data fitted electrochemical parameters.

Electrode	pH	R_s ($\Omega\cdot cm^2$)	Q_{dl} ($\mu F\cdot cm^{-2}$)	n_1	R_{ct} ($\Omega\cdot cm^2$)	Q_f($mF\cdot cm^{-2}$)	n_2	R_f ($\Omega\cdot cm^2$)	Chi-Squared
	3.10	1.50	0.213	0.814	63.9	-	-	-	2.22×10^{-3}
Bare	5.15	1.04	0.382	0.776	230	-	-	-	5.14×10^{-3}
	7.18	1.70	2.04	0.570	351	-	-	-	4.17×10^{-3}
	3.10	1.68	5.65	0.731	4.02	0.490	0.485	21.7	4.85×10^{-4}
S8-coated	5.15	1.55	7.81	0.825	8.46	0.174	0.413	37.3	3.08×10^{-3}
	7.18	1.85	1.17	0.739	25.74	0.058	0.458	49.8	2.08×10^{-3}

3.3. Potential and Current Distribution

To further study the localized corrosion behavior under S8 deposition, a WBE was used to reveal the corrosion difference between the bare steel electrode and the S8-coated electrode. Tan et al. [26] first adopted the WBE to study localized corrosion and obtained localized potential and current information. A new parameter, the localized corrosion intensity index (*LCII*), is proposed to quantify the degree of localized corrosion using the following equation:

$$LCII = \frac{i_{max}}{i_{tot}} \tag{1}$$

where i_{max} is the maximum anode current, and i_{tot} is the positive galvanic current density of the WBE. Tan et al. [26] presumed that localized corrosion was not serious when the *LCII* index was lower than 0.1.

When sulfuric acid and sodium hydroxide were added to the medium solution as a simple pH modifier to adjust the pH of the solution, the corrosion mechanism changed and the corrosion distribution became highly different, as reflected by the color and scale. For the bare electrode, as shown in Figure 3, all the point current densities were positive, and the potential distribution was uneven and random. The potential difference was relatively small, indicating that the localized corrosion tendency of the WBE was relatively weak. The maximum anodic currents of 0.607, 0.279, and 0.279 A/cm^2 were present at the electrodes (8, 4), (6, 6), and (6, 6), respectively (X-column, Y-row), at pH of 3.10–7.18.

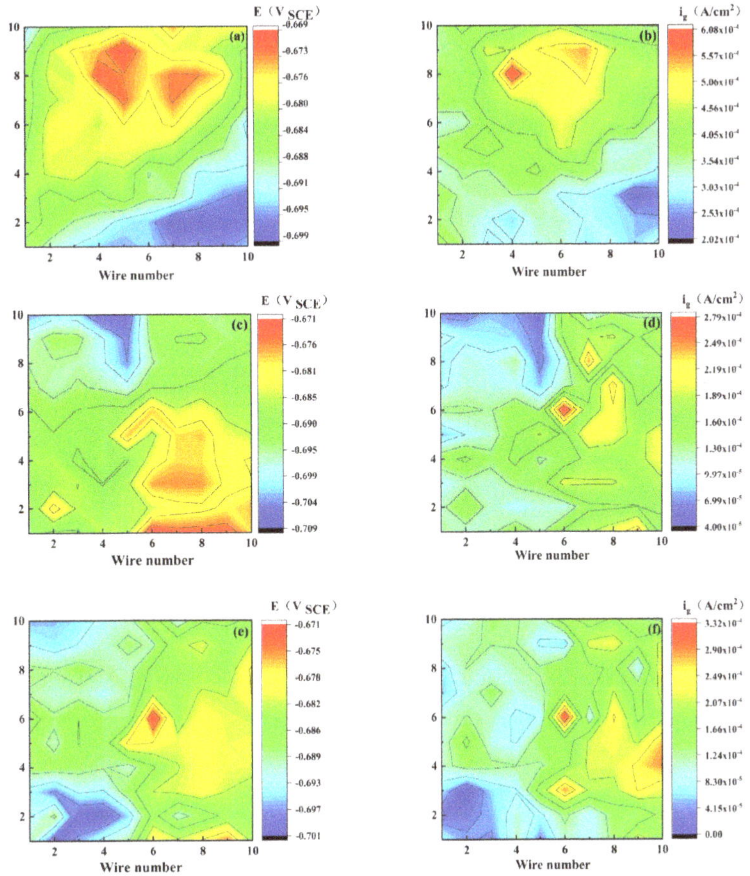

Figure 3. Potential (E) and galvanic current density (ig) distribution maps of WBE bare electrode at different pH values ((**a**,**b**): pH 3.10), ((**c**,**d**): pH 5.15), ((**e**,**f**): pH 7.18).

Figure 4 shows the coupling potential and current density distribution maps of the WBE coated with S8. It was found that with the increase in pH, the anode region gradually enlarged, and its position changed. The non-uniformity of the potential distribution reflects the heterogeneity of the corrosion. As shown in Figure 4e, the upper half of the electrode potential is small. In general, the anode position causes the OCP of the electrode to move negatively; therefore, the electrode in the upper half region is used as the anode region, and the corresponding Figure 4f shows a high position current value, which can also be reflected by the color. The other electrode potentials presented similar results.

In the pH range of 3.10–7.18, Figure 5 shows the calculation results of some key parameters (i_{tot}, i_{max}, *LCII*, and N_A) based on Figure 3, where N_A is the number of anodes. When the pH was adjusted from 3.10 to 7.18, both the i_{tot} and i_{max} decreased, all N_A values were 100, and a small *LCII* value (*LCII* < 0.1) was obtained, indicating that all bare electrodes experienced homogeneous corrosion, and localized corrosion did not occur. As shown in Figure 6c, as the pH increased, the surface of the S8-coated electrode produced new anode sites, and the N_A value increased from 57 to 99; therefore, the anode current density increased significantly, and i_{max} also increased. This also means that the current distribution became less concentrated on a small number of anode sites, and then generated more active corrosion activities, and corrosion on the electrode surface became more homogeneous. After the pH increased from acidic to neutral, the corrosion pattern changed significantly.

The distribution of the anode and cathode sites became random, and the *LCII* value also decreased rapidly from 0.3061 to 0.0295. Obviously, higher *LCII* values correspond to concentrated anode regions and intensive corrosion locations. As shown in Figure 5c, the *LCII* (*LCII* = 0.3061 > 0.1) of the electrode coated with S8 was higher than that of the bare electrode at pH = 3.10, which indicates that the localized corrosion trend of the electrode coated with S8 was enhanced. This result agrees with the polarization curves and EIS results.

An optical picture of the WBE is shown in Figure 7. For the bare electrode (Figure 7a), the current density in the anode region was extremely small, and the corresponding electrode was covered with a negligible amount of black corrosion products, causing mild corrosion. For the S8-coated electrode (Figure 7b), part of the electrode was covered with black corrosion products, and the other was still bright. S8 accelerated the corrosion of the electrode, and the electrode suffered from localized corrosion, which agrees with the calculated results shown in Figures 5 and 7.

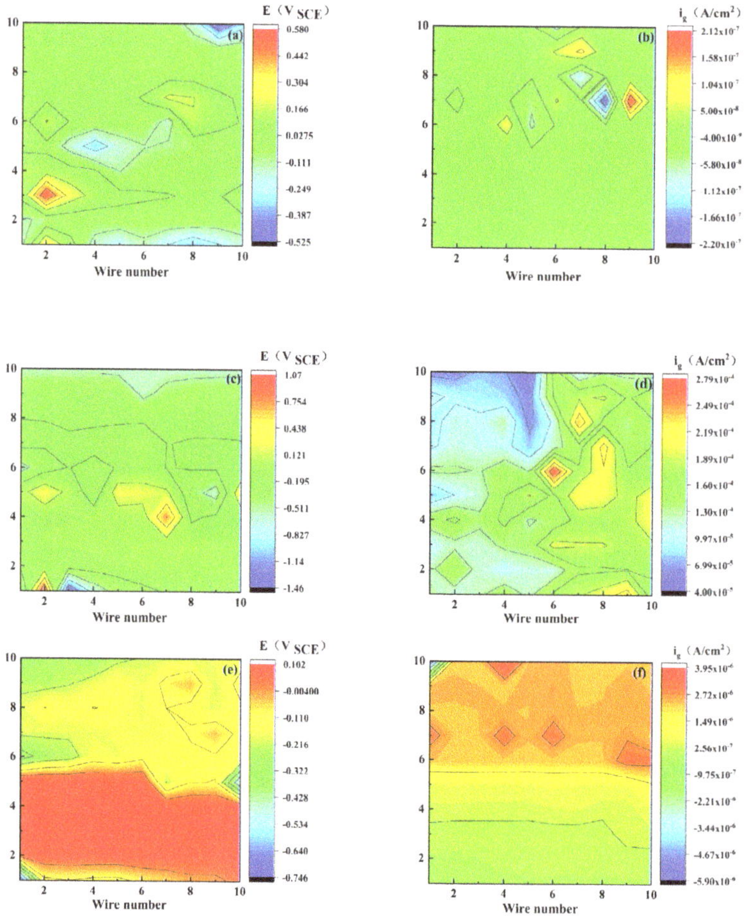

Figure 4. Potential (*E*) and galvanic current density (i_g) distribution maps of WBE with S8-coating at different pH values ((**a**,**b**): pH 3.10), ((**c**,**d**): pH 5.15), ((**e**,**f**): pH 7.18).

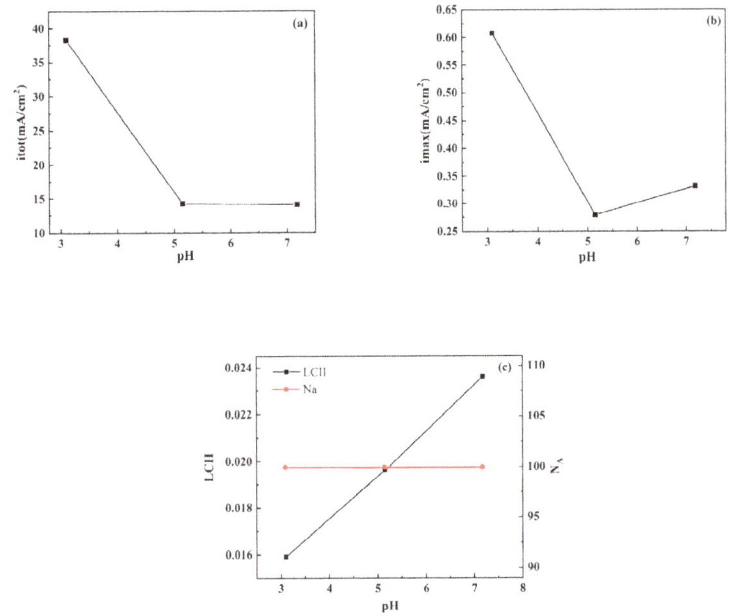

Figure 5. Bare electrode statistical and calculated results for (**a**) the sum of anode current density i_{tot}, (**b**) the maximum anodic current density i_{max}, (**c**) the localized corrosion intensity index *LCII*, and the number of anodes N_A.

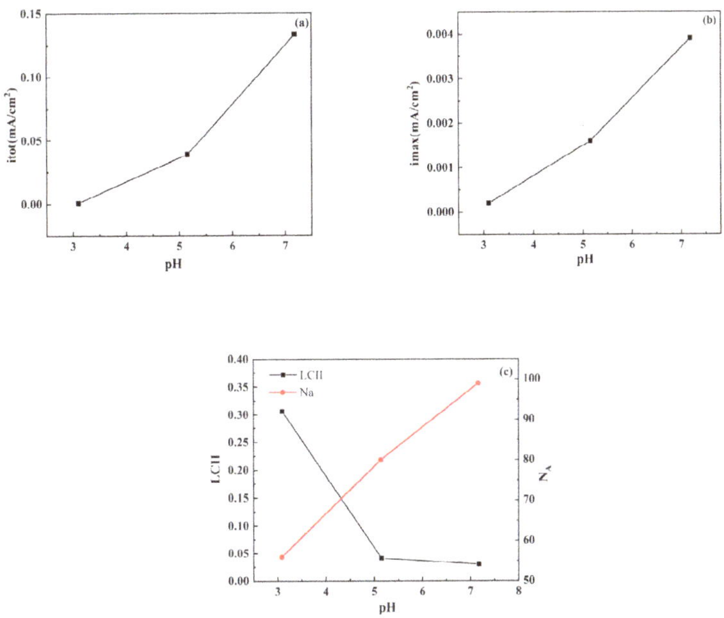

Figure 6. S8-coated electrode statistical and calculated results for (**a**) the sum of anode current density i_{tot}, (**b**) the maximum anodic current density i_{max}, (**c**) the localized corrosion intensity index *LCII*, and the number of anodes N_A.

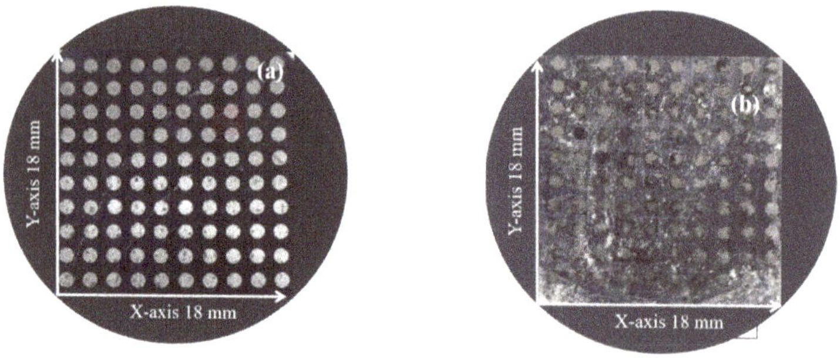

Figure 7. Optical picture of WBE at pH = 3.10: (**a**) bare electrode, (**b**) S8−coated electrode.

3.4. Surface and Component Characterization

The micromorphology of the corrosion products formed on the bare steel and S8-coated steel at different pH values with an autoclave is shown in Figure 8. The entire surface was uniformly covered by a thick and dense corrosion scale (Figure 8a–c), and there was no obvious localized corrosion. The dense corrosion scale provided protection for the substrate by hindering the corrosive species from penetrating the interface between the corrosion scale and the steel. The surface of the S8-coated steel was covered with an incomplete and loose corrosion scale and dispersed solid particles, and some small holes and recessed areas were observed at pH = 3.10, indicating that the steel suffered from serious localized corrosion attacks. Therefore, the presence of S8 changed the structure of the corrosion scale, leading to a reduction in the resistance of corrosion scale to the corrosive medium. The corrosion of the S8-coated steel was more serious than that of the bare steel, which was consistent with the results of potential polarization.

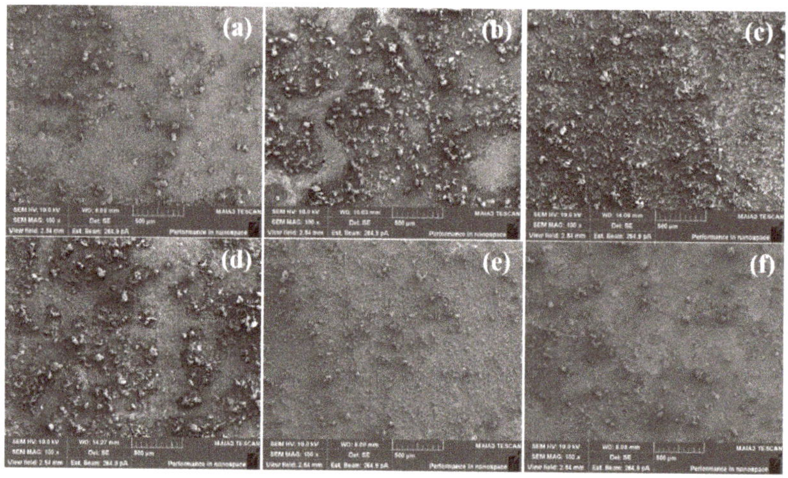

Figure 8. SEM surface morphologies of the samples at different pH values: (**a**) bare steel, 3.10, (**b**) bare steel, 5.15, (**c**) bare steel, 7.18, (**d**) S8−coated steel, 3.10, (**e**) S8−coated steel, 5.15, (**f**) S8−coated steel, 7.18.

To further evaluate the corrosion difference between the bare electrode and S8-coated electrode at pH = 3.10, an OLYMPUS DSX500 optical digital microscope was used to observe and calculate the maximum corrosion depth of the metal substrate after removing the corrosion scale in the bright field (BF) mode. Representative 3D images are shown in Figure 9. To comprehend the severity of the localized corrosion, a pitting factor was proposed to characterize the severity of pitting corrosion [15]:

$$PF = \frac{p}{d} \tag{2}$$

where *PF* is the pitting factor, p is the deepest metal penetration (μm), and d is the average metal penetration (μm). The *PF* value was positively correlated with the destructiveness of pitting, and the pitting corrosion threat was obvious at $PF > 5$. According to the results in Table 3, the *PF* value of the S8-coated electrode was above 5, whereas that of the bare electrode was at a low level.

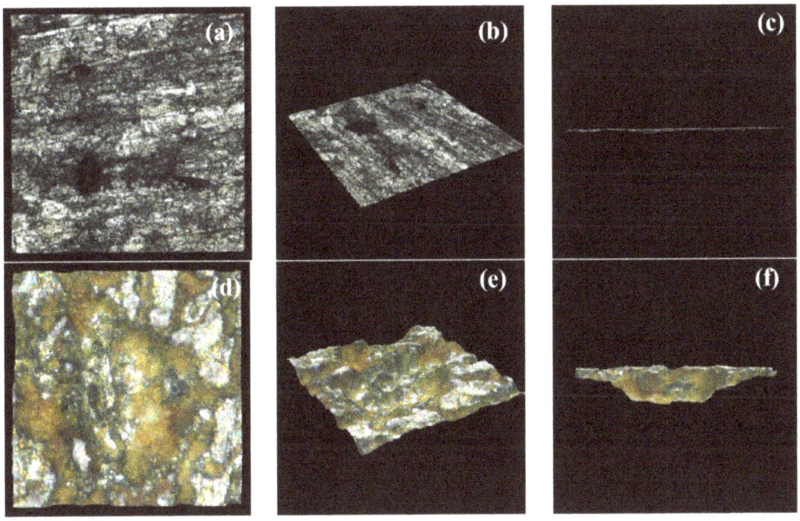

Figure 9. Three-dimensional profilometry measurement of corroded samples after removing corrosion products at pH = 3.10 for (**a**–**c**) the bare electrode, (**d**–**f**) S8–coated electrode.

Table 3. Three–dimensional profilometry measurement depth at pH = 3.10.

Parameter	p/μm	d/μm	*PF*
Bare electrode	25.84	21.02	1.23
S8-coated electrode	137.18	22.61	6.07

Only a small number of pores and scratches were observed on the bare electrode's surface, and the substrate was relatively flat. Many large pores were found on the surface of the S8 deposition electrode, and the metal substrate was uneven. Therefore, the characteristics of deeper holes and surface roughness indicated that the corrosion of the S8-coated electrode was more serious, which further confirmed the results mentioned above.

Furthermore, Figure 10 shows the corresponding chemical composition of the corrosion scale. The main elements in the deposits were C, O, Fe, and S. Figure 11 illustrates the XRD spectra of the corrosion products of the S8-coated and bare electrodes at pH 3.10. The corrosion scale on the bare electrode was composed of $FeCO_3$, which is a typical product in CO_2 systems. $FeCO_3$ is deposited on the electrode surface when the concentration of ferrous ions and carbonate ions surpasses the solubility product of $FeCO_3$ at a certain temperature. The product on the S8-coated electrode was FeS. Therefore, the sediments on the substrate were mainly ferrous carbonate and ferrous sulfide.

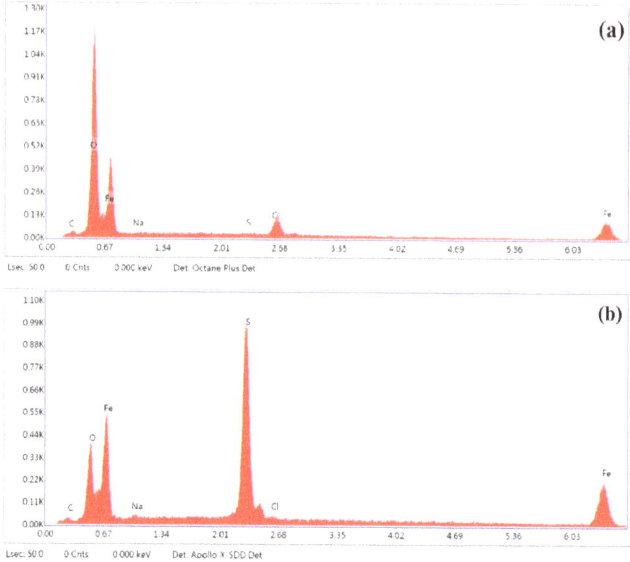

Figure 10. Elemental analysis (EDS) spectra: (**a**) bare electrode, (**b**) S8-coated electrode.

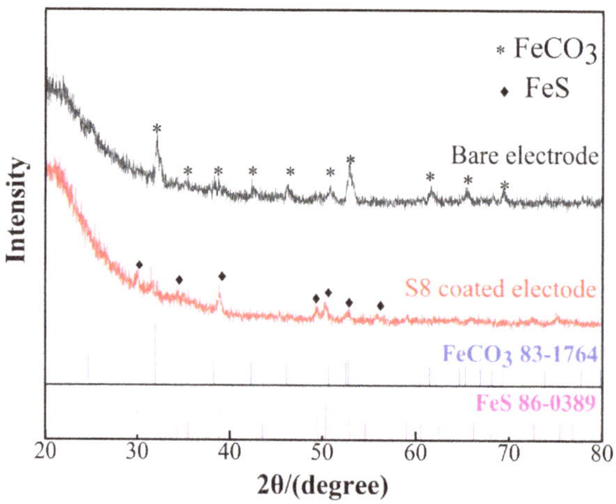

Figure 11. XRD patterns of the bare electrode and S8-coated electrode at pH = 3.10.

4. Corrosion Mechanism

The presence of S8 and CO_2 accelerated corrosion by reacting directly with the metal substrate or forming acid by hydrolysis to further promote corrosion. The corrosion process under S8 deposition mainly included the hydrolysis of S8, formation of the corrosion product (ferrous sulfide), and further catalytic corrosion caused by ferrous sulfide in addition to Cl^- [27]. The following equations are proposed to explain the corrosion formation of steel caused by S8 deposition on the electrode surface.

Cathodic reaction:

$$2H_2CO_3(aq) + 2e^- \rightarrow H_2(g) + 2HCO_3^-(aq) \quad (3)$$

$$2HCO_3^-(aq) + 2e^- \rightarrow H_2(g) + 2CO_3^{2-}(aq) \quad (4)$$

$$2H_2O(l) + 2e^- \rightarrow H_2(g) + 2OH^-(aq) \quad (5)$$

$$S_{y-1}{\cdot}S^{2-} + 2(x-1)\,e^- + 2xH^+ \overset{FeS}{\rightarrow} xH_2S + S_{y-x} \quad (6)$$

$$H^+ + e^- \rightarrow H_{ads} \rightarrow 1/2H_2 \quad (7)$$

Anodic reaction process:

$$(x-1)Fe \rightarrow (x-1)Fe^{2+} + 2(x-1)e^- \quad (8)$$

$$(x-1)Fe^{2+} + (x-1)H_2S \rightarrow (x-1)FeS + 2(x-1)H^+ \quad (9)$$

$$Fe^{2+}(aq) + CO_3{}^{2-}(aq) \rightarrow FeCO_3(s) \quad (10)$$

Overall corrosion reaction:

$$S_8 + 8H_2O \rightarrow 6H_2S + 2H_2SO_4 \quad (11)$$

$$H_2S \Leftrightarrow H^+ + HS^- \quad (12)$$

$$HS^- \Leftrightarrow H^+ + S^{2-} \quad (13)$$

$$(x-1)Fe + S_{y-1} \cdot S^{2-} + 2H^+ \overset{FeS}{\rightarrow} (x-1)FeS + H_2S + S_{y-x} \quad (14)$$

In the above reaction equations, Fe^{2+} generated by the substrate entered the solution, and initially combined with $CO_3{}^{2-}$ to form $FeCO_3$ or with S^{2-} and $SO_4{}^{2-}$ to form FeS and $FeSO_4$. In the CO_2-saturated acidic solution, the direct reduction of carbonic acid was the main cathodic reaction, and other possible cathodic reactions, including H^+ reduction, were ignored. $S_{y-1}{\cdot}S^{2-}$, as a reduced substance under the catalysis of ferrous sulfide, was considered to be formed by the chemical adsorption of sulfide ions on the surfaces of S8 particles, and corrosive substances might also involve adsorbed polysulfate ions, which may be formed by the reaction of S8 with water [14]. It has been reported that acidification occurs in sulfur-containing aqueous systems, and H_2SO_2, H_2SO_3, H_2SO_4, H_2S, and polysulfides are species for the hydrolysis products of S8.

At the beginning of corrosion, the surface of the metal substrate was covered by S8, and the corrosion process was a self-catalytic process. The direct contact between the metal substrate and S8 was a necessary factor for corrosion, which would lead to the corrosion potential moving in the positive direction, and it was easy to induce pitting corrosion with defect areas [27]. Then, the Cl^- and acid solution contacted the substrate, further aggravating the corrosion. However, some scholars believe that, in the presence of S8, the hydrolysis of S8 to generate acid is the main factor controlling corrosion, and sulfide-catalyzed cathodic sulfur reduction is considered to be the most important step [28]. To better demonstrate the corrosion mechanism of the S8-coated electrode, a pattern was proposed, as shown in Figure 12.

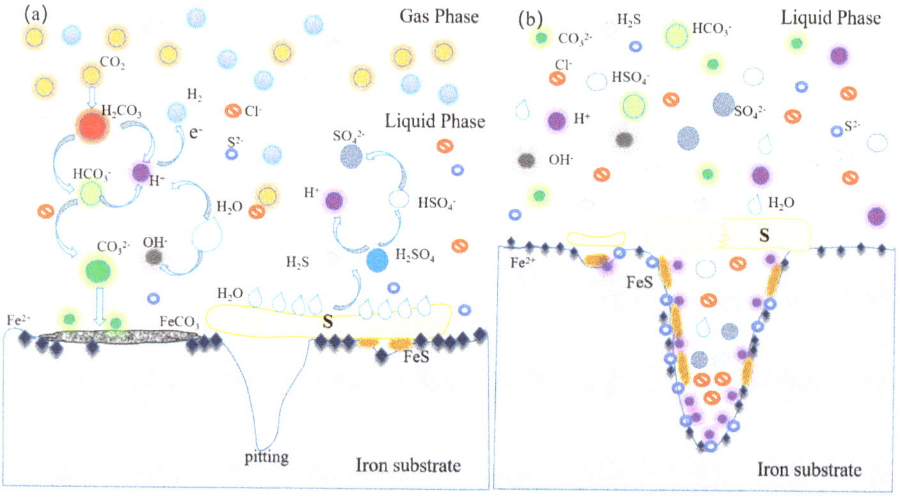

Figure 12. Description of corrosion mechanism of L360 pipeline steel under S8 deposition in NaCl solution containing CO_2: (**a**) S8 non-uniform deposition, (**b**) under S8 deposition.

The deposition of S8 on the metal surface was non-uniform for the electrode coated without S8, which directly contacted the NaCl solution containing CO_2, mainly resulting in the anodic dissolution of the metal and the reduction of H_2CO_3. The typical corrosion scale of $FeCO_3$ was formed, covering the surface of the metal and preventing Cl^- from contacting the metal substrate, which is consistent with the results of XRD and SEM. In addition, with an increase in pH, the solubility of CO_2 in the solution decreased, and the stability of the corrosion product $FeCO_3$ was also enhanced. Finally, the corrosion rate slowed with an increase in pH, as shown in the left half of Figure 12a. S8 reacted with the metal substrate and formed the corresponding sulfide, such as FeS (Equation (9)). At the same time, hydrolysis of S8 occurred, and the solution was acidified. With the passage of time, the hydrolysis of S8 and the dissolution of H_2CO_3 occurred slowly, and the local environment between the S8 deposition layer and metal substrate changed, which was mainly manifested in the excessive pH shifting from neutral to acidic and activated the dissolution of metal. In addition, the hydrolysis products of S8, H_2SO_4 and H_2S, contacted the metal and combined with Fe^{2+} to form FeS and $FeSO_4$, as shown in Figure 11.

Crevice corrosion was also formed due to the non-uniform deposition of S8, and localized corrosion occurred first at the edge of the deposition of S8. There was a potential difference between the steel substrate and the corrosion products with electronic conductivity, such as FeS and $FeCO_3$, and the substrate covered by S8. Therefore, the potential of the electrode covered by S8 was significantly lower than that of the bare electrode or corrosion products. The metal under S8 deposition was used as the anode corrosion area, and the bare electrode or corrosion products were used as the cathodic protection area, forming a large cathode and a small anode, which caused galvanic corrosion and further accelerated the corrosion under deposits [29,30], as shown in Figure 12a.

At the same time, the environment under the S8 deposition was different from that of the surface of the bare electrode, and the corrosion behavior changed. The self-catalytic cathodic reaction accelerated the corrosion of the substrate during S8 deposition. A self-catalytic cathodic reaction occurred when the electrode was covered with S8 in an aqueous solution. Ferrous sulfide is an electronic conductor with structural defects and low hydrogen overpotential, so it has a catalytic effect on the cathodic reaction [14], as shown in Figure 12b.

In addition, the chloride concentration had a significant influence on the overall corrosion rate (Equations (13) and (14)). When the chloride content was high, the concentration of dissolved Fe^{2+} ions was much higher than the solubility constant of FeS [28].

Anodic metal dissolution in the presence of Cl^-:

$$Fe + 2Cl^- + H_2O \rightarrow [Fe(OH)^+ + Cl^-] + HCl + 2e^- \quad (15)$$

Cathodic sulfur reduction in the presence of Cl^-:

$$[FeS_{x+1}] + 2e^- + 2Na^+ + H_2O \rightarrow [FeS_x] + (Na^+ + SH^-) + NaOH \quad (16)$$

With a decrease in pH, especially at the tips of pits, higher concentrations of H^+ (lower pH) accumulated under S8 deposition due to the hydration of Fe^{2+}, forming an occluded corrosion chamber, which aggravated pitting corrosion. Cl^- with a negative charge in the solution was adsorbed into the corrosion chamber with a positive charge to maintain the charge balance, which accelerated the anodic dissolution of the metal and negatively shifted the cathode potential, further deteriorating the environment under the S8 deposition layer. Furthermore, pitting began to expand outward, increasing the range of pitting and further accelerating the corrosion of the L360 pipeline steels.

5. Conclusions

In this work, the effect of S8 deposition on the corrosion behavior of L360 pipeline steel in CO_2-Cl^- medium at different pH values at 60 °C was studied using electrochemical measurements and surface analysis.

(1) The corrosion current densities of the electrode coated with S8 were larger, and the *R*ct value was 1/27 that of the bare electrode. Pitting corrosion cavities were observed on the surface of the electrode under S8 deposition in acidic conditions, and the bare electrode suffered uniform corrosion.

(2) The *LCII* and morphology further proved that the deposition of S8 enhanced localized corrosion at pH = 3.10.

(3) The deposition of S8 not only acidized the solution but also increased the difference in the corrosion environment inside and outside of the S8 deposition layer, resulting in serious pitting corrosion under S8 deposition.

Author Contributions: Formal analysis, C.Q. and F.C.; Investigation, S.Z.; Resources, T.Y. and B.Y.; Supervision, Y.L.; writing—original draft preparation, F.W. and J.L. All authors have read and agreed to the published version of the manuscript.

Funding: This work was funded by the National Natural Science Foundation of China (21808182, 51974245). Key Laboratory Scientific Research Project of Shaanxi Provincial Department of Education (18JS088).

Institutional Review Board Statement: Not applicable.

Informed Consent Statement: Not applicable.

Data Availability Statement: Not applicable.

Conflicts of Interest: The authors declare no conflict of interest.

References

1. Obot, I.B. Under Deposit Corrosion on Steel Pipeline Surfaces: Mechanism, Mitigation and Current Challenges. *J. Bio-Tribo-Corros.* **2021**, *7*, 1–14. [CrossRef]
2. Tan, Y.; Fwu, Y.; Bhardwaj, K. Electrochemical evaluation of under deposit corrosion and its inhibition using the wire beam electrode method. *Corros. Sci.* **2011**, *53*, 1254–1261. [CrossRef]
3. Katerina, K.; Gubner, R. Development of standard test method for investigation of under deposit corrosion in carbon dioxide environment and its application in oil and gas industry. In Proceedings of the CORROSION 2010, San Antonio, TX, USA, 14–18 March 2010.

4. Hoshowski, J.; Pineiro, R.P.; Nordvik, T.; Barnes, P.; Jenkins, A. The Development of Novel Corrosion Inhibitors for High Temperature Sour Gas Environments. In Proceedings of the CORROSION 2020, Online, 14–18 June 2020.
5. Cao, M.; Liu, L.; Yu, Z.; Fan, L.; Li, Y.; Wang, F. Electrochemical corrosion behavior of 2A02 Al alloy under an accelerated simulation marine atmospheric environment. *J. Mater. Sci. Technol.* **2019**, *35*, 651–659. [CrossRef]
6. Zheng, Y.; Brown, B.; Nešić, S. Electrochemical study and modeling of H_2S corrosion of mild steel. *Corrosion* **2014**, *70*, 351–365. [CrossRef]
7. Wen, X.; Bai, P.; Luo, B.; Zheng, S.; Chen, C. Review of recent progress in the study of corrosion products of steels in a hydrogen sulphide environment. *Corros. Sci.* **2018**, *139*, 124–140. [CrossRef]
8. Wang, Q.; Wu, W.; Li, Q.; Zhang, D.; Yu, Y.; Cao, B.; Liu, Z. Under deposit corrosion of tubing served for injection and production wells of CO_2 flooding. *Eng. Fail. Anal.* **2021**, *127*, 105540–105552. [CrossRef]
9. Chang, Y.N.; Wei, F.I. High temperature chlorine corrosion of metals and alloys. *J. Mater. Sci.* **1991**, *26*, 3693–3698. [CrossRef]
10. Zhang, S.; Hou, L.; Du, H.; Wei, H.; Liu, B.; Wei, Y. A study on the interaction between chloride ions and CO_2 towards carbon steel corrosion. *Corros. Sci.* **2020**, *167*, 108531–108542. [CrossRef]
11. Yang, L.; Zhu, S.; Lu, T. Study on the mechanism of elemental sulfur hydrolysis. *Chem. Res. Appl.* **2016**, *28*, 390–395. (In Chinese)
12. Zhang, Y.N.; Wang, T.L.; Han, X.; Wang, Z.M.; Zhang, J. Corrosion of artificial rock layer covered steel electrodes in a CO_2 environment: The influence of permeability. *Corros. Sci.* **2016**, *105*, 190–201. [CrossRef]
13. Gong, K.; Wu, M.; Liu, G. Stress corrosion cracking behavior of rusted X100 steel under the combined action of Cl^- and HSO_3^- in a wet dry cycle environment. *Corros. Sci.* **2020**, *165*, 108382–108396. [CrossRef]
14. MacDonald, D.D.; Roberts, B.; Hyne, J.B. The corrosion of carbon steel by wet elemental sulphur. *Corros. Sci.* **1978**, *18*, 411–425. [CrossRef]
15. Gong, Q.; Xiang, Y.; Zhang, J.; Wang, R.; Qin, D. Influence of elemental sulphur on the corrosion mechanism of X80 steel in supercritical CO_2 saturated aqueous phase environment. *J. Supercrit. Fluids* **2021**, *176*, 105320–105332. [CrossRef]
16. Zhang, G.A.; Yu, N.; Yang, L.Y.; Guo, X.P. Galvanic corrosion behavior of deposit covered and uncovered carbon steel. *Corros. Sci.* **2014**, *86*, 202–212. [CrossRef]
17. Xu, Y.; Liu, L.; Zhou, Q.; Wang, X.; Huang, Y. Understanding the influences of pre-corrosion on the erosion corrosion performance of pipeline steel. *Wear* **2020**, *442*, 203151. [CrossRef]
18. Pan, C.q.; Zhong, Q.d.; Yang, J.; Frank Cheng, Y.; Chen, C. Investigating crevice corrosion of copper and copper alloys using wire beam electrode. *Corros. Eng. Sci. Technol.* **2021**, *56*, 407–418. [CrossRef]
19. Tan, Y. Monitoring localized corrosion processes and estimating localized corrosion rates using a wire beam electrode. *Corrosion* **1998**, *54*, 403–413. [CrossRef]
20. Wu, Y.L.; Zhang, D.P.; Cai, G.Y.; Zhang, X.X.; Dong, Z.H. Effects of temperature on polarity reversal of under deposit corrosion of mild steel in oilfield produced water. *Corros. Eng. Sci. Technol.* **2020**, *55*, 708–720. [CrossRef]
21. Chen, J.; Wu, J.; Wang, P.; Zhang, D.; Chen, S.; Tan, F. Corrosion of 907 steel influenced by sulfate reducing bacteria. *J. Mater. Eng. Perform.* **2019**, *28*, 1469–1479. [CrossRef]
22. Marcus, P. Surface science approach of corrosion phenomena. *Electrochim. Acta* **1998**, *43*, 109–118. [CrossRef]
23. Bommersbach, P.; Alemany-Dumont, C.; Millet, J.P.; Normand, B. Formation and behaviour study of an environment friendly corrosion inhibitor by electrochemical methods. *Electrochim. Acta* **2005**, *51*, 1076–1084. [CrossRef]
24. Jin, Z.; Ge, H.; Lin, W.; Zong, Y.; Liu, S.; Shi, J. Corrosion behaviour of 316L stainless steel and anti-corrosion materials in a high acidified chloride solution. *Appl. Surf. Sci.* **2014**, *322*, 47–56. [CrossRef]
25. Cheng, Q.; Tao, B.; Song, L.; Zhang, W.; Liu, X.; Li, W.; Hou, B.; Liu, Q. Corrosion behaviour of Q235B carbon steel in sediment water from crude oil. *Corros. Sci.* **2016**, *111*, 61–71. [CrossRef]
26. Tan, Y.; Aung, N.N.; Liu, T. Evaluating localised corrosion intensity using the wire beam electrode. *Corros. Sci.* **2012**, *63*, 379–386. [CrossRef]
27. Fang, H.; Young, D.; Nesic, S. Corrosion of mild steel in the presence of elemental sulfur. *Corrosion* **2018**. 08, Paper no.08637, OnePetro: 2008, NACE-08637.
28. Schmitt, G. Effect of elemental sulfur on corrosion in sour gas systems. *Corrosion* **1991**, *47*, 285–308. [CrossRef]
29. Zhao, Q.; Zhao, J.; Cheng, X.; Huang, Y.; Lu, L.; Li, X. Galvanic corrosion of the anodized 7050 aluminum alloy coupled with the low hydrogen embrittlement CdTi plated 300M steel in an industrial marine atmospheric environment. *Surf. Coat. Technol.* **2020**, *382*, 125171. [CrossRef]
30. Yang, W.; Liu, Z.; Huang, H. Galvanic corrosion behavior between AZ91D magnesium alloy and copper in distilled water. *Corros. Sci.* **2021**, *188*, 109562. [CrossRef]

metals

MDPI

Article

Influence of Anodization on the Fatigue and Corrosion-Fatigue Behaviors of the AZ31B Magnesium Alloy

Leandro Antonio de Oliveira [1], Silvano Leal dos Santos [2], Vinicius Antonio de Oliveira [3] and Renato Altobelli Antunes [1,*]

[1] Centro de Engenharia, Modelagem e Ciências Sociais Aplicadas (CECS), Federal University of ABC (UFABC), Santo André 09210-580, SP, Brazil; leandro.oliveira@ufabc.edu.br
[2] Laboratório de Processamento e Caracterização de Materiais, Faculdade de Tecnologia de São Paulo (FATEC-SP), São Paulo 01124-060, SP, Brazil; silvanoleal@fatecsp.br
[3] Departamento de Engenharia de Materiais (DEMa), Universidade Federal de São Carlos (UFSCar), São Carlos 13565-905, SP, Brazil; vinicius.oliveira@estudante.ufscar.br
* Correspondence: renato.antunes@ufabc.edu.br; Tel.: +55-11996-8241

Abstract: The aim of the present work was to investigate the effect of anodization on the fatigue and corrosion-fatigue behavior of the AZ31B magnesium alloy. Samples were anodized in constant current density mode at 20 mA cm^{-2} for 5 min at room temperature, in an environmentally friendly electrolyte consisting of a mixture of sodium hydroxide and sodium silicate. Fatigue tests were conducted in air and in phosphate buffer solution (PBS) at room temperature in the tension-tension mode, at a frequency of 5 Hz and stress ratio of 0.1. S-*N* curves were obtained for polished and anodized samples. Fracture surface morphology was examined by optical stereo-microscopy and scanning electron microscopy. Results indicated that the fatigue limit was reduced approximately 60% at 10^6 cycles for the anodized specimens, either for the fatigue tests conducted in air or PBS solution. Anodization had a remarkable effect on the fatigue behavior of the AZ31B alloy. The effect of the corrosive environment, in turn, was not significant.

Keywords: AZ31B magnesium alloy; anodizing treatment; fatigue resistance; corrosion fatigue

Citation: de Oliveira, L.A.; dos Santos, S.L.; de Oliveira, V.A.; Antunes, R.A. Influence of Anodization on the Fatigue and Corrosion-Fatigue Behaviors of the AZ31B Magnesium Alloy. *Metals* **2021**, *11*, 1573. https://doi.org/10.3390/met11101573

Academic Editor: Sebastian Feliú, Jr.

Received: 3 September 2021
Accepted: 27 September 2021
Published: 1 October 2021

1. Introduction

The interest in magnesium alloys for applications demanding high strength-to-weight ratio is growing rapidly, mainly driven by the inherent low density of these materials. Additional attributes, such as high damping capacity, good castability, and machinability, are also attractive for industrial purposes [1,2]. One further, but no less important, aspect of magnesium alloys is their well-known low corrosion resistance in aqueous environments [3]. The naturally formed magnesium hydroxide layer is non-protective in chloride-containing electrolytes, making the material susceptible to pitting corrosion [4]. Corrosion control is, therefore, a serious concern for magnesium alloys, as its high chemical reactivity limits a widespread use in several applications [5].

In fact, the low corrosion resistance of magnesium alloys is undesirable for most commercial applications. Nevertheless, it may be advantageous if one considers biodegradable implant materials. Due to its intrinsic biocompatibility, magnesium can be employed in temporary implants for fracture fixation [6,7]. Research on magnesium alloys as temporary orthopedic devices has gained huge interest in the past few years. Recently, Sezer et al. [8] reviewed the main aspects of biodegradable Mg-based implants. The most important feature of temporary fixation devices is to withstand the mechanical loads to which the implant is subject during its use while the fracture heals. In order to meet this goal, the inherent high chemical reactivity of magnesium alloys must be controlled to avoid premature failure of the fixation device [9].

Several methods for improving the corrosion resistance of magnesium have been reported in the literature, such as alloying and surface treatments. In the case of alloying, microstructural control is pursued to promote the formation of precipitates and/or grain refining, thus improving the corrosion properties of the alloy by reducing the propensity to the formation of local microgalvanic cells [10]. Surface modification methods for the corrosion control of magnesium alloys include, but are not limited to, chemical conversion coatings, ion implantation, microarc oxidation, physical vapor deposition, plasma spraying, and anodization [11].

Anodization is a traditional way of improving the corrosion resistance of magnesium alloys. Many studies are devoted to investigating the effect of electrical parameters on the corrosion protection ability of the anodic film or the electrolyte type and concentration of chemical species in the formation of a compact and protective anodized layer [12–15]. The most recent trends in this research field are focused on anodizing treatments based on environmentally friendly electrolytes. Silicate-containing alkaline baths play a prominent role in this scenario. Salami et al. [16] have shown that dense and uniform anodic films could be produced on the AZ31B alloy by controlling the concentration of sodium silicate in the electrolyte, favoring the formation of Mg_2SiO_4 in the coating layer. Due to its non-toxic character, silicate-based electrolytes are also envisaged as good options for the surface treatment of magnesium implants [17].

Fatigue resistance plays a central role in the service life of biomedical alloys. Corrosion fatigue is recognized as a serious problem for different metallic biomaterials, being associated with a great part of the mechanical failures of implantable devices [18]. Raman et al. [19] highlighted the critical aspects of corrosion fatigue of magnesium implants, emphasizing that it is not frequently addressed in the current literature. Nonetheless, despite the scarcity of studies in this area, some reports can be found. Liu et al. [20] studied the corrosion fatigue behavior of a biocompatible Mg-Zn-Y-Nd alloy in simulated body fluid (SBF). The fatigue resistance decreased in SBF in comparison with the fatigue tests conducted in air. Another interesting aspect was related to the source of fatigue cracks. In SBF, multiple cracks were generated, while only one crack source was found in the absence of the corrosion medium. Gu et al. [21] have also reported a deleterious effect of the corrosive physiological environment in the fatigue resistance of the AZ91D alloy. Corrosion pits were associated with the preferential sites for fatigue crack nucleation. Bian et al. [22] studied the corrosion fatigue of Mg-Ca and Mg-Zn-Ca alloys in SBF. A significant decrease of the fatigue properties was reported when compared to the tests conducted in air.

In spite of the relevant findings accumulated so far on the corrosion fatigue behavior of magnesium-based biomaterials, the concomitant effect of the presence of an anodized layer and a corrosive environment on the fatigue response is not currently found in the literature. In one hand, anodization can be an interesting method for the corrosion control of temporary fixation devices, as it allows one to tailor the morphology and composition of the oxide layer to produce a dense, uniform, and biocompatible anodic film. On the other hand, the presence of the anodic film may affect the fatigue properties of the anodized alloy. According to Eifert et al. [23], anodization influences both the crack initiation and propagation processes during cyclic loading of magnesium alloys. Khan et al. [24] reported that the fatigue resistance of anodized AM60 alloy (Mg-Zn-Mn) decreased due to the porous nature of the anodic film. Hence, the morphology of the anodized layer influences the fatigue behavior of the material. Nemcova et al. [25] reported a reduction of 56% for the fatigue limit of a microarc oxidized AZ61 alloy in NaCl solution due to the presence of the oxide layer. Although these reports provide relevant information on the interplay between anodization and corrosion fatigue of structural magnesium alloys, similar information is not found if one envisages their application as temporary fixation devices for biomedical purposes.

In light of this scenario, the present work aims to fill this gap by investigating the effect of anodization on the corrosion fatigue of the AZ31B alloy in phosphate buffered solution. Anodization was carried out in an environmentally friendly sodium silicate-based electrolyte. The fatigue and corrosion fatigue tests were conducted in the tension-tension mode.

2. Materials and Methods

2.1. Material and Specimen Preparation

The material was a hot rolled AZ31B magnesium alloy sheet (composition in wt.% Al 2.54%, Zn 1.08%, Mn 0.38% and Mg balance), supplied by Xi'an Yuechen Metal Products Co. Ltd. (Xi'an, China). The tensile and fatigue samples were machined by laser-cutting along the rolling direction following ASTM E8M-16a [26] and ASTM E466-15 [27], respectively. Figure 1 shows the shape and size of the specimens employed for tensile and fatigue tests. The dog-bone shaped specimens were sequentially ground using waterproof silicon carbide paper (from #220 to #4000), and polished using diamond paste slurry (diameter 3 μm and 1 μm). Right after, they were cleaned using deionized water and ethanol, being dried in a warm air stream provided by a conventional heat gun.

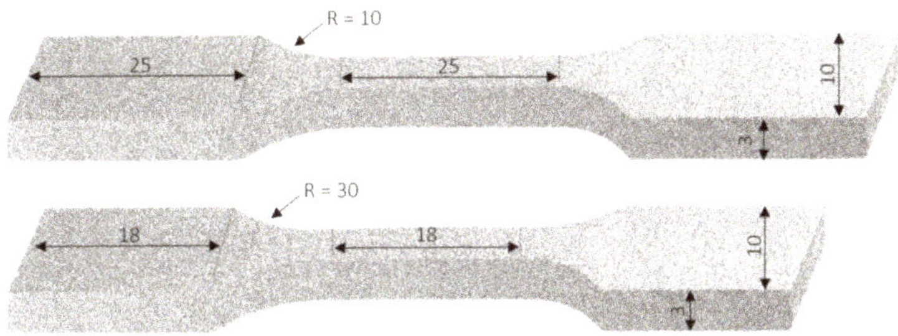

Figure 1. Shape and dimensions (in mm) of (**a**) tensile and (**b**) fatigue specimens.

For the corrosion test, the AZ31B magnesium alloy was cut using a cut-off saw in a square section with area of 100 mm^2 and thickness of 3.5 mm. The AZ31B alloy pieces were connected to a copper wire at the rear side using a conductive colloidal silver paste, being subsequently embedded in epoxy resin. Next, the surfaces were ground using waterproof silicon carbide paper (from #220 to #2400 grit size), polished using diamond paste (diameter 1 μm), washed using deionized water, and dried in a warm air stream provided by a conventional heat gun.

2.2. Anodizing Treatment

Anodization was performed in an aqueous solution consisting of a mixture of 1.0 M NaOH and 0.5 M Na_2SiO_3 at a constant current density of 20 mA cm^{-2} for 5 min at room temperature. Details about the anodizing experiments can be found in our previous work [28].

2.3. Coating Morphology and Corrosion Test

The surface and cross-section morphologies of the anodized sample were acquired using scanning electron microscopy (SEM, JSM-6010LA, JEOL, Tokyo, Japan).

The corrosion behavior was assessed in a phosphate-buffered saline (PBS) solution consisting of 0.355 g L^{-1} $NaH_2PO_4 \cdot H_2O$, 8.2 g L^{-1} NaCl, and 0.105 g L^{-1} Na_2HPO_4 (anhydrous). The electrolyte was prepared with deionized water (18.2 MΩ cm) and analytical grade reagents. The measurements were carried out using a potentiostat/galvanostat (M101, Metrohm Autolab B.V., Utrecht, The Netherlands) in a conventional three-electrode cell configuration. The test cell consisted of a platinum wire as the counter-electrode, Ag/AgCl (3 M, KCl) as the reference and the investigated AZ31B alloy as the working electrode. Firstly, the open potential circuit (OCP) was monitored for 1 h. Next, potentiodynamic polarization tests were conducted in a potential range from −0.50 vs. OCP to 0 V vs. Ag/AgCl/(3 M, KCl), at a scanning rate of 1 mV s^{-1}. The experiments were performed at room temperature and in triplicate.

The porosity of the anodized layer was calculated using an electrochemical relationship based on the assessment of polarization resistance (Rp) values obtained from the potentiodynamic curves. This relationship is based on the variation of the corrosion potential ($\Delta Ecorr = Ecorr, substrate - Ecorr, substrate + coating$) incited by the presence of the coating and from individual measurements of the polarization resistance (Rp) of the polished and anodized AZ31B, according to Equation (1) [28,29].

$$P = \left(\frac{R_{p,s}}{R_p}\right) \times 10^{\frac{|\Delta E_{corr}|}{b_a}} \quad (1)$$

where Rp,s indicates the polarization resistance of the polished AZ31B alloy and Rp is related to the polarization resistance of the anodized AZ31B alloy, b_a is the anodic Tafel slope of the bare material. Rp,s and b_a are determined from separate analysis of the polished substrate.

2.4. Tensile Test

The tensile tests were carried out following the ASTM E8M-16a [26] at a displacement rate of 12.7 mm/min at room temperature on a universal material testing machine (MTS Exceed E45, Eden Prairie, MN, USA). Five measurements were taken for the AZ31B alloy in the polished and anodized conditions.

2.5. Fatigue and Corrosion Fatigue Tests

Axial fatigue tests were conducted using a computer-controlled servo-hydraulic testing machine (MTS Landmark 370, Eden Prairie, MN, USA) with a sinusoidal loading control. The stress ratio was 0.1 (tension-tension mode) and the test frequency was 5 Hz. The fatigue test was continued until fracture or until the specimen did not fail up to 10^6 cycles. The procedure was defined according to ASTM F1801-97 [30], which is specific for temporary metal-based implants. Thus, the maximum stress at which the sample has not failed at 10^6 cycles is defined as a fatigue limit in this work. At every test load condition at least three specimens were evaluated.

The electrolyte employed in the corrosion fatigue tests was phosphate-buffered saline (PBS) solution, which contains: 0.355 g L^{-1} $NaH_2PO_4{\cdot}H_2O$, 8.2 g L^{-1} NaCl, 0.105 g L^{-1} Na_2HPO_4 (anhydrous). The solution was prepared with deionized water and analytical grade reagents. An acrylic chamber was designed and mounted on MTS Landmark 370 servo-hydraulic machine, in order to safeguard the gage length of the specimen was immersed in the PBS during the fatigue test. The electrolyte was subject to room temperature, open to air, and static during the testing procedure. A new fresh solution was employed for each test.

2.6. Fracture Surface Analysis

The fatigue fractured specimens were immersed in 10 g L^{-1} CrO_3 for 5 min in order to remove corrosion products, in agreement to ASTM G1-90 [31]. Next, they were rinsed in deionized water, and dried in warm air stream provided by a conventional heat gun prior to fractographic analysis.

The fractured specimen surface was examined using scanning electron microscopy (SEM, JEOL JSM-6010LA, Tokyo, Japan) and stereomicroscopy (Olympus SZ61, Tokyo, Japan).

3. Results and Discussion

3.1. Anodized Layer Morphology and Corrosion Test

SEM micrographs of the top surface and cross-section of the AZ31B magnesium alloy in the anodized condition are shown in Figure 2. As seen in Figure 2a, the artificial oxide layer produced by the anodization treatment covered the whole substrate. Several cracks and carved regions are unevenly distributed over the surface. The cross section of the anodized specimen is showed in Figure 2b. The interface anodized layer/substrate displays an irregular thickness. A roughened interface greatly affects the design of the implantable devices, once it plays an important role behaving as a stress concentrator [32], thus limiting the fatigue resistance of the component.

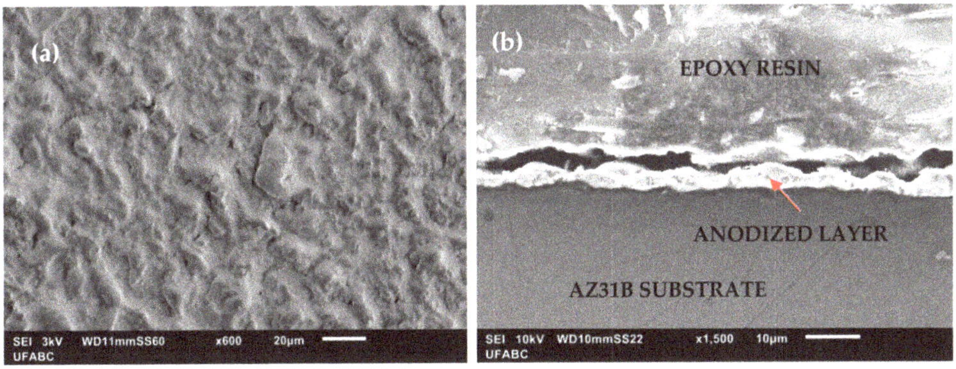

Figure 2. SEM images of the anodized layer (**a**) top surface and (**b**) cross-section.

Potentiodynamic polarization curves of the polished and anodized samples tested in PBS solution at room temperature are presented in Figure 3. The values of corrosion potential (E_{corr}) and corrosion current densities (j_{corr}) were determined from the curves using the Tafel extrapolation method. The results are shown in Table 1. The j_{corr} values were significantly affected by the presence of the anodic film. The dissolution rate of the anodized sample was reduced by one order of magnitude. The anodized layer acted as a barrier layer against electrolyte penetration. However, as seen in Table 1, the produced oxide layer presents an inherent porosity, which could be related to the cracks and cavities observed from the SEM micrographs (Figure 2).

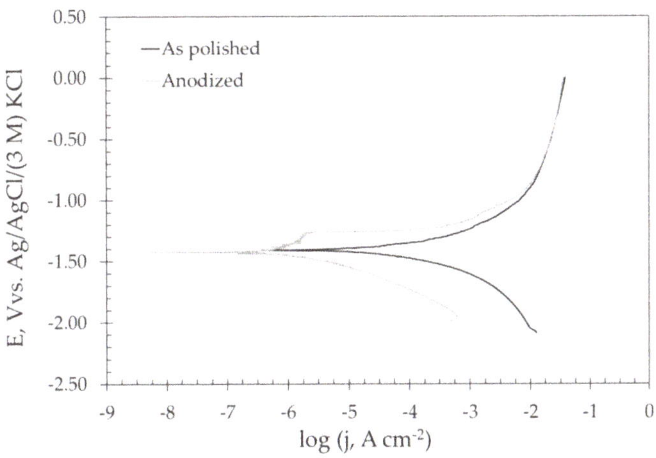

Figure 3. Potentiodynamic polarization curves obtained for the as polished and anodized samples immersed in PBS solution at room temperature.

Table 1. Electrochemical parameters obtained from the potentiodynamic polarization curves displayed in Figure 2. The last column indicates the porosity percentage of the anodized sample.

Condition	E_{corr}	j_{corr}	R_p	ba^{AZ31B}	$\mid bc \mid$	P
	(V vs. Ag/AgCl/(3 M) KCl)	(μA cm^{-2})	(kΩ cm^2)	(mV dec^{-1})	(mV dec^{-1})	(%)
As polished	-1.407 ± 0.004	30.4 ± 8.5	0.71	100	130	-
Anodized	-1.470 ± 0.041	2.0 ± 0.7	51.2	-	160	0.84

In the cathodic branches, the value of the cathodic Tafel slope (bc) was increased after anodization, as shown in Table 1. According to Rahman et al. [33], this effect could be due to the formation of $Mg(OH)_2$ inside the pores and cracks of the anodized layer, as shown in Equations (2) and (3).

$$2H_2O + 2e^- \rightarrow H_2 + 2OH^- \quad (2)$$

$$Mg^{2+} + 2OH^- \rightarrow Mg(OH)_2 \quad (3)$$

When film breakdown occurs in the anodic part of the polarization curve, the possible sequence of reactions is depicted in Equations (4) and (5). These processes (anodic and cathodic reactions) are accompanied by the formation of a corrosion product layer, following the same reaction shown in Equation (3). Furthermore, due to the presence of chloride ions in the PBS solution, $Mg(OH)_2$ could further react according to Equation (6).

$$Mg \rightarrow Mg^{2+} + 2e^- \text{ (anodic dissolution)} \quad (4)$$

$$2H_2O + 2e^- \rightarrow H_2 + 2OH^- \text{ (cathodic reaction)} \quad (5)$$

$$Mg(OH)_2 + 2Cl^- \rightarrow MgCl_2 + 2OH^- \quad (6)$$

3.2. Tensile Properties

Figure 4 shows the stress–strain curves of the AZ31B magnesium alloy in the as received and anodized conditions. The average values of yield strength, ultimate strength, and elongation at break of the alloy in the as received and anodized conditions are displayed in Table 2. It is clear that the ultimate tensile strength and yield strength decreased after anodization. According to the literature [34], this effect is due to the presence of pores and defects in the oxide layer that act as crack nucleation sites, reducing the mechanical strength of the anodized substrate.

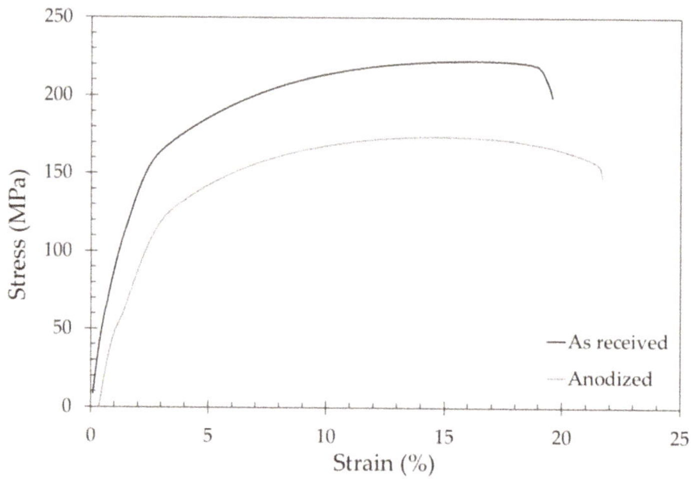

Figure 4. Stress-strain curves in uniaxial tensile tests at room temperature.

Table 2. Stress-strain curves in uniaxial tension tests at room temperature for anodized and as received specimens.

Specimens	Ultimate Tensile Strength (MPa)	Yield Strength (MPa)	Elongation (%)
As received	223 ± 5	165 ± 5	20 ± 3
Anodized	175 ± 5	120 ± 5	20 ± 3

3.3. Fatigue and Corrosion Fatigue Behavior

3.3.1. Influence of the Corrosive Environment

Figure 5 presents the S-*N* curves for the polished AZ31B alloy exposed to the air and to the PBS solution. Comparatively, a decrease on the maximum applied stress is noticed when the tests were performed in the PBS solution. Considering 10^6 cycles to failure, according to ASTM F1801, the fatigue limit was reduced from 142.5 MPa when the tests were conducted in the air, to 137.5 MPa when the tests were conducted in PBS. However, at an intermediate number of cycles to failure, the influence of the testing environment is even more evident. It can be seen that an applied stress of 140 MPa led to the fatigue life of approximately 69,000 cycles, while the fatigue life was extended to 10^6 cycles at a compatible applied stress when the tests were conducted in the air. These results indicate that there is an abrupt change in the fatigue life for a relatively low variation of the applied stress in the corrosion-fatigue condition. A similar trend was observed by Nan et al. during fatigue tests of the AZ31 alloy [35]. A slight loading variation was sufficient to drastically reduce the fatigue life of this material. In another work, Nan et al. have also investigated the influence of the 3.5 wt.% NaCl on the corrosion fatigue behavior of the AZ31 alloy [36]. A remarkable decrease in fatigue resistance was noticed when the tests were performed in the chloride solution. Authors have pointed out the combination of cyclic stress and pit growth as responsible for this behavior [37,38].

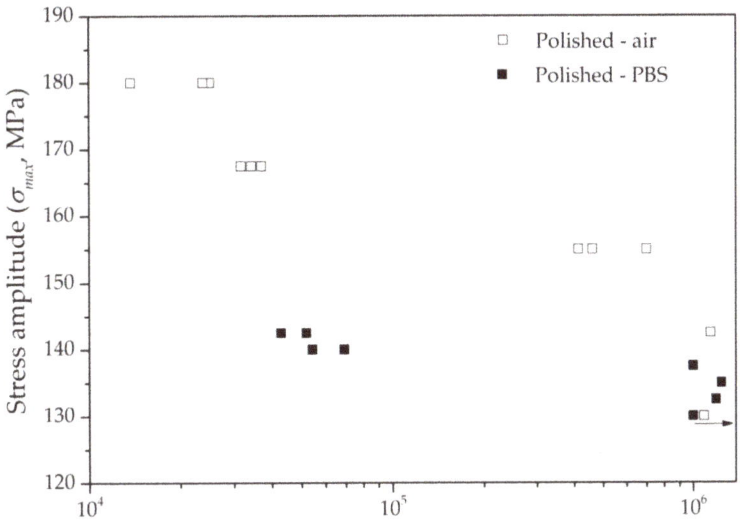

Figure 5. S-*N* curves for the polished AZ31B alloy specimens in air and in PBS solution at room temperature.

The S-*N* curves for the anodized specimens exposed to air and PBS solution are shown in Figure 6. Corrosion was less detrimental for the surface treated samples than the polished ones. As displayed in Figure 5, the fatigue life was little affected by the corrosive environment for stress levels lower than 130 MPa.

In order to give a more quantitative interpretation on the influence of corrosion on the fatigue resistance of the AZ31B alloy, we employed Equations (7) and (8) to calculate the reduction rate ($R\sigma_N$) of the maximum fatigue strength at 10^6 cycles, as proposed by He et al. [35]. Based on this procedure, fitting equations shown in Table 3 were obtained by a logarithmic transformation of the experimental data points shown in Figures 4 and 5.

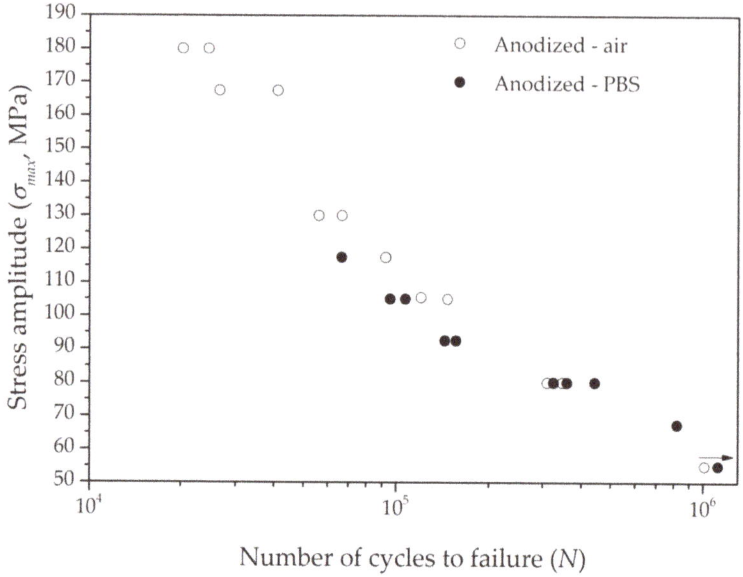

Figure 6. S-*N* curves for the anodized AZ31B alloy specimens in air and in PBS solution at room temperature.

$$y = k + bx \tag{7}$$

$$R\sigma_N = \left(\frac{\sigma_{air} - \sigma_{PBS}}{\sigma_{air}}\right) \times 100 \tag{8}$$

Table 3. Fitting equations for each experimental condition.

Specimen Conditions and Environment	Equations
Polished–air	$\log(\sigma_{max}) = -7894\ln N + 254.61$
Polished – PBS solution	$\log(\sigma_{max}) = -1336\ln N + 155.86$
Anodized–air	$\log(\sigma_{max}) = -32{,}96\ln N + 501.26$
Anodized–PBS solution	$\log(\sigma_{max}) = -20{,}39\ln N + 340.47$

In Equation (2), y and x are log (σ_{max}) and log N, respectively. The maximum applied stress is represented by σ_{max}, while N is the number of cycles to failure at a certain stress level. The parameters k and b are coefficients. In Equation (3), $R\sigma_N$ indicates the reduction rate of the maximum applied stress at specific fatigue cycles in air and in PBS solution. The maximum applied stress that specimens can resist at specific fatigue cycles are denoted by σ_{air} and σ_{PBS}. Table 4 displays the values of these parameters, along with the corresponding $R\sigma_N$ at 10^6 cycles.

Table 4. Reduction rate of the maximum applied stress for the AZ31B alloy in the polished and anodized conditions.

Conditions	σ_{air} (MPa)	σ_{PBS} (MPa)	$R\sigma_N$ (%)
Polished	144.6	137.4	5.6
Anodized	45.9	58.8	−28.0

As seen in Table 4, the reduction rate of the polished AZ31B alloy was 5.6% when immersed in the PBS solution, while the anodized condition, in turn, showed an increment. Therefore, the effect of the PBS solution was less harmful to the fatigue strength of the anodized samples. This behavior is probably associated with the protection ability promoted by the anodization treatment. Generally, the main mechanism related to the corrosion fatigue failure of magnesium and its alloys is pit nucleation and growth [19,32,39]. In this case, the presence of the oxide layer possibly provided a higher local dissolution resistance for the AZ31B alloy.

3.3.2. Influence of Anodization

Figures 7 and 8 show the S-*N* curves for the polished and anodized samples exposed to air and PBS solution, respectively. As seen in these figures, there was a remarkable reduction of the fatigue resistance for the anodized specimens compared to the polished condition regardless of the environment. In both scenarios, the reduction of the fatigue limit at 10^6 cycles was approximately 60% for the anodized samples.

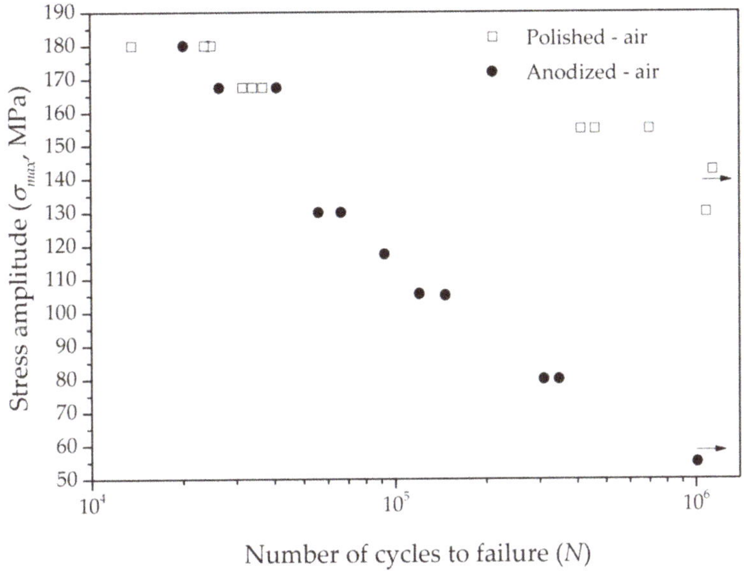

Figure 7. S-*N* curves for the polished and anodized AZ31B alloy specimens in air.

The reduction of the fatigue strength for the anodized specimens was much more intense at low stress levels, independently of the testing environment. This behavior has been associated to a significant influence of the crack nucleation than the crack propagation, which causes the premature failure of the material [40]. In other words, the anodic layer facilitates fatigue crack nucleation. Khan et al. [41] observed a distinguished effect of the crack nucleation of anodized magnesium alloy when subjected to fatigue tests. The authors concluded that the crack nucleation step, which is defined by the material surface conditions in a crucial manner, exerts more influence than the crack growth and propagation stages. As the anodized layer greatly affects the surface characteristics, this means that fatigue failure would easily happen when a certain number of cycles are able to nucleate the crack. Besides that, other adverse effects concerned with the fatigue behavior of anodized magnesium alloys have been reported in the current literature. Yerochin et al. [42] mentioned that the plasma micro-discharges which occur during the oxidation process led to strain distortions of the metal subsurface layers. Some other works have pointed out the presence of a microcrack network [39] and disordered porous structure [24,32] of the coating layer. Furthermore, the compactness and uniformity of the oxide layer, its adhesion to the substrate, and relevant aspects of roughness on the interface of coating and the material may also affect the fatigue properties of anodized alloys [43,44].

The results obtained in the present work highlighted that the presence of the oxide layer caused a hostile effect to the AZ31B magnesium alloy regardless of the environment conditions.

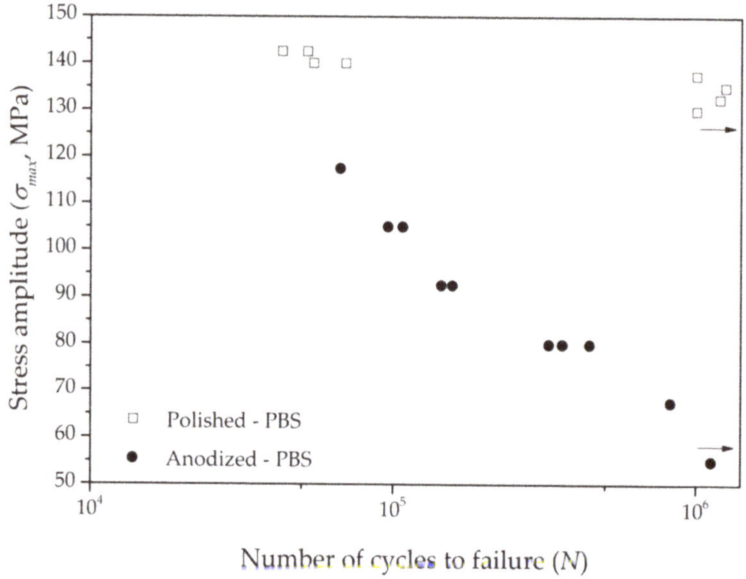

Figure 8. S-N curves for the polished and anodized AZ31B alloy specimens in the PBS solution at room temperature.

3.4. *Fractographic Analysis*

Optical stereo-micrographs of the fracture surfaces of polished and anodized AZ31B specimens after fatigue tests in air and PBS solution are shown in Figure 9. The specimens for optical stereoscopic analysis were chosen based on the number of cycles to failure. The aim was to identify the microfractographic features in the initial hours of the test, as it would lead to the most relevant information regarding the nucleation of fatigue cracks. Hence, specimens that failed in the first hours of test, at different applied stresses but with similar fatigue lives were chosen in order to investigate the effect of the PBS solution on the fractographic aspects of the AZ31B alloy. Table 5 displays the applied stress and corresponding number of cycles to failure of the polished and anodized specimens selected for fractographic analysis.

Table 5. Specimens chosen for fracture analysis after the corrosion fatigue tests.

Corrosion Fatigue Results	Polished		Anodized	
	Air	PBS Solution	Air	PBS Solution
Applied stress (MPa)	167.5	140	105	92.5
Number of cycles to failure	34,200	54,355	146,754	143,681

The fatigue fracture surface may display distinguishable regions, denoting three different steps of fatigue failure: (i) crack nucleation; (ii) crack growth and propagation; (iii) catastrophic failure (final fracture) [45,46]. These regions are indicated in Figure 9 for the AZ31B specimens in the polished and anodized conditions. Fatigue fracture is often initiated in the surface or subsurface region [47]. However, identifying the exact point of crack nucleation is not an easy task [48]. These sites are indicated as A1 and A2 in Figure 9. Growth region generally expands radially for the surface of the specimen (region B in

Figure 9). The transition from region B to final fracture (region C) is accompanied by a change of the fractographic features to a darker, uneven region. Details regarding each one of these regions were obtained by SEM analysis.

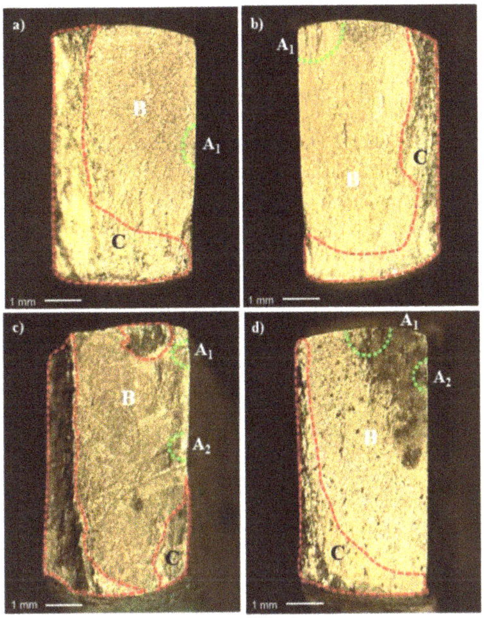

Figure 9. Optical stereo-micrographs of the AZ31B alloy: (**a**) polished (air; 167.5 MPa); (**b**) anodized (air—105 MPa); (**c**) polished (PBS solution—140 MPa); and (**d**) anodized (PBS solution—92.5 MPa). Marks on the micrographs represent the following fracture regions: (A_1 and A_2) crack initiation; (B) crack propagation; (C) catastrophic failure.

Figure 10 shows the SEM micrographs corresponding to regions A (Figure 10A), B (Figure 10B) and C (Figure 10C) of Figure 9a. Green arrows point to possible crack nucleation sites in Figure 10A, denoted by discontinuities and small particles at the bottom right of the micrographs. However, the nucleation site was not clearly distinguished. A different aspect is seen in the crack growth region (Figure 10B). Faceted regions and shallow dimples are encountered over this region, suggesting an unstable crack growth mechanism was predominant. Conversely, a typical ductile dimpled fracture surface is seen in Figure 10C, suggesting that plastic deformation occurred in this region, before final fracture. Similar features of fractographic aspects of fatigue fracture surfaces of magnesium alloys were reported by other authors [48–51].

Regions A–C of the fracture surface of the air-tested anodized specimen shown in Figure 9b were further explored by SEM analysis. The corresponding micrographs are displayed in Figure 11. The nucleation region (A1 in Figure 9b) is shown in Figure 11A. Green arrows point to the interface between the anodized layer and the bulk AZ31B alloy, revealing different roughness at these sites. Moreover, subsurface particles are also observed and may have contributed to crack nucleation, as observed by other authors [24,45]. Yet, a gradual transition in the surface appearance from a smooth to a rough aspect is seen at the middle of the micrograph, indicating that crack propagation may have started at this site [52]. In Figure 11B, the crack growth region is further detailed. Its aspect resembles that shown in Figure 9b for the polished sample tested in air, being characterized by faceted regions and shallow dimples, typical of unstable crack growth. In the final fracture region (Figure 11C), the fracture surface is dominated by an intense dimpled structure, indicating plastic deformation at these sites, as also observed in Figure 10C for the polished sample.

It is clear, therefore, that anodization did not alter the microfractographic features of the AZ31B alloy subject to fatigue testing in air.

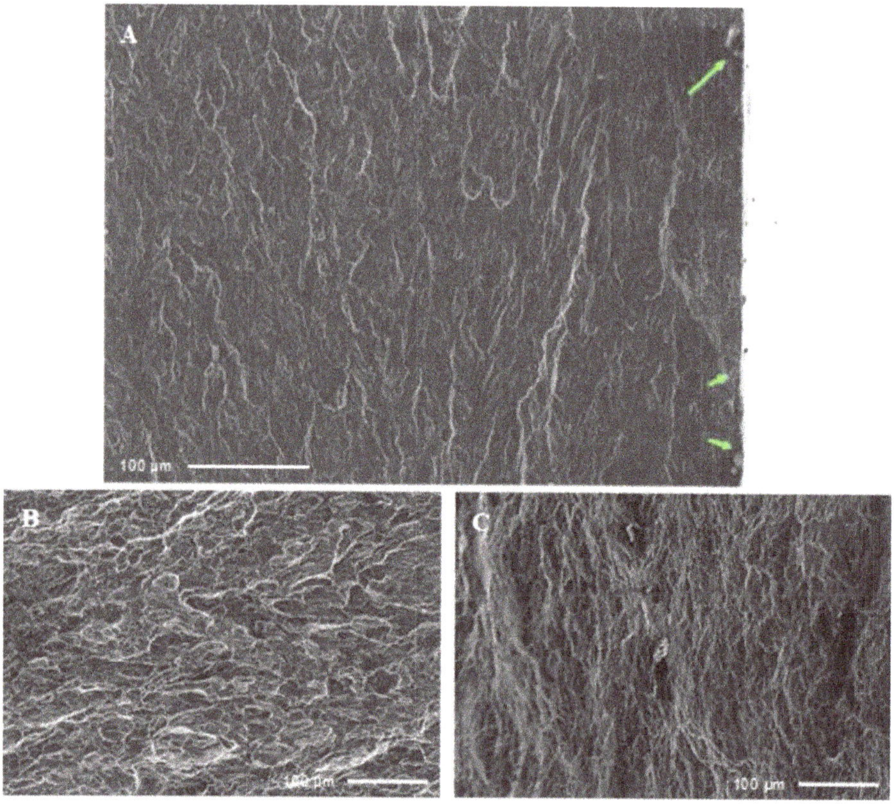

Figure 10. Fatigue fracture morphologies of the polished specimen tested in air at 167.5 MPa. The micrographs (**A**–**C**) correspond to the three different regions pointed out in Figure 9a.

SEM micrographs of regions A1, B, and C (Figure 9c) for the polished sample tested in PBS solution are displayed in Figure 12. The most probable crack nucleation sites (region A1, Figure 9c) are shown in Figure 11A where potential stress risers are pointed by the green arrows, such as discontinuities in the $MgO/Mg(OH)_2$ layer spontaneously formed in the electrolyte and subsurface inclusions that may have facilitated fatigue crack nucleation. The transition between the smooth and rough aspects of the fracture surface is remarkable, as seen in the left part of the micrograph, in contrast with that of the polished samples tested in air (Figure 10A). According to the literature [23,53], this suggests a faster transition from the nucleation step to crack growth. Cracks would appear simultaneously at different sites, triggered by surface porosity or subsurface cracks. Hence, material plasticity is reduced, increasing the crack propagation rate, and leading to rough aspect of the fracture surface. It is likely that the concomitant action of corrosive environment and cyclic loading is responsible for such an enhanced propagation rate. In spite of the surface cleaning step before fractographic analysis, some oxide inclusions still remained in the fracture surface, as indicated by the arrows in Figure 12B,C. The final fracture region (Figure 12C) is similar to those of the specimens tested in air (Figures 10C and 11C), showing a dimpled structure.

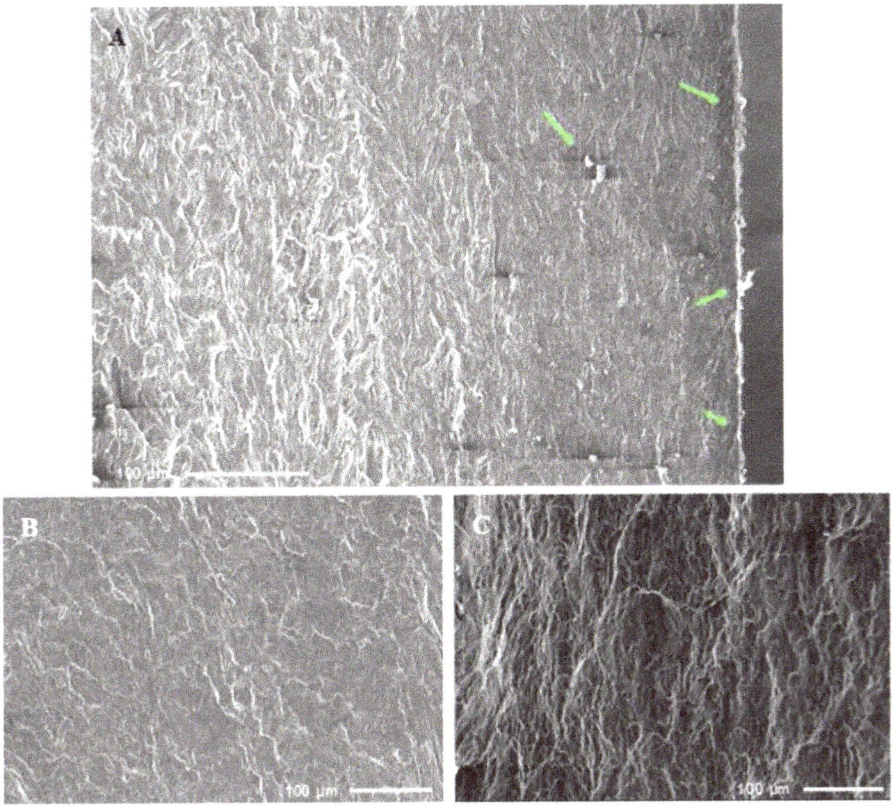

Figure 11. Fatigue fracture morphologies of the anodized specimen tested in air at 105 MPa. The micrographs (**A–C**) correspond to the three different regions pointed out in Figure 9b.

SEM micrographs of the fracture surface of the anodized specimen tested in PBS solution are shown in Figure 13. In Figure 13A (region A1 in Figure 9d), green arrows indicate pores, subsurface defects between the substrate and the anodized layer, and inclusions that may be related to crack nucleation by acting as stress risers during fatigue loading. In the crack propagation region (Figure 13B), the fracture surface is quite flat and some micropores are indicated by green arrows. The aspect of the final fracture region (Figure 13C) is quite different from that of the polished sample fractured in PBS (Figure 12C). Instead of a dimpled structure, the surface is flat, indicating that plastic deformation was not as marked as observed for the polished sample. In this respect, it is evident that the presence of the anodized layer reduced plasticity at the final fracture step.

The main observations with respect to the microfractographic features of the fracture surfaces for the AZ31B alloy specimens described in Table 5 are synthesized in Table 6, along with possible causes of the final failure.

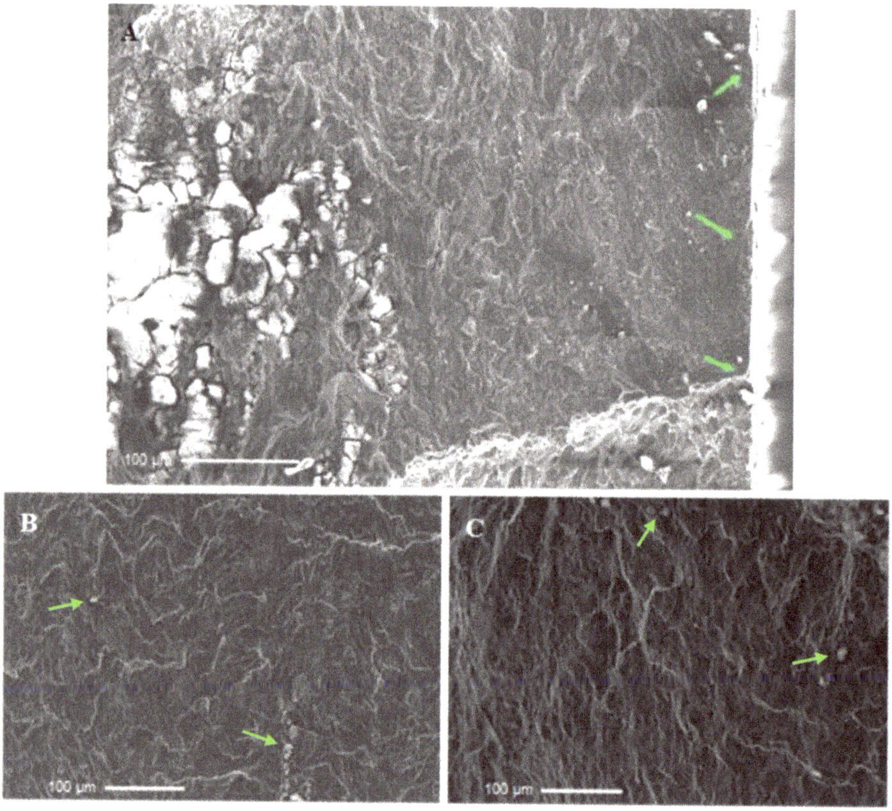

Figure 12. Fatigue fracture morphologies of the polished specimen tested in PBS solution at 140 MPa. The micrographs (**A–C**) correspond to the three different regions pointed out in the Figure 9c.

Table 6. Summary of characteristics of the fractured specimens submitted to the corrosion fatigue tests.

Conditions	Crack Status	Possible Causes	Fracture Appearance
Polished—Air Figure 8	Single crack nucleation	Surface and subsurface discontinuities	Brittle with radial pattern
Anodized—Air Figure 8	Single crack nucleation	Rough or uneven anodized layer	Brittle with radial pattern
Polished—PBS solution Figure 8	Multiple crack nucleation	Surface and subsurface discontinuities combined with local corrosion	Brittle and hackly types
Anodized—PBS solution Figure 8	Multiple crack nucleation	Rough or uneven anodized layer	Brittle and quasi-cleavage types

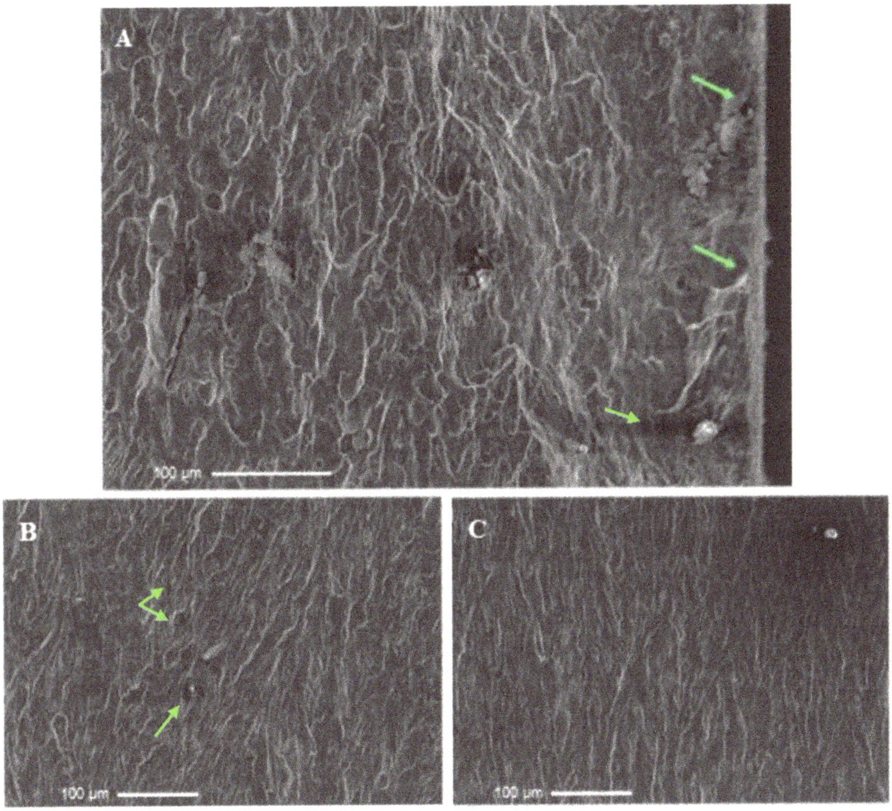

Figure 13. Fatigue fracture morphologies of the anodized specimen tested in the PBS solution at 92.5 MPa at room temperature. The images (**A–C**) correspond to the three different regions pointed out in the Figure 9d.

4. Conclusions

Anodization displayed a profound effect on the fatigue strength of the AZ31B alloy. S-*N* curves obtained in the tension-tension mode, in air and PBS solution, indicated that the fatigue limit was reduced by the corrosive electrolyte when the magnesium alloy was tested in the polished condition. Conversely, it increased for the anodized alloy tested in PBS in comparison to the tests conducted in air. However, a reduction of approximately 60% of the fatigue strength was observed for the anodized samples with respect to the polished ones, either in air or PBS solution. The presence of the anodized layer entails remarkable surface changes that affect the fatigue behavior of the AZ31 alloy, mostly related to surface porosity and roughness heterogeneities at the anodized layer/substrate interface. The fracture surface displayed a more brittle character for the anodized alloy fractured in PBS solution when compared to air. Oxide particles are likely to act as stress risers, increasing the crack propagation rate in PBS solution.

Author Contributions: Conceptualization, R.A.A. and L.A.d.O.; Methodlogy, R.A.A. and L.A.d.O.; investigation, L.A.d.O., S.L.d.S. and V.A.d.O.; Writing—original draft preparation, L.A.d.O. and R.A.A.; Supervision, R.A.A.; funding acquisition, R.A.A. All authors have read and agreed to the published version of the manuscript.

Funding: This research was funded by the São Paulo Research Foundation (FAPESP), process number 2015/22921-6 and CAPES (Finance Code 001).

Data Availability Statement: The data obtained in this study are available from the corresponding author on reasonable request.

Acknowledgments: FATEC-SP, DEMa-UFSCar and the Multiuser Experimental Facilities (CEM-UFABC) are acknowledged for the experimental support.

Conflicts of Interest: The authors declare no conflict of interest.

References

1. Esmaily, M.; Svensson, J.E.; Fajardo, S.; Birbilis, N.; Frankel, G.S.; Virtanen, S.; Arrabal, R.; Thomas, S.; Johansson, L.G. Fundamentals and advances in magnesium alloy corrosion. *Prog. Mater. Sci.* **2017**, *89*, 92–193. [CrossRef]
2. Si, Y.; Xiong, Z.; Zheng, X.; Li, M.; Yang, Q. Improving the anti-corrosion ability of anodization film of AZ31B magnesium alloy by addition of NH4VO3 in the electrolyte. *Int. J. Electrochem. Sci.* **2016**, *11*, 3261–3268. [CrossRef]
3. Song, G.L.; Shi, Z. Corrosion mechanism and evaluation of anodized magnesium alloys. *Corros. Sci.* **2014**, *85*, 126–140. [CrossRef]
4. Cui, L.; Liu, Z.; Hu, P.; Shao, J.; Li, X.; Du, C.; Jiang, B. The corrosion behavior of AZ91D magnesium alloy in simulated haze aqueous solution. *Materials* **2018**, *11*, 970. [CrossRef] [PubMed]
5. Mousa, H.M.; Hussein, K.H.; Woo, H.M.; Park, C.H.; Kim, C.S. One-step anodization deposition of anticorrosive bioceramic compounds on AZ31B magnesium alloy for biomedical application. *Ceram. Int.* **2015**, *41*, 10861–10870. [CrossRef]
6. Wang, B.J.; Wang, S.D.; Xu, D.K.; Han, E.H. Recent progress in fatigue behavior of Mg alloys in air and aqueous media: A review. *J. Mater. Sci. Technol.* **2017**, *33*, 1075–1086. [CrossRef]
7. Yang, Y.; He, C.; Dianyu, E.; Yang, W.; Qi, F.; Xie, D.; Shen, L.; Peng, S.; Shuai, C. Mg bone implant: Features, developments and perspectives. *Mater. Des.* **2020**, *185*, 108259. [CrossRef]
8. Sezer, N.; Evis, Z.; Kayhan, S.M.; Tahmasebifar, A.; Koç, M. Review of magnesium-based biomaterials and their applications. *J. Magnes. Alloy.* **2018**, *6*, 23–43. [CrossRef]
9. Radha, R.; Sreekanth, D. Insight of magnesium alloys and composites for orthopedic implant applications—A review. *J. Magnes. Alloy.* **2017**, *5*, 286–312. [CrossRef]
10. Nassif, N.; Ghayad, I. Corrosion protection and surface treatment of magnesium alloys used for orthopedic applications. *Adv. Mater. Sci. Eng.* **2013**, *2013*, 532896. [CrossRef]
11. Uddin, M.S.; Hall, C.; Murphy, P. Surface treatments for controlling corrosion rate of biodegradable Mg and Mg-based alloy implants. *Sci. Technol. Adv. Mater.* **2015**, *16*, 53501. [CrossRef]
12. Cheng, I.C.; Fu, E.G.; Liu, L.D.; Lee, C.Y.; Lin, C.S. Effect of Fluorine Anions on Anodizing Behavior of AZ91 Magnesium Alloy in Alkaline Solutions. *J. Electrochem. Soc.* **2008**, *155*, C219–C225. [CrossRef]
13. Chai, L.; Yu, X.; Yang, Z.; Wang, Y.; Okido, M. Anodizing of magnesium alloy AZ31 in alkaline solutions with silicate under continuous sparking. *Corros. Sci.* **2008**, *50*, 3274–3279. [CrossRef]
14. Li, L.L.; Cheng, Y.L.; Wang, H.M.; Zhang, Z. Anodization of AZ91 magnesium alloy in alkaline solution containing silicate and corrosion properties of anodized films. *Trans. Nonferrous Met. Soc. China* **2008**, *18*, 722–727. [CrossRef]
15. Girón, L.; Aperador, W.; Tirado, L.; Franco, F.; Caicedo, J.C. Electrochemical Performance Estimation of Anodized AZ31B Magnesium Alloy as Function of Change in the Current Density. *J. Mater. Eng. Perform.* **2017**, *26*, 3710–3718. [CrossRef]
16. Salami, B.; Afshar, A.; Mazaheri, A. The effect of sodium silicate concentration on microstructure and corrosion properties of MAO-coated magnesium alloy AZ31 in simulated body fluid. *J. Magnes. Alloy.* **2014**, *2*, 72–77. [CrossRef]
17. De Oliveira, L.A.; Silva, R.M.P.; Antunes, R.A. Scanning Electrochemical Microscopy (SECM) Study of the Electrochemical Behavior of Anodized AZ31B Magnesium Alloy in Simulated Body Fluid. *Mater. Res.* **2019**, *22*, 1–8. [CrossRef]
18. Niinomi, M. Fatigue characteristics of metallic biomaterials. *Int. J. Fatigue* **2007**, *29*, 992–1000. [CrossRef]
19. Jafari, S.; Singh Raman, R.K.; Davies, C.H.J. Corrosion fatigue of a magnesium alloy in modified simulated body fluid. *Eng. Fract. Mech.* **2015**, *137*, 2–11. [CrossRef]
20. Liu, M.; Wang, J.; Zhu, S.; Zhang, Y.; Sun, Y.; Wang, L.; Guan, S. Corrosion fatigue of the extruded Mg–Zn–Y–Nd alloy in simulated body fluid. *J. Magnes. Alloy.* **2020**, *8*, 231–240. [CrossRef]
21. Gu, X.N.; Zhou, W.R.; Zheng, Y.F.; Cheng, Y.; Wei, S.C.; Zhong, S.P.; Xi, T.F.; Chen, L.J. Corrosion fatigue behaviors of two biomedical Mg alloys—AZ91D and WE43—In simulated body fluid. *Acta Biomater.* **2010**, *6*, 4605–4613. [CrossRef]
22. Bian, D.; Zhou, W.; Liu, Y.; Li, N.; Zheng, Y.; Sun, Z. Fatigue behaviors of HP-Mg, Mg–Ca and Mg–Zn–Ca biodegradable metals in air and simulated body fluid. *Acta Biomater.* **2016**, *41*, 351–360. [CrossRef]
23. Eifert, A.J.; Thomas, J.P.; Rateick, R.G. Influence of anodization on the fatigue life of WE43A-T6 magnesium. *Scr. Mater.* **1999**, *40*, 929–935. [CrossRef]
24. Khan, S.A.; Miyashita, Y.; Mutoh, Y. Corrosion fatigue behavior of AM60 magnesium alloy with anodizing layer and chemical-conversion-coating layer. *Mater. Corros.* **2015**, *66*, 940–948. [CrossRef]
25. Němcová, A.; Skeldon, P.; Thompson, G.E.; Morse, S.; Čížek, J.; Pacal, B. Influence of plasma electrolytic oxidation on fatigue performance of AZ61 magnesium alloy. *Corros. Sci.* **2014**, *82*, 58–66. [CrossRef]
26. ASTM E8M-16a. Standard test methods for tension testing of metallic materials 1. *ASTM Stand.* **2010**, *4*, 1–27. [CrossRef]
27. ASTM E466-15. Standard Practice for Conducting Force Controlled Constant Amplitude Axial Fatigue Tests of Metallic Materials. *ASTM Stand.* **2002**, *3*, 4–8. [CrossRef]

28. De Oliveira, L.A.; Silva, R.M.P.; Rodas, A.C.D.; Souto, R.M.; Antunes, R.A. Surface chemistry, film morphology, local electrochemical behavior and cytotoxic response of anodized AZ31B magnesium alloy. *J. Mater. Res. Technol.* **2020**, *9*, 14754–14770. [CrossRef]
29. De Oliveira, M.C.L.; Pereira, V.S.M.; Correa, O.V.; Antunes, R.A. Corrosion performance of anodized AZ91D magnesium alloy: Effect of the anodizing potential on the film structure and corrosion behavior. *J. Mater. Eng. Perform.* **2014**, *23*, 593–603. [CrossRef]
30. ASTM F1801-97. Standard Practice for Corrosion Fatigue Testing of Metallic Implant Materials 1. *ASTM Stand.* **2014**, *3*, 5–10. [CrossRef]
31. ASTM G1-90. Standard Practice for Preparing, Cleaning, and Evaluation Corrosion Test Specimens. *ASTM Stand.* **2003**, *8*, 1–9. [CrossRef]
32. Khan, S.A.; Miyashita, Y.; Mutoh, Y.; Koike, T. Effect of anodized layer thickness on fatigue behavior of magnesium alloy. *Mater. Sci. Eng. A* **2008**, *474*, 261–269. [CrossRef]
33. Rahman, Z.U.; Deen, K.M.; Haider, W. Controlling corrosiion kinetics of magnesium alloys by electrochemical anodization and investigation of film mechanical properties. *Appl. Surf. Sci.* **2019**, *484*, 906–916. [CrossRef]
34. Barjaktarevic, D.; Medjo, B.; Gubeljak, N.; Cvijovic-Alagic, I.; Stefane, P.; Djokic, V.; Rakin, M. Experimental and numerical analysis of tensile properties of Ti-13Nb-13Zr alloy and determination of influence of anodization process. *Proc. Struct. Integr.* **2020**, *28*, 2187–2194.
35. Nan, Z.Y.; Ishihara, S.; McEvily, A.J.; Shibata, H.; Komano, K. On the sharp bend of the S-N curve and the crack propagation behavior of extruded magnesium alloy. *Scr. Mater.* **2007**, *56*, 649–652. [CrossRef]
36. Nan, Z.Y.; Ishihara, S.; Goshima, T. Corrosion fatigue behavior of extruded magnesium alloy AZ31 in sodium chloride solution. *Int. J. Fatigue* **2008**, *30*, 1181–1188. [CrossRef]
37. He, X.L.; Wei, Y.H.; Hou, L.F.; Yan, Z.F.; Guo, C.L. High-frequency corrosion fatigue behavior of AZ31 magnesium alloy in different environments. *Proc. Inst. Mech. Eng. Part C J. Mech. Eng. Sci.* **2013**, *228*, 1645–1657. [CrossRef]
38. Bhuiyan, M.S.; Mutoh, Y.; Murai, T.; Iwakami, S. Corrosion fatigue behavior of extruded magnesium alloy AZ61 under three different corrosive environments. *Int. J. Fatigue* **2008**, *30*, 1756–1765. [CrossRef]
39. Raman, S.R.K.; Jafari, S.; Harandi, S.E. Corrosion fatigue fracture of magnesium alloys in bioimplant applications: A review. *Eng. Fract. Mech.* **2015**, *137*, 97–108. [CrossRef]
40. Hilpert, M.; Wagner, L. Corrosion fatigue behavior of the high-strength magnesium alloy AZ 80. *J. Mater. Eng. Perform.* **2000**, *9*, 402–407. [CrossRef]
41. Khan, S.A.; Miyashita, Y.; Mutoh, Y.; Koike, T. Fatigue behavior of anodized AM60 magnesium alloy under humid environment. *Mater. Sci. Eng. A* **2008**, *498*, 377–383. [CrossRef]
42. Yerokhin, A.L.; Shatrov, A.; Samsonov, V.; Shashkov, P.; Leyland, A.; Matthews, A. Fatigue properties of Keronite® coatings on a magnesium alloy. *Surf. Coat. Technol.* **2004**, *182*, 78–84. [CrossRef]
43. Ceschini, L.; Morri, A.; Angelini, V.; Messieri, S. Fatigue behavior of the rare earth rich EV31A Mg alloy: Influence of plasma electrolytic oxidation. *Metals* **2017**, *7*, 212. [CrossRef]
44. Lonyuk, B.; Apachitei, I.; Duszczyk, J. The effect of oxide coatings on fatigue properties of 7475-T6 aluminium alloy. *Surf. Coat. Technol.* **2007**, *201*, 8688–8694. [CrossRef]
45. Wang, B.J.; Xu, D.K.; Wang, S.D.; Sheng, L.Y.; Zeng, R.C.; Han, E. hou Influence of solution treatment on the corrosion fatigue behavior of an as-forged Mg-Zn-Y-Zr alloy. *Int. J. Fatigue* **2019**, *120*, 46–55. [CrossRef]
46. Wang, S.D.; Xu, D.K.; Wang, B.J.; Han, E.H.; Dong, C. Effect of corrosion attack on the fatigue behavior of an as-cast Mg-7%Gd-5%Y-1%Nd-0.5%Zr alloy. *Mater. Des.* **2015**, *84*, 185–193. [CrossRef]
47. Xu, D.K.; Liu, L.; Xu, Y.B.; Han, E.H. The crack initiation mechanism of the forged Mg-Zn-Y-Zr alloy in the super-long fatigue life regime. *Scr. Mater.* **2007**, *56*, 1–4. [CrossRef]
48. Yang, Y.; Liu, Y.B. High cycle fatigue characterization of two die-cast magnesium alloys. *Mater. Charact.* **2008**, *59*, 567–570. [CrossRef]
49. Lv, F.; Yang, F.; Duan, Q.Q.; Yang, Y.S.; Wu, S.D.; Li, S.X.; Zhang, Z.F. Fatigue properties of rolled magnesium alloy (AZ31) sheet: Influence of specimen orientation. *Int. J. Fatigue* **2011**, *33*, 672–682. [CrossRef]
50. Ochi, Y.; Masaki, K.; Hirasawa, T.; Wu, X.; Matsumura, T.; Takigawa, Y.; Higashi, K. High cycle fatigue property and micro crack propagation behavior in extruded AZ31 magnesium alloys. *Mater. Trans.* **2006**, *47*, 989–994. [CrossRef]
51. Tokaji, K.; Kamakura, M.; Ishiizumi, Y.; Hasegawa, N. Fatigue behaviour and fracture mechanism of a rolled AZ31 magnesium alloy. *Int. J. Fatigue* **2004**, *26*, 1217–1224. [CrossRef]
52. Rivers, G. Cyclic Fatigue Behaviour of Wrought AZ80 Magnesium Alloy from Forged Automotive Wheel. Ph.D. Thesis, McMaster University, Hamilton, ON, Canada, 2011.
53. Fouad, Y. Fatigue behavior of a rolled AZ31 magnesium alloy after surface treatment by EP and BB conditions. *Alex. Eng. J.* **2011**, *50*, 23–27. [CrossRef]

metals

MDPI

Article

Corrosion Behavior of Copper Bearing Steels and the Derived In-Situ Coating

Na Li [1,2,*], Ling Yan [2,*], Shaodong Wang [1], Changshun Wang [2], Hongmei Zhang [1], Fangfang Ai [2] and Zhengyi Jiang [1,3,*]

1 School of Materials and Metallurgy, University of Science and Technology Liaoning, Anshan 114051, China; pnmx521@126.com (S.W.); zhanghm@ustl.edu.cn (H.Z.)
2 State Key Laboratory of Metal Material for Marine Equipment and Application, Anshan 114009, China; changshun1101@sina.com (C.W.); aifangfang@163.com (F.A.)
3 School of Mechanical, Materials and Mechatronic Engineering, University of Wollongong, Wollongong, NSW 2522, Australia
* Correspondence: lina@ustl.edu.cn (N.L.); yanling_1101@126.com (L.Y.); jiang@uow.edu.au (Z.J.)

Abstract: Using a period immersion wet/dry cyclic corrosion test, in-situ copper-coated steels prepared by corroding copper-bearing steels were investigated in this study. The steel with a higher copper content (>3%) has a higher initial corrosion rate due to its obvious two-phase microstructure. The corrosion rates of all copper bearing steels tend to be stable after a certain time of corrosion. A copper-rich layer is formed between the matrix and the rust layer, which is due to the diffusion of copper from the rust layer to the metal surface. The copper's stability under this corrosion condition led to the formation of a thin copper-rich film, which was uncovered after removing the rust by choosing appropriate descaling reagents. The copper coating was generated from the matrix itself during the corrosion process at 25 °C, which provided a new approach for producing in-situ composite materials without any bonding defect. It is found that the corrosion rate, corrosion time, and copper content in steel all affect the formation of copper-rich layer. In addition to the noble copper surface, the electrochemical corrosion test results show that the corrosion resistance of copper-coated steel has been significantly improved.

Keywords: copper; steel; coating; descaling; corrosion; composite

Citation: Li, N.; Yan, L.; Wang, S.; Wang, C.; Zhang, H.; Ai, F.; Jiang, Z. Corrosion Behavior of Copper Bearing Steels and the Derived In-Situ Coating. *Metals* **2021**, *11*, 1462. https://doi.org/10.3390/met11091462

Academic Editor: Renato Altobelli Antunes

Received: 1 August 2021
Accepted: 13 September 2021
Published: 15 September 2021

1. Introduction

Copper accumulates in steel when steel scrap is reused, because it is difficult to remove copper in steelmaking processes [1]. Therefore, with the increase in the amount of scrap steel being used, it is imperative to study the application of copper in copper-bearing steels. Meanwhile, much research has been done to produce copper cladding on a steel surface using various methods. Copper, with its unique physical, chemical, mechanical, and physiological properties, has an important role in the lives of humans, and it also has an important application in many industrial and technical fields. Copper has been used both as an intermediate or undercoating, and as a top coat material [2].

Steel/copper clad metals are widely studied because they have applications ranging from functional coatings in the electrical and electronics industries, to adhesion promoting coatings, intermediate coatings for improving brightness and leveling effects in decorative/corrosion-protective coating systems, and purely decorative uses as a top coating [3,4]. Electroplating and electroless plating [2,5], cladding [6,7], cold sprayed coating [8–10], and casting [11] approaches have been widely used to produce copper-clad steel. Composite materials are more difficult to produce than single materials because of the diversity of their material properties. According to the characteristics that copper in copper-bearing steels, such as weathering steels, are easy to segregate on the surface of steels, accelerated corrosion tests of copper-bearing steels at 25 °C (room temperature) were carried out in this study in order to make copper-clad steel composites. A nearly perfect in-situ copper coating was observed after the rusts was removed from the corroded copper-bearing steels. As one of the basic properties of cladding material, the corrosion resistance of the copper-coated steels was tested by electrochemical corrosion test.

2. Materials and Methods

2.1. Sample Preparation

In order to eliminate the influence of other alloying elements, ultra-low carbon steel was used as a raw material in this study. Copper-bearing steels were prepared in a high-frequency vacuum induction furnace.

Samples with a side length of about 50 mm and a thickness of around 10 mm were cut from each ingot. The surface of the samples were polished smooth, and multipoint ($\geq$3) composition tests were carried out by direct reading spectrometer (ARL 3460, Thermo Fisher, New York, America), and the average value of the test values was taken as the alloy composition. The chemical compositions (in wt%) of the test samples are listed in Table 1, where sample No.0 is a raw steel. At present, the high content of copper in commercial copper steel is about 3%. In this experiment, two samples with higher content of copper were specially added for comparative study.

Table 1. Chemical composition of test steel (in wt%).

Steel No.	C	Cu	Si	Mn	P	S	Fe
0	0.005	0	0.030	0.046	0.010	0.007	Bal
1	0.006	1.900	0.020	0.045	0.010	0.006	Bal
2	0.001	2.380	0.018	0.035	0.010	0.007	Bal
3	0.001	6.000	0.030	0.037	0.008	0.007	Bal
4	0.001	12.640	0.056	0.064	0.009	0.007	Bal

Rectangular specimens with the approximate dimensions of 25 mm × 25 mm × 3 mm (with a small hole to hang the sample during the corrosion test) were cut from the ingots, and then mechanically ground using SiC papers in succession up to 800 grit. Afterwards, they were treated by descaling, cleaning, and rinsing in acetone, and then drying.

2.2. Wet-Dry Cyclic Accelerated Test

Dry-wet cycle accelerated test is a conventional method to test the corrosion resistance of steels at 25 °C [12]. This method is used to test the corrosion resistance of test steels with different copper content, and to corrode iron on the surfaces of steels to obtain enriched copper.

Before and after each corrosion test, the specimens were weighed using an analysis balance with an accuracy of 1 mg. The samples were subjected to a periodic immersion wet/dry cyclic corrosion test, which was an accelerated corrosion test under an artificially simulated atmospheric environment. Each cycle of this wet/dry cyclic corrosion test lasted 24 h and included three stages: (1) Put the dried samples one by one on the filter paper in the high-precision balance, and handle the samples gently to prevent the rust layer from falling off, (2) immersing the specimens into a 5% sodium chloride solution with a pH value of 6.5–7.2 (as measured using a digital pH tester) at 25 °C for 3 s, and (3) drying the specimens immediately by hanging them in a test chamber with a constant humidity of RH60% (±5%) for 24 h. The total corrosion test lasted 288 h. No direct connection between the different metals occurred during the entire testing process.

2.3. Descaling and Observation

For all the corroded samples, the rust was removed with citric acid (30 g) and sodium dihydrogen phosphate anhydrous (50 g), which were dissolved in 1 L of deionized water [13]. The pH value of the descaling solution was approximately 1.6–1.7, and the descaling process lasted 24 h at 25 °C.

The samples were mounted with a slope, because pure copper is softer than steel and is easy to wear off. The microstructures of the mounted samples were observed using an optical microscope (OM, Axioskop 2 MAT, Carl Zeiss AG, Jena, Germany). A scanning electron microscope (SEM, JSM-6500, JEOL, Tokyo, Japan) equipped with an energy dispersive X-ray spectroscopy (EDS) feature (Quanta FEG250 FEI, Hillsboro, OR, USA) was used to obtain the elemental distribution at different areas of the coating microstructure. X-ray diffraction (XRD D/max-2500) was used to determine the surface phase compositions of the descaled samples. The working voltage is 40 KV, the working current is 40 mA, the radiation target is CuKa, and the scanning speed is 6°/min. Scanning ranges from 10° to 90° (2θ). JCPDS (Joint Committee on Powder Diffraction Standards) cards are used as reference to identify crystal phases in XRD patterns. The JCPDS card numbers used in this paper are: Iron 01-089-7194, Copper 01-070-3038, Fe_3O_4 00-003-0863, Fe_2O_3 00-004-0755, and FeOOH 01-081-0463.

2.4. Electrochemical Corrosion Test

The steel samples (with and without Cu coating) were rectangular in shape ($10 \times 8 \times 3$ mm^3). All samples were suitably embedded in epoxy resin, and the exposed surface was 0.8 cm^2. In this study, polarization curves were measured by Autolab electrochemical station (ECO CHEMIE BV BST7276). The samples with an electric contact were used as working electrodes in a standard three-electrode cell, i.e., platinum counter electrode, silver/silver chloride (Ag/AgCl/sat. KCl) reference electrode, and a working electrode. The latter was placed in a capillary in such a way that it remained outside the cell at 25 °C, while the capillary tip was immersed in the testing solution close to the working electrode surface.

Before experiment, the working electrode surface was degreased with acetone followed by rinsing with deionized water, and a 3.5% sodium chloride solution with pH value of 6.5–7.0 was used at 25 °C. Potentiodynamic polarization were conducted from a potential of −1.0 V to −0.1 V versus Ag/AgCl with a scan rate of 0.05 V/min after achieving a stationary value for the open circuit potential (potential variation not higher than 0.1 mV/s).

3. Results

3.1. Microstructure

Figure 1 shows the optical microstructures of the as-cast Cu-bearing steels. According to Fe-Cu phase diagram, at room temperature, the solid solubility of copper in iron is about 0.6%, while that of iron in copper is almost zero. Therefore, the copper-bearing steel is mainly composed of solid solution ferrite phase and copper-rich phase close to pure copper. Therefore, the main phases in Figure 1 should be α ferrite and a small amount of copper-rich phase.

For steels 3 and 4, there are obvious bright particle precipitates distributed along the grain boundaries and in the grains, and the grains are remarkably refined as a result of the precipitation of these phases. No bright precipitate can be observed in steels 1 and 2, which confirm that the bright phase is Cu-rich phase. The EDS analysis of steel 4 shown in Figure 2 shows that the bright precipitates are copper-rich phases with a small amount of iron in solid solution. Meanwhile, about 2.8–7.4 mass% of copper is distributed in α ferrite matrix. According to these results, there should be little of the copper-rich phase precipitated in steels 1 and 2, which agrees well with the optical observations shown in Figure 1. All the relatively larger white, gray, and black phases are α ferrite, which are shown in different colors because of their different crystal orientations.

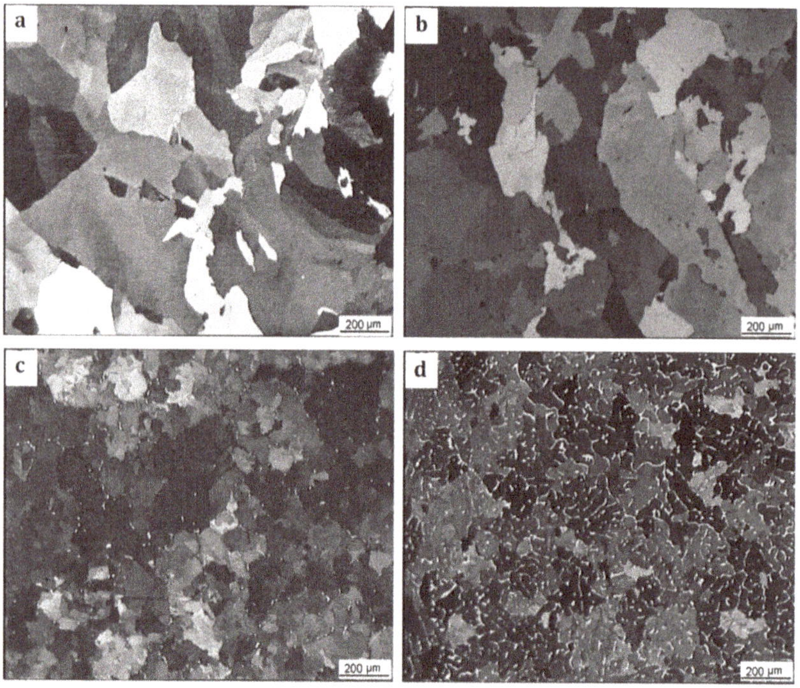

Figure 1. Optical micrographs of Cu-bearing steel 1 (**a**), steel 2 (**b**), steel 3 (**c**), and steel 4 (**d**). The etchant was 4% nital.

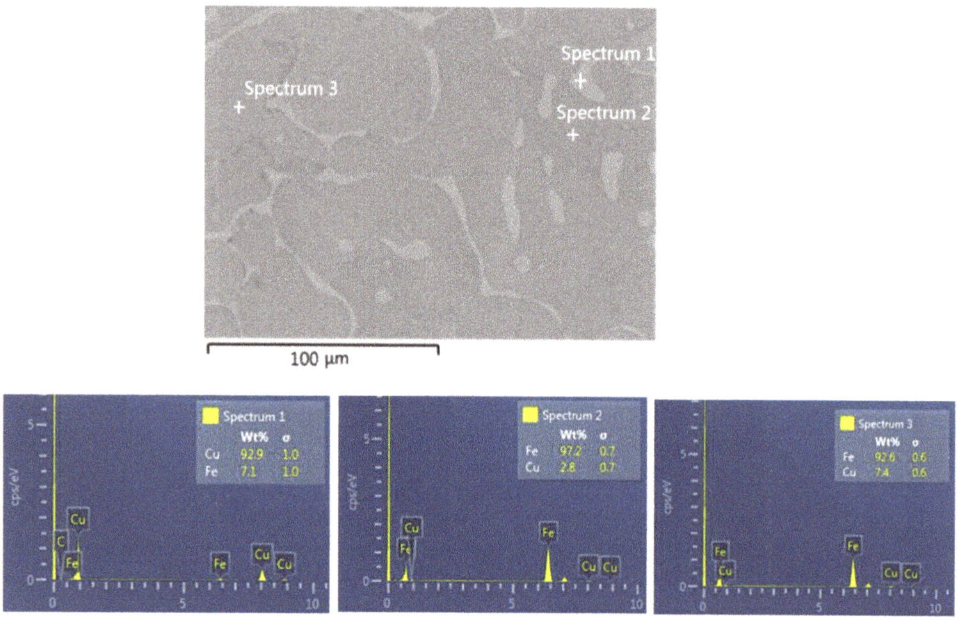

Figure 2. SEM (scanning electron microscope) image and EDS (energy dispersive X-ray spectroscopy) results of steel 4.

The actual solid solubility of metals may be slightly higher than that shown in the phase diagram. Therefore, copper beyond the solid solubility content exists as copper atoms or as near pure copper phase in steel. With an increase of the copper content, the grain sizes of the steel became fine and uniform, as shown in Figure 1c,d, indicating that copper precipitated in excess of the solid solubility is more likely to accumulate at grain boundaries and hinder their migration.

3.2. Corrosion Kinetics

Copper is widely employed as an outdoor building material for fabricating statues, sculptures, and monuments because of its aesthetic qualities, and good mechanical and antibacterial properties [14,15]. It is noted for its excellent corrosion resistance. Therefore, the corrosion behavior of copper-bearing steel is an important performance factor. The corrosion properties of Cu-bearing steels are shown in Figure 3, where steel 0 is copper free as a benchmark.

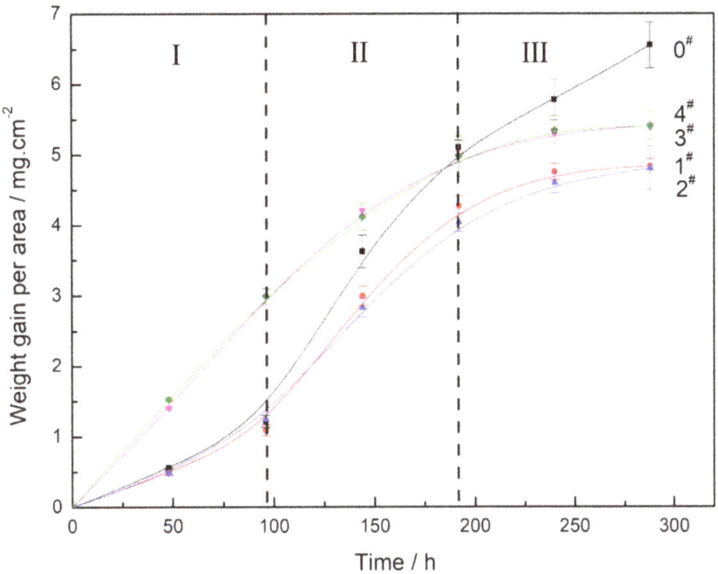

Figure 3. Corrosion behavior (weight gain per unit area in early (I), middle (II), and late (III) corrosion stage) of Cu-bearing steels.

The entire corrosion process (up to 288 h) for all the test steels can be clearly divided into three stages, as shown in Figure 3. Each stage shows distinctive slope behavior for all the samples.

In stage I, both steels 3 and 4 show higher corrosion rates. The corrosion is much more aggressive as the copper and iron begin the process of galvanic coupling at the very beginning of the corrosion process. In contrast, steels 0, 1, and 2 all show lower corrosion rates in stage I, which correspond to no obvious Cu-rich phases in the steels, and these relatively lower corrosion rates may be attributed to the formation of oxides in the beginning [16]. Therefore, in the initial stage of corrosion, steels 0, 1, and 2 exhibit surface corrosion of ferrite samples with relatively low corrosion rate, while steels 3 and 4 exhibit two-phase galvanic corrosion with relatively high corrosion rate.

The subsequent gain in weight in stage II becomes gradual for steels 0, 1, and 2. The growth rate of the corrosion rate of steel 0 is much greater than those of steels 1 and 2, which are even greater than those of steels 3 and 4, as shown in Figure 3. This phenomenon indicates that the surface corrosion products in the first stage are not dense, and the oxygen penetrate through the open holes of the oxide layer to enter the metal matrix from the surface in stage II, and the matrix is directly corroded rapidly. Therefore, in the corrosion stages up to about 192 h, although the corrosion mechanism of steel changes when the copper content of steel reaches or exceeds 6.8%, a certain amount of copper in steel has the effect of reducing and stabilizing the corrosion rate. This phenomenon should be related to the corrosion products of copper-bearing steels.

When the corrosion time exceeds 192 h, the corrosion process enters the third stage. Although the copper-free steel (steel 0) was still corroded at a higher rate, the corrosion rate of all the experimental steels was reduced compared with that of the second stage. The decrease of corrosion rate of steel 0 should be related to the increase of thickness of corrosion products on steel surface. The corrosion rate of copper-bearing steel is obviously lower than that of the previous stage, especially the steel with higher copper content, which indicates that a protective layer is formed on the corrosion surface, which is related to copper in the steel.

3.3. Corrosion Products

It is well known that copper can accumulate at the surface of steel after a certain period of corrosion for copper-bearing weathering steels. Figure 4 show optical microscope pictures of the inclined cross section of test steels after an accelerated corrosion test at 25 °C for 288 h.

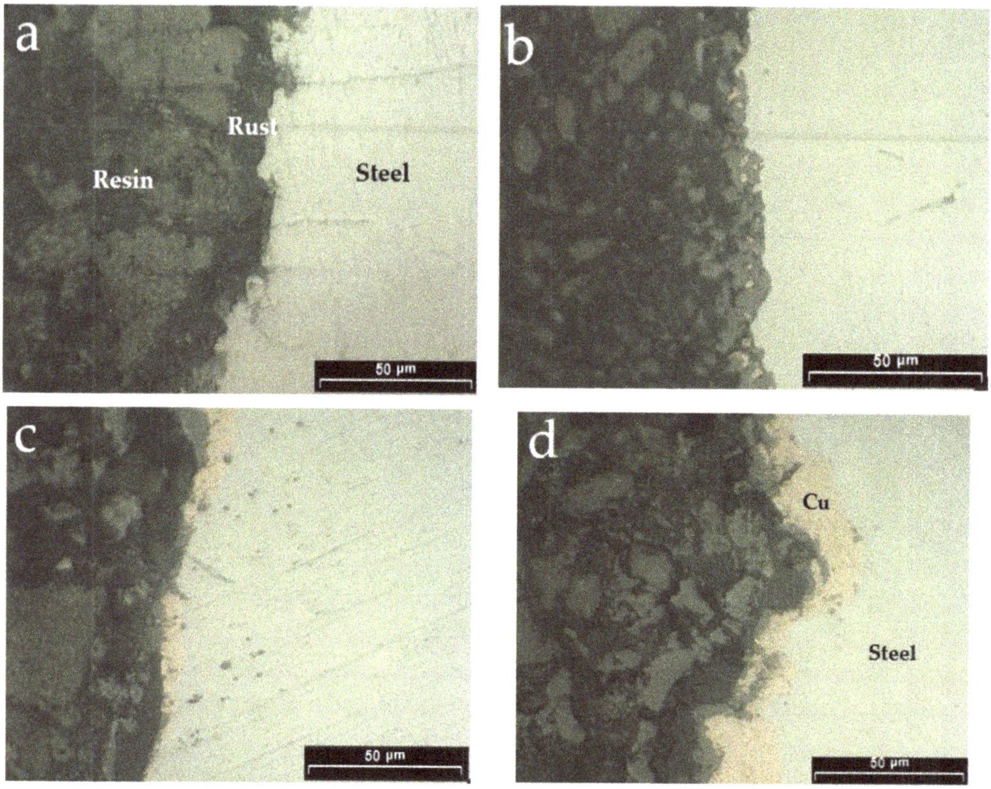

Figure 4. Optical microscope picture of inclined cross section of steel 1 (**a**), 2 (**b**), 3 (**c**), and 4 (**d**) after corrosion.

All samples were heat mounted to keep the rust layer from falling off. There is no evidence of copper segregation on the surface of steel 1 with less copper content (Figure 4a). With the increase of copper content in steel, sporadic copper segregation can be observed in the rust layer on the surface of steel 2 (Figure 4b), and the copper segregated on the surface of steel 3 begin to connect with each other locally (Figure 4d). The copper layer between the matrix and rust was obvious as shown in Figure 4d. The similar parts, which are much brighter in the back-scattered electron image are mainly pure copper of greater than 95% according to the EDS results. The copper layer is integrated in steel 4. However, this copper layer is easily ground away because pure copper is much softer than steel. It can be calculated that the thickness of the copper layer varies from a few microns to a dozen microns according to the tilt angle of the sample.

The oxides appear in black in Figure 4. These should primarily consist of iron oxide because copper is more stable than iron. After forming the copper layer between the rust and the matrix, the oxidation speed greatly slowed down before halting, and then the copper on the surface would oxidize slowly if the corrosion tests continued.

The bright gray part shown in Figure 4 is the matrix, which is an iron solid solution in which copper is the main constituent.

3.4. Rust Removal from Corroded Steel

By choosing an appropriate descaling reagent and descaling condition, the rust could be removed while keeping the copper layers, which could be determined by the naked eye because of copper's distinctive color. Macrographs of the test steels before and after descaling are shown in Figure 5.

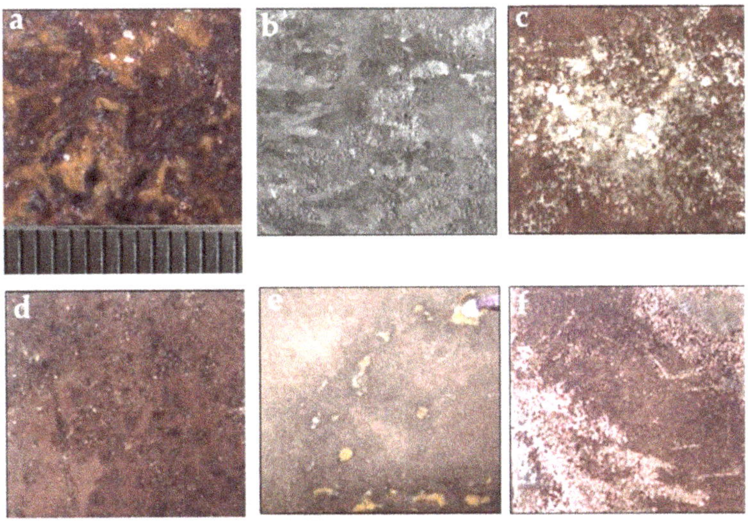

Figure 5. Macro-photograph of corrosion samples before (**a**) and after descaling (**b**) for steel 0, (**c**) for steel 1, (**d**) for steel 2, (**e**) for steel 3, and (**f**) for steel 4.

Before descaling, there is a thick rust layer on the steel surface, and a typical macro-photograph is shown in Figure 5a. After descaling, the rusts had obviously been removed, as shown in Figure 5b–f. With the increasing of copper content in steel, the copper on the steel surface becomes more and more obvious after rust removal. In order to show the copper layer on the surface of steel 4, the corners of rust removal surface were partially ground with sandpaper as shown in Figure 5f, which shows that the copper layer on the steel surface is dense and closely combined with the matrix.

In order to confirm the presence of a pure metallic Cu coating, the XRD analysis results for steel 2 at different stages are shown in Figure 6. Before corrosion, the original material is iron (Figure 6a), where copper should act as solid solution atoms. After corrosion, the surface of steel 2 is covered with a thick rust layer consisting of Fe_2O_3, Fe_3O_4, Fe, and FeO(OH), which may include α-FeOOH, γ-FeOOH, and amorphous ferric oxyhydroxide ($FeOx(OH)_{3-2x}$, x = 0–1) [17,18]. There is little copper or copper oxide present, because these cannot be detected by XRD, as shown in Figure 6b. After descaling, there is approximately 80% Cu and 20% Fe with a small amount of Fe_2O_3 left on the surface of steel 2, as shown in Figure 6c. These results agree well with that shown in Figure 4.

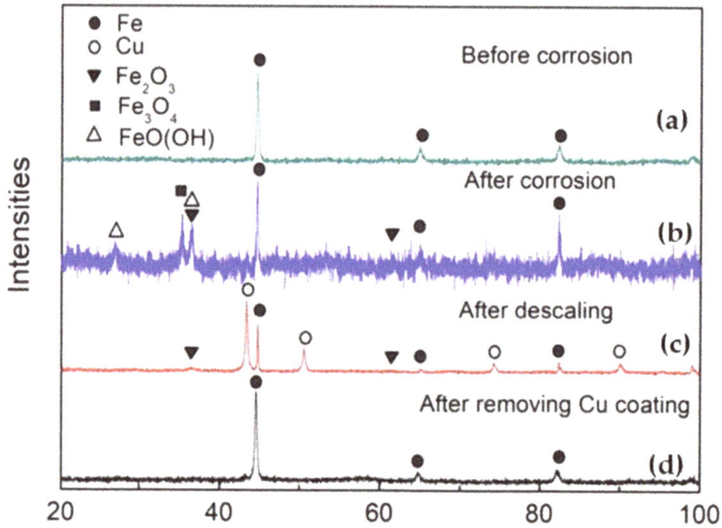

Figure 6. XRD patterns on the surface of steel 2 under different conditions (before corrosion (**a**), after corrosion (**b**), after descaling (**c**) and after removing Cu coating (**d**)).

According to XRD results of all the descaled steel samples, the main contents of the descaled surfaces are pure copper and pure iron, with a small amount of Fe_2O_3 left. For steel 2, the copper content in the bulk steel is much smaller than that in steel 4, while the composition of the surface after descaling is similar. The higher iron content in the XRD pattern of steel 2 shown in Figure 6c may correspond to a relatively thinner copper coating.

The microstructures of the steel surfaces after descaling were observed using SEM, as shown in Figure 7. Except for steel 1, almost all other steel surfaces were covered with more than 95% copper according to EDS results of steels 2, 3, and 4, which is nearly pure copper, as shown in the XRD results, and the greater amount of iron recognized by the XRD should come from the matrix, as previously mentioned. However, there were some oxides left on the descaled surface according to the XRD results shown in Figure 6. Nevertheless, they were only found in small quantities. The carbon detected by EDS may have come from the microscope chamber, as did some of the oxygen in the spectra, because the EDS tends to over-measure the levels of light elements.

It can also be seen that the copper coating is not very dense for steel 2, as shown in Figure 7b. Thus, further oxidation of the matrix might occur when the corrosion test continues. For steel 4, the copper coating is denser than that of other steels, as shown in Figure 7. The "grain boundary" morphology shown in Figure 7d should come from the copper-rich phase precipitated in the matrix, as shown in Figure 1d. Therefore, the copper content of the "grain boundary" may be a little higher than that of the "grain".

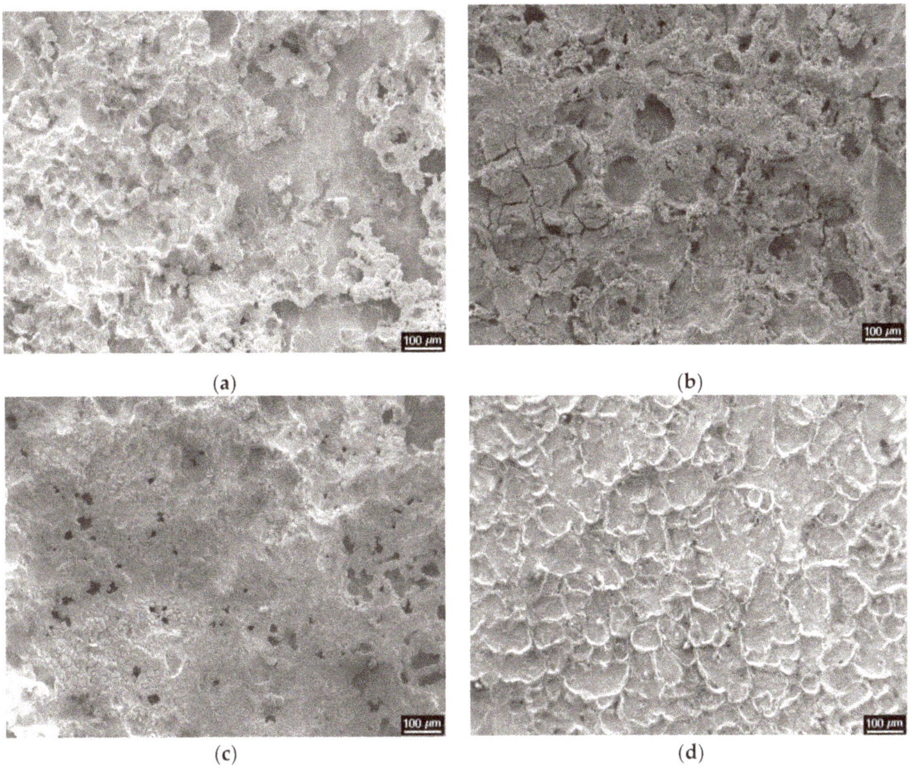

Figure 7. SEM (a) for steel 1, (b) for steel 2, (c) for steel 3, and (d) for steel 4 micrographs on steel surface after descaling.

3.5. Copper Enrichment Mechanisms

3.5.1. Movement of Copper Atoms

It is well known that the oxidation and corrosion rate of copper is relatively low compared to that of iron. Kondo [19] examined the distribution of copper in the scale of oxidized steel containing copper. The schematic diagram of corrosion and oxidation process of copper-bearing steel is shown in Figure 8. In the corrosion process, oxygen and other atoms tend to combine with iron atoms to form Fe_2O_3, Fe_3O_4 and other non-metallic corrosion products, while the combination of oxygen and copper requires higher energy and longer time. The combination of oxygen and iron atoms results in the increase of sample weight [20]. Meanwhile, the volume of the oxide layer expands with the increase of material weight, which indicates that the distance between the compound molecules in the oxide layer and the metal atoms in the matrix increases as shown in Figure 8b. As a result, the binding forces between iron atoms in the oxide molecule and those in the matrix are weakened, sometimes accompanied by some defects [21]. The appearance of defects is more conducive to the oxidation and movement of metal atoms.

Because of different bonds, the bonding force between metallic atoms is stronger than that between metal and oxide. With the formation of oxides on the surface of steel, the binding force between the copper atoms and the surrounding iron oxide molecules is weakened, whereas the bonding force between the copper atoms in the oxide layer and the iron (and/or copper) atoms in the matrix becomes stronger. The metallic atoms would move closer to each other. In Figure 8b, FexOy is used to represent corrosion products. As a result, the copper atoms move continuously from the inner oxide layer to the interface between the oxide and the metal matrix under the action of the resultant force, as shown in Figure 8c. After a period of corrosion, the continuous oxidation of iron atoms and the continuous movement of copper atoms lead to the formation of a certain number of copper atoms on the steel surface, which is macroscopically shown as a copper rich layer (Figure 8d).

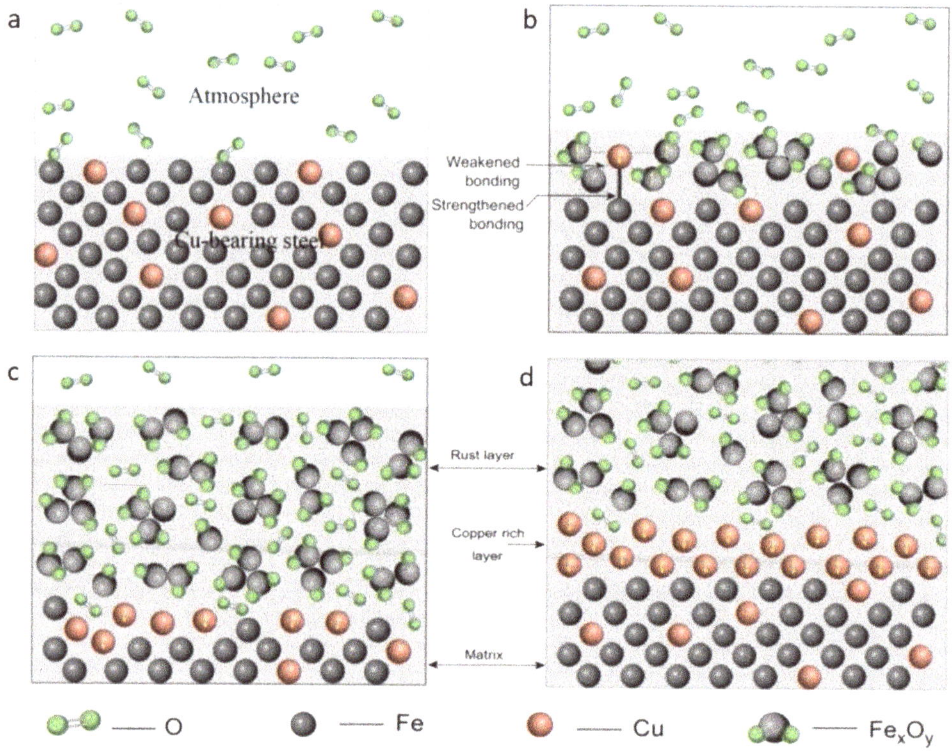

Figure 8. Schematic diagram of copper enrichment mechanism. (**a**) Before corrosion, (**b**) initial stage of corrosion, (**c**) after a short period of corrosion, and (**d**) after a long period of corrosion.

In the oxidation process, the iron atoms in the lattice positions of the matrix crystal deviate from their original positions because they combine with oxygen to form ferric oxide. The migration of surrounding iron atoms and the increase of the microcosmic distance due to the weakening of the bonding force between the metal atoms and non-metal molecules is favorable for the directional movement of copper atoms. The oxidation of the matrix surface and diffusion of un-oxidized copper atoms occur at the same time, which form the quasi-layered microstructures of the matrix, copper enrichment layer, and upper rust layer after a period. Therefore, it can be deduced that the copper enrichment layer in the structure originates from the movement of copper atoms in the oxidized surface layer.

When steel is corroded, both iron atoms and solid solution copper atoms in the oxide layer deviate from their atomic lattice position. Under the experimental conditions, most copper atoms have time to move towards the matrix and combine to form a copper-rich phase. From the macroscopic point of view, a layer of "copper coating" has been formed, and the coating is generated by the matrix itself.

The force between the copper atoms moving from the oxide layer to the interface and the iron atoms in the matrix is a metal bond. Thus, there is no bonding problem.

It should be noticed that in the process of corrosion, some defects on the surface of the substrate, such as vacancies, will affect the concentration of copper, and some copper atoms will be immersed in the scale.

3.5.2. Influence Factors

The formation of the quasi-layered structure may be affected by the following factors: the corrosion rate, corrosion time, and copper content in the steel.

The corrosion rate should correspond to the movement speed of the copper atoms in the oxide layer, although the values of these two are difficult to measure at present. If the corrosion rate is very high, the copper atoms are not able to move to the interface, because all the iron atoms around them are quickly oxidized and the copper atoms lose the driving force. Thus, no copper enrichment layer will be formed. However, if the corrosion rate is rather low, the moving velocity of the copper atoms toward the metal matrix may also be very low. In this process, the copper atoms themselves may be oxidized before they form the copper enrichment layer, and the copper atoms inside the metal matrix have no reason to move to the surface at 25 °C. Thus, no copper enrichment layer will be formed in this case either.

The effect of the oxidation time is related to the copper content in the steel. As previously mentioned, the formation of a copper enrichment layer is a process in which copper atoms in the oxide layer keep moving and accumulate on the surface of the matrix. Therefore, the continuous copper enrichment layer on the matrix surface can be formed in a relatively short time if the copper content is high. In contrast, if the copper content is low, the corrosion time will be longer, and the thickness of the oxide layer will be greater, allowing sufficient copper atoms to form a continuous layer of copper, which can explain the early corrosion deceleration stage of steels 3 and 4 in the corrosion weight gain curve shown in Figure 3.

3.6. Electrochemical Corrosion Test

Figure 9 shows the polarization curves of all test steels (a) and steel 4 (b) before corrosion, without Cu coating, and after corrosion and descaling, with Cu coating (under the experimental conditions, only steel 4 shows a complete copper coating), in 3.5% NaCl solution.

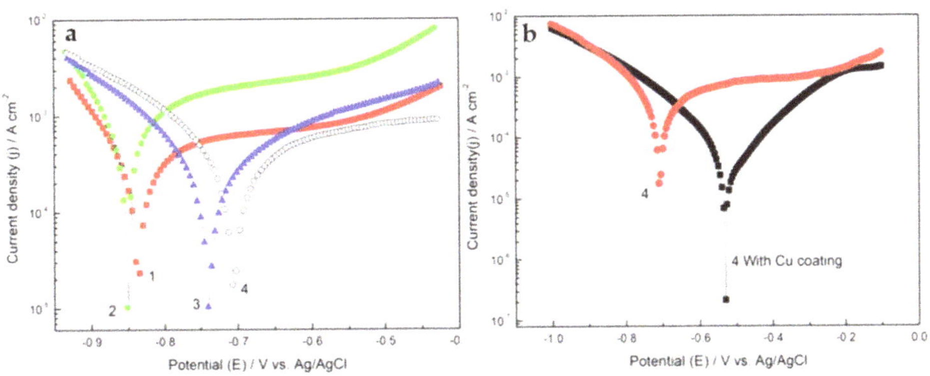

Figure 9. Potentiodynamic polarization curves of all test steels (**a**) and steel No.4 with Cu coating (**b**) in 3.5% NaCl solution.

The electrochemical parameters E_{corr}, I_{corr}, j_{corr}, and v were obtained by fitting the curve are listed in Table 2. With the increase of copper content in steel, E_{corr} tends to move positively, which implies a better corrosion resistance property. At the same time, when the copper content is high enough, the corrosion current density and corrosion rate of the experimental steel are reduced, and the passivation interval has been extended, which also shows that the corrosion resistance of the steel is improved. Relevant research results also reported that high-copper-bearing steel has excellent corrosion resistance [22].

Table 2. Electrochemical parameters of test steels.

Steel No.	Corrosion Current I_{corr}/A	Corrosion Current Density j_{corr}/(A·cm^{-2})	Corrosion Potential E_{corr}/V	Corrosion Rate v/(g·m^2·h^{-1})
1	2.13×10^{-4}	2.66×10^{-4}	−0.83	1.8534
2	3.26×10^{-4}	4.08×10^{-4}	−0.85	2.8376
3	1.16×10^{-4}	1.44×10^{-4}	−0.66	1.0056
4	1.67×10^{-4}	2.09×10^{-4}	−0.70	1.4542
4 with Cu coating	1.90×10^{-5}	2.38×10^{-5}	−0.53	0.1324

E_{corr} of steel 4 with Cu coating is more positive than that of steel 4 without Cu coating. The OCP of steel 4 with Cu coating shifted significantly to a positive (anodic) direction, suggesting the formation of an efficient barrier for the aggressive medium [23,24].

The polarization potential of the steel treated by corrosion and descaling shifts positively and the corrosion current per unit area is reduced by nearly one order of magnitude, which is similar to the result of polyaniline coating on mild steel [25]. Therefore, the corrosion resistance of Cu-bearing steel can be significantly improved by the copper coating after corrosion and descaling treatment at 25 °C.

4. Conclusions

The corrosion behaviors of copper bearing steels were studied, and the accumulation of copper on the surface of corroded steels after the removal of rust was observed and analyzed. The following conclusions can be made:

(1) Steel with a high copper content (>3%) had a high initial corrosion rate due to its obvious two-phase microstructure. The corrosion rates of all copper bearing steels tended to be smooth after a certain period of corrosion.

(2) A complete copper coating was obtained on the surfaces of the steels after a periodic immersion wet/dry cyclic corrosion test. The oxidation of the matrix and the diffusion of copper atoms at the interface formed a quasi-layered structure consisting of the matrix, copper coating, and upper rust layer.

(3) The copper coating was generated from the matrix itself during the corrosion process at 25 °C. A lower corrosion rate, longer corrosion time, and higher copper content were favorable for the formation of a copper enriched layer.

(4) Copper-bearing steel with copper coating shows better corrosion resistance property.

Author Contributions: Conceptualization, N.L.; methodology, N.L.; validation, L.Y., S.W. and H.Z.; formal analysis, N.L., Z.J.; investigation, H.Z.; resources, L.Y., C.W., H.Z. and F.A.; data curation, L.Y. and C.W.; writing—original draft preparation, N.L.; writing—review and editing, N.L. and Z.J.; supervision, H.Z. and Z.J.; project administration, H.Z.; funding acquisition, N.L., L.Y., C.W. and H.Z. All authors have read and agreed to the published version of the manuscript.

Funding: This research was funded by the State Key Laboratory of Metal Material for Marine Equipment and Application-School of Material and Metallurgy, University of Science and Technology Liaoning co-project, grant number No. SKLMEA-USTL 2017010 and HGSKL-USTLN 201905.

Institutional Review Board Statement: Not applicable.

Informed Consent Statement: Not applicable.

Data Availability Statement: Data is contained within the article.

Acknowledgments: The authors thank Wei Sha from Queen's University Belfast, UK, for some helpful discussions.

Conflicts of Interest: The authors declare no conflict of interest.

References

1. Ramadan, A.; Shash, A.Y.; El-Mahallawi, I.S.; Senk, D.; Mattar, T. Identification of copper precipitates in scrap based recycled low carbon rebar steel. *Mater. Des.* **2017**, *120*, 157–169. [CrossRef]
2. Zhang, X.; Wang, J.; Tang, J.; Tian, Z. Unraveling the multilayer structure formation mechanism of copper coating for the metallization of poly para-phenylene terephthalamide fibers. *Colloids Surf. A* **2021**, *618*, 126450. [CrossRef]
3. Kumar, R.R.; Rajalakshmy, P.; Saranya, M.D.; Kirubakaran, S.; Elwin, J.G.R.; Marichamy, S. Process automation through internet of things on copper coating process of stainless steel. *Mater. Today Proc.* **2021**, *45*, 2354–2356. [CrossRef]
4. Baiocco, G.; Rubino, G.; Ucciardello, N. Pretreatments effects on mechanical and morphological features of copper coatings. *Surf. Interfaces* **2020**, *20*, 100625. [CrossRef]
5. Lv, J.; Wang, Q.; Zhao, J.; Liu, W.; Chen, P.; Liu, H. The difference in the improvement of electrochemical hydrogen storage performance between two methods of coating copper on the surface of Co_2B alloy. *Chem. Phys. Lett.* **2020**, *754*, 137697. [CrossRef]
6. Arai, S.; Iwashita, R.; Shimizu, M.; Inoue, J.; Horita, M.; Nagaoka, T.; Itabashi, M. Fabrication of roughened electrodeposited copper coating on steel for dissimilar joining of steel and thermoplastic resin. *Metals* **2021**, *11*, 591. [CrossRef]
7. Paul, H.; Chulist, R.; Mania, I. Structural properties of interfacial layers in tantalum to stainless steel clad with copper interlayer produced by explosive welding. *Metals* **2020**, *10*, 969. [CrossRef]
8. Tam, J.; Li, W.; Yu, B.; Poirier, D.; Legoux, J.G.; Lin, P.; Palumbo, G.; Giallonardo, J.D.; Erb, U. Reducing complex microstructural heterogeneity in electrodeposited and cold sprayed copper coating junctions. *Surf. Coat. Technol.* **2020**, *404*, 126479. [CrossRef]
9. Silva, F.S.; Cinca, N.; Dosta, S.; Cano, I.G.; Guilemany, J.M.; Caires, C.S.A.; Lima, A.R.; Silva, C.M.; Oliveira, S.L.; Caires, A.R.L.; et al. Corrosion resistance and antibacterial properties of copper coating deposited by cold gas spray. *Surf. Coat. Technol.* **2019**, *361*, 292–301. [CrossRef]
10. Singh, S.; Singh, H.; Chaudhary, S.; Buddu, R.K. Effect of substrate surface roughness on properties of cold-sprayed copper coatings on SS316L steel. *Surf. Coat. Technol.* **2020**, *389*, 125619. [CrossRef]
11. Dong, T.; Liu, M.; Feng, Y.; Li, G.; Li, X. Microstructure and properties of a wear resistant Al–25Si–4Cu–1Mg coating prepared by supersonic plasma spraying. *Int. J. Min. Met. Mater.* **2020**, *27*, 1287–1294. [CrossRef]
12. Chen, Y.Y.; Tzeng, H.J.; Wei, L.I.; Wang, L.H.; Oung, J.C.; Shih, H.C. Corrosion resistance and mechanical properties of low-alloy steels under atmospheric conditions. *Corros. Sci.* **2005**, *47*, 1001–1021. [CrossRef]
13. Zeng, L.; Xu, Y.; Zou, Z. Study on normal temperature neutral rust remover for iron and steel. *Electroplat. Pollut. Control* **1996**, *16*, 18–20. (In Chinese)
14. Pan, C.; Lv, W.; Wang, Z.; Su, W.; Wang, C.; Liu, S. Atmospheric corrosion of copper exposed in a simulated coastal-industrial atmosphere. *J. Mater. Sci. Technol.* **2017**, *33*, 587–595. [CrossRef]
15. Li, P.; Zhao, Y.; Liu, Y.; Zhao, Y.; Xu, D.; Yang, C.; Zhang, T.; Gu, T.; Yang, K. Effect of Cu addition to 2205 duplex stainless steel on the resistance against pitting corrosion by the pseudomonas aeruginosa biofilm. *J. Mater. Sci. Technol.* **2017**, *33*, 723–727. [CrossRef]
16. Pilling, N.B.; Bedworth, R.E. The oxidation of metals at high temperatures. *J. Inst. Met.* **1923**, *29*, 529–591.
17. Kim, K.T.; Tsuchiya, H.; Hanaki, K.; Yamashita, M.; Fujimoto, S. Effect of cations on protective properties of rust layer formed on carbon steel during wet/dry cyclic corrosion. *Mater. Trans.* **2020**, *61*, 506–514. [CrossRef]
18. Zhang, Q.C.; Wu, J.S.; Wang, J.J.; Zheng, W.L.; Chen, J.G.; Li, A.B. Corrosion behavior of weathering steel in marine atmosphere. *Mater. Chem. Phys.* **2002**, *77*, 603–608. [CrossRef]
19. Yasumitsu, K. Behaviour of copper during high temperature oxidation of steel containing copper. *ISIJ Int.* **2004**, *44*, 1576–1580.
20. Li, N.; Jia, R.; Zhang, H.; Sha, W.; Li, Y.; Jiang, Z. In-situ Cu coating on steel surface after oxidizing at high temperature. *Materials* **2019**, *12*, 3536. [CrossRef]
21. Yang, F.; Choi, Y.M.; Liu, P.; Stacchiola, D.; Hrbek, J.; Rodriguez, J.A. Identification of 5-7 defects in a copper oxide surface. *J. Am. Chem. Soc.* **2011**, *133*, 11474–11477. [CrossRef]
22. Liu, H.; Teng, Y.; Guo, J.; Li, N.; Wang, J.; Zhou, Z.; Li, S. Corrosion resistance and corrosion behavior of high-copper-bearing steel in marine environments. *Mater. Corros.* **2021**, *72*, 816–828. [CrossRef]
23. Shabani-Nooshabadi, M.; Allahyary, E.; Jafari, Y. Electrosynthesis of poly(ortho-phenetidine) coatings on steel and investigation of their corrosion protection properties. *Prot. Met. Phys. Chem. Surf.* **2018**, *54*, 104–112. [CrossRef]
24. Aigbodion, V.S. Morphological and electrochemical behavior of epoxy –activated orange juice functionalized rice husk nanoparticles composite coating on mild steel. *Chem. Data Collect.* **2020**, *29*, 100526. [CrossRef]
25. Gupta, D.; Neupane, S.; Singh, S.; Karki, N.; Yadav, A. The effect of electrolytes on the coating of polyaniline on mild steel by electrochemical methods and its corrosion behavior. *Prog. Org. Coat.* **2021**, *152*, 106127. [CrossRef]

metals

Article

The Effect of Immersion Corrosion Time on Electrochemical Corrosion Behavior and the Corrosion Mechanism of EH47 Ship Steel in Seawater

Hongmei Zhang [1,2,*], Ling Yan [2,*], Yangyang Zhu [1], Fangfang Ai [2], Hongnan Li [1], Yan Li [2] and Zhengyi Jiang [3]

[1] School of Material and Metallurgy, University of Science and Technology Liaoning, Anshan 114051, China; zhu123yangyang@126.com (Y.Z.); lihongnan7270@ustl.edu.cn (H.L.)
[2] State Key Laboratory of Metal Material for Marine Equipment and Application, Anshan 114009, China; aifangfang@163.com (F.A.); 2323liyan@sina.com (Y.L.)
[3] School of Mechanical, Materials, Mechatronic and Biomedical Engineering, University of Wollongong, Wollongong, NSW 2522, Australia; jiang@uow.edu.au
* Correspondence: zhanghm@ustl.edu.cn (H.Z.); yanling_1101@126.com (L.Y.); Tel.: +86-138-0792-7151 (H.Z.); +86-139-4122-0530 (L.Y.)

Abstract: In this paper, electrochemical corrosion tests and full immersion corrosion experiments were conducted in seawater at room temperature to investigate the electrochemical corrosion behavior and the corrosion mechanism of high-strength EH47. The polarization curve, EIS (electrochemical impedance spectroscopy), SEM (scanning electron microscope), and EDS analyses were employed to analyze the results of the electrochemical corrosion process. The electrochemical corrosion experiments showed that the open circuit potential of EH47 decreases and then increases with an increase in total immersion time, with the minimum value obtained at 28 days. With an increase in immersion time, the corrosion current density (I_{corr}) of EH47 steel first decreases and then increases, with the minimum at about 28 days. This 28-day sample also showed the maximum capacitance arc radius, the maximum impedance and the minimum corrosion rate. In the seawater immersion test in the laboratory, the corrosion mechanism of EH47 steel in the initial stage of corrosion is mainly pitting corrosion, accompanied by a small amount of crevice corrosion with increased corrosion time. The corrosion products of EH47 steel after immersion in seawater for 30 days are mainly composed of FeOOH, Fe_3O_4 and Fe_2O_3.

Keywords: electrochemical corrosion behavior; corrosion mechanism; the corrosion current density; Nyquist diagram

Citation: Zhang, H.; Yan, L.; Zhu, Y.; Ai, F.; Li, H.; Li, Y.; Jiang, Z. The Effect of Immersion Corrosion Time on Electrochemical Corrosion Behavior and the Corrosion Mechanism of EH47 Ship Steel in Seawater. *Metals* **2021**, *11*, 1317. https://doi.org/10.3390/met11081317

Academic Editor: Renato Altobelli Antunes

Received: 29 June 2021
Accepted: 16 August 2021
Published: 20 August 2021

1. Introduction

For marine ships, the hull steel must withstand the impact of waves, the action of huge bending moments formed by surging waves, temperature changes in cold winters and hot summers, and the corrosion of seawater. Therefore, in addition to sufficient mechanical properties and good technological properties, hull steel should also have good corrosion resistance. Corrosion will reduce the strength of the ship structure and shorten the service life. It will also endanger the safety of navigation and reduce its operational performance. Therefore, corrosion prevention of the hull structure has always been an important issue in the design, construction and use of ships [1,2].

In the marine environment, the corrosion of marine equipment and marine engineering steel is electrochemical corrosion. Electrochemical corrosion refers to the corrosion of metals in the electrolyte due to the action of microcells on its surface [3,4]. The main reason for corrosion is the presence of dissolved oxygen. In the marine environment, the surface of steel and iron will form anode and cathode regions due to inhomogeneity. In the positive and negative regions, the formation of local batteries will cause a certain degree of corrosion [5,6]. A series of reports have shown that the surface of carbon steel does not form a stable passivating corrosion film, but a porous rust layer, which makes it difficult to prevent further corrosion under marine conditions [7,8]. The passivation film of low-alloy steels has low stability and is easily broken and corroded in seawater. Chen et al. [9] showed that the passivating film could be formed on the surface of carbon steel. When immersed in 3% NaCl solution or seawater, the weak parts of the passivating film would crack quickly and induce corrosion. The corrosion of carbon steel and low-alloy steel is uneven in a seawater environment, and local pits are formed. Furthermore, a relatively closed microenvironment and occlusive corrosion cell can be formed under the rust layer, resulting in serious local corrosion. The strength of the occlusive corrosion cell directly affects the expansion of the pit and the local corrosion resistance of carbon steel and low-alloy steel. The pH of seawater is about 8.0 to 8.2, and seawater contains a large amount of chloride ions. With a decrease in Cl^- content in the solution, the thickness of the passivation film decreases and its protection ability increases due to the failure of the passivation layer on the steel surface at the threshold of Cl^- concentration [10,11]. In addition, with an increase in Cl^- concentration, a porous and non-protective product is formed on the steel surface. Zhang studied the corrosion behavior of 304 stainless steel under different Cl^- concentrations and soaking times. The results showed that with the increase in Cl^- concentration, the stability of 304 stainless steel decreases with the passivation film [12]. The effect of Cl^- on the passivation film and pitting corrosion is particularly serious with long-term use. Yang [13] used the potentiometric polarization method to measure pitting potential in the corrosion process. The results showed that with the increase in temperature and Cl^- concentration, the size and number of corrosion pits increase.

Xia studied the erosion and corrosion of hull steel under different flow rates in the seawater environment of the East China Sea [14]. Jia et al. [15] analyzed the corrosion resistance of F690, F460 and Q235B steel in 3.5% NaCl solution. Melchers [16] proposed various models to explain the corrosion mechanism. Zayed et al. [17] and Panayotova et al. [18] investigated the main corrosion mechanisms of ship steels.

Pitting corrosion is a common form of corrosion of carbon steel and low-alloy steel in the marine environment [19,20]. It is uneven and comprehensive, and so is often called pitting corrosion, which is different from the typical blunt metal pitting corrosion. Pitting corrosion forms because of electrochemical inhomogeneity on the steel surface, which leads to the formation of local corrosion microcells. Typical forms of corrosion of carbon and low-alloy ship steels in the marine environment include uniform corrosion, pitting corrosion, oxygen concentration difference corrosion, etc. [21,22].

Electrochemical technologies and hanging plate tests, such as the corrosion potential measurement, the polarization test, linear polarization technology, electrochemical impedance spectroscopy, and the simulated block cell test, have been widely used in the study of the corrosion behavior of ship steels and the development of products [23–26]. The main advantage of the electrochemical test method is that it is fast and can obtain instantaneous corrosion information, but the main disadvantage is the lack of accuracy. In contrast to the electrochemical test, the results of the hanging test and field test are accurate, but the required time is longer and the process is more complicated [27].

In this study, due to the disadvantages of the long test period and significant influence of environmental factors, the seawater hanging sheet test was carried out in the laboratory, and the seawater was changed every seven days during the experiment, which overcomes the shortcoming that the medium changes due to the limitation of the amount of medium and the corrosion and dissolution of metal as the test progresses. In this paper, the electrochemical corrosion behavior of ship plate steel EH47 was studied by electrochemical techniques such as corrosion potential measurement, polarization curve and electrochemical impedance spectroscopy. At the same time, the corrosion mechanism of low-alloy and high-strength ship plate steel in seawater was investigated by actual seawater hanging plate test.

2. Experimental

2.1. Materials

The chemical compositions of the samples were measured by a direct reading spectrometer (ARL-3460L, Thermoelectric Corporation of America, Chicago, IL, USA): C: 0.07, Si: 0.20, Mn: 1.42, Mo: 0.22, Ni: 0.79, Cr: 0.07, Ti: 0.013, Nb: 0.04, V: 0.006, Al: 0.03, P: 0.006, S: 0.001, Fe balanced.

2.2. Electrochemical Test

In order to ensure the accuracy of the experiments, the electrochemical experiments were conducted in triplicate, meanwhile, the specimens were 4 × 3 = 12, every set needed 4 specimens, and there were 12 specimens in total.

The experimental samples were polished with W28, W20, W14 and W10 sandpaper, then polished with a metallographic polishing machine, and finally wiped with ethanol and acetone cotton balls. After welding the copper wire on the non-working surface, the non-working surface was sealed with epoxy resin, and a 10 mm × 10 mm electrode working surface was reserved.

Electrochemical experiments were carried out on a weekly basis after 7, 14, 28 and 49 days of total immersion corrosion in seawater.

All electrochemical tests were performed using a CHI760E electrochemical workstation produced by Shanghai ChenHua instrument Co., LTD (Shanghai, China). The sample was used as the working electrode, the saturated calomel electrode (SCE) was used as the reference electrode, and the platinum wire around the working electrode was used as the auxiliary electrode.

EIS (electrochemical impedance spectroscopy) measurements were performed at the open circuit potential. The open circuit potential was monitored for 60 s before electrochemical measurement. The frequency range was 100 kHz to 100 mHz and the voltage amplitude was 10 mV. Potentiodynamic polarization curves were obtained in the scanning range of −600 mV to 1400 mV with respect to the Open Circuit Potential (OCP) at a scanning rate of 0.5 mV/s.

The corrosion morphology was detected by scanning electron microscope (SEM, EV018, Carl Zeiss AG, (Oberkochen, Germany).

X-ray diffraction (XRD X'Pert Powder, Malvern Panalytical, Malvern, UK) was used to determine the surface phase compositions of the samples. The working voltage was 40 kV, the working current was 40 mA, the radiation target was Cu Ka, and the scanning speed was 6/min. The scanning ranged from 10° to 90° (2θ). The numbers of the JCPDS (Joint Committee on Powder Diffraction Standards) cards are used in this paper as references to identify the crystalline phases in the XRD patterns. The JCPDS of Fe_3O_4 is 98-015-8506, Fe_2O_3 is 98-010-890, and FeOOH of JCPDS is 96-100-8763.

2.3. Immersion Test

In order to ensure the accuracy of the experiments, the immersion experiments were conducted in duplicate, meanwhile, the specimens were 4 × 2 = 8, every set needed 4 specimens, and there were 8 specimens in total.

Considering the short electrochemical experiment period, the total immersion corrosion experiment lengthens the experiment period. The immersion test experiments were carried out in natural seawater for different total immersion times in the laboratory. The samples were immersed in natural seawater for 20, 40, and 80 days at room temperature.

3. Results

3.1. Energy Spectrum Analysis of Steel

Figure 1 is the surface EDS analysis and optical micrograph of the original EH47. It can be seen from Figure 1b that the surface of the original steel EH47 is composed of three main elements: Fe, Mn and C, indicating that Fe, Mn and C are the primary elements. The optical micrograph of the original EH47 is presented in Figure 1c. The microstructure is mainly composed of ferrite and bainite.

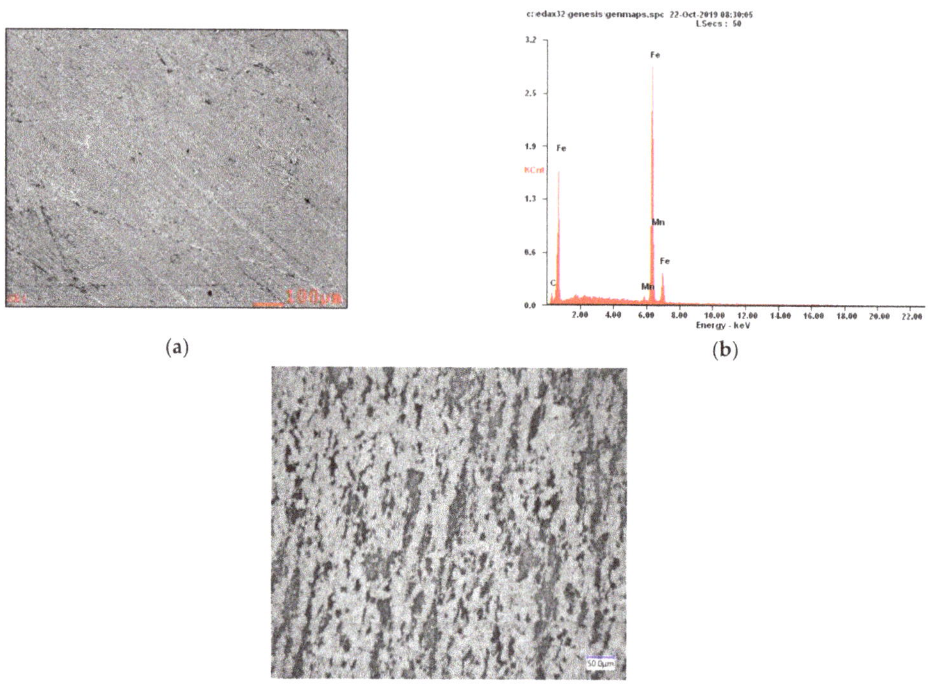

(a) (b) (c)

Figure 1. (**a**) The surface of the original steel EH47; (**b**) surface EDS analysis of EH47 steel; (**c**) the optical micrograph before the immersion test of EH47 steel.

3.2. *Electrochemical Characteristics of EH47 in Seawater*

EH47 samples were immersed in seawater for 0, 14, 28 and 49 days for electrochemical experiments. The open circuit potential, polarization curve and Nyquist diagram of electrochemical impedance spectroscopy were measured.

3.2.1. Open Circuit Potential of EH47

The change in open circuit potential can indicate the corrosion state and corrosion behavior of the material surface [27]. Figure 2 shows the open circuit potential curves of EH47 steel samples immersed in seawater for different durations (0, 14, 28, and 49 days) at room temperature. It can be seen from Figure 2 that the corrosion potential changes significantly after total immersion in seawater for 14 days, and the minimum value is obtained after 28 days of immersion. The variation in the open circuit potential from 14 days to 28 days is lower than that from 0 days to 14 days, which indicates that the corrosion rate of EH47 steel is relatively slow with the increase in corrosion time from 14 days to 28 days. With further extension of the corrosion time, the open circuit potential moves in the positive direction and the corrosion rate decreases gradually.

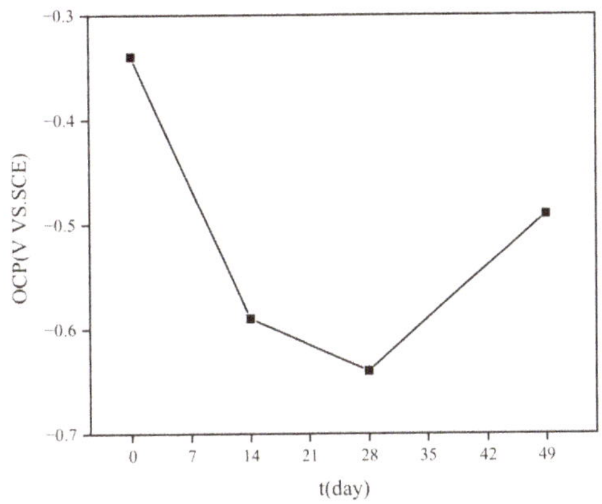

Figure 2. Open circuit potential of the EH47 steel sample immersed in seawater for different durations.

3.2.2. Potentiodynamic Polarization Tests

Figure 3 shows the potentiodynamic polarization curves of the EH47 samples immersed in seawater for different immersion durations (0, 14, 28, and 49 days) at room temperature. It can be seen from Figure 3 that the cathodic branch at day 0 immersion indicates the limiting diffusion control characteristics of the reduction reaction of dissolved oxygen. The diffusion of dissolved oxygen cannot be more effectively inhibited because there is no rust layer on the surface of the sample under this condition. With the increase in immersion time, the cathodic branch at day 0 immersion is not significantly different from 14, 28 and 49 days. The relatively smooth curves indicate that oxygen reduction can occur but it is not diffusion-controlled. The limit control characteristics of dissolved oxygen reduction on the surface of the steel with a rust layer have disappeared and changed to the charge transfer control transformation dominated by the reduction of corrosion products in the whole process of corrosion. The anode branch has a certain passivation characteristic, and the passivation range is about 0.6 V (from −0.9 V to −0.2 V). The experimental steel underwent a cathodic electrode reaction under the limit diffusion control of dissolved oxygen; meanwhile, the anode electrode reaction underwent electrochemically active dissolution under the control of charge transfer (current). After forming a certain corrosion layer thickness at the early stage of corrosion, the reductive corrosion product γ-FeOOH appears in the corrosion layer [28]. As the corrosion reaction continues, the reduction reaction ($Fe^{3+} + e^{-} \rightarrow Fe^{2+}$) of the corrosion layer mainly occurs in the cathode region. With the continuous reaction, Fe_3O_4 and β-FeOOH were formed with high stability in the corrosion layer, which played a certain protective role, resulting in a relatively slow corrosion rate and a relatively low corrosion current [29].

A comparison of the corrosion current density (I_{corr}) of samples at different corrosion durations in seawater, according to the data in Table 1, shows that the maximum value of I_{corr} is 1.444×10^{-4} A/cm^2 on day 0, because there is no rust layer on the surface of the day 0 immersion sample, and that the oxygen on the surface of the sample is sufficient, the anodic dissolved Fe^{2+} can diffuse rapidly, and the corrosion rate is the fastest [30]. With the prolongation of immersion time, the I_{corr} of the EH47 steel sample first decreases and then increases, and reaches the minimum value of 8.093×10^{-5}A/cm^2 at day 28 immersion. Because of the dense and tight corrosion layer formed on the surface of the sample under this condition, which plays a prominent role in protecting the matrix and hinders the diffusion of Fe^{2+} generated by the anodic reaction on the steel surface in seawater, the corrosion resistance of the sample is relatively good at day 28 immersion. With a further increase in immersion time, the rust layer formed on the surface of the sample is easily removed after long-term immersion in seawater, which provides a channel for the diffusion of iron ions into seawater under the condition of sufficient O_2 [31]. Furthermore, the corrosion products become loose and porous, and the protective effect of the corrosion layer on the matrix is weakened, which makes the I_{corr} of the EH47 steel sample increase after 49 days of immersion [5,32,33].

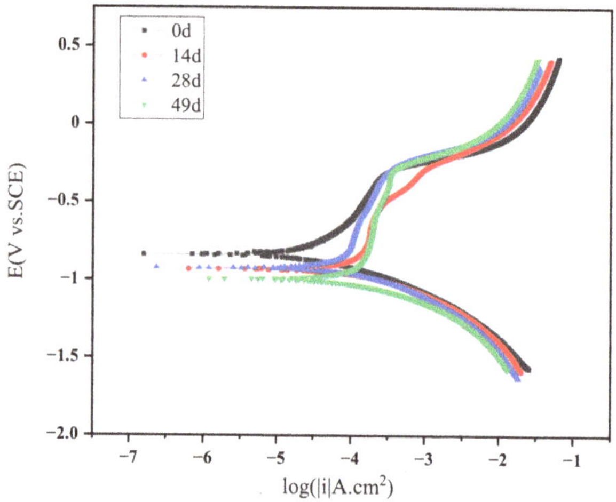

Figure 3. Potentiodynamic polarization curves of the EH47 samples immersed in seawater for different durations.

Table 1. Results of fitting polarization curves of the EH47 sample immersed in seawater for different corrosion durations.

Corrosion Time (d)	Corrosion Potential, vs. SCE	Corrosion Current Density (A/cm^2)	Passive Domain, vs. SCE
0	−0.915	1.444×10^{-4}	−0.8~−0.2
14	−0.934	1.107×10^{-4}	−0.8~−0.5
28	−0.924	8.093×10^{-5}	−0.8~−0.2
49	−0.995	1.030×10^{-4}	−0.9~−0.2

3.2.3. EIS Measurements

Electrochemical impedance spectroscopy is an effective method in the field of corrosion electrochemistry [34]. Figure 4 shows the Nyquist diagrams of EH47 steel corroded in seawater for different immersion durations (0, 14, 28, and 49 days). It can be seen from Figure 4 that the obvious capacitive arc is an incomplete semicircle. The radius of the semicircle arc increases gradually. With the extension of full immersion time, the radius of the semicircle arc decreases, which may be caused by the thickening of the rust layer affecting the charging and discharging process. Among them, the arc radius of the day 28 immersion sample is the largest, which indicates that the impedance is the largest, the corrosion rate is the smallest, and the corrosion resistance is good. The arc radius of the day 0 immersion sample is the smallest, which indicates that its impedance is the smallest and the corrosion rate is the largest, which is consistent with the results of the polarization curve analysis above.

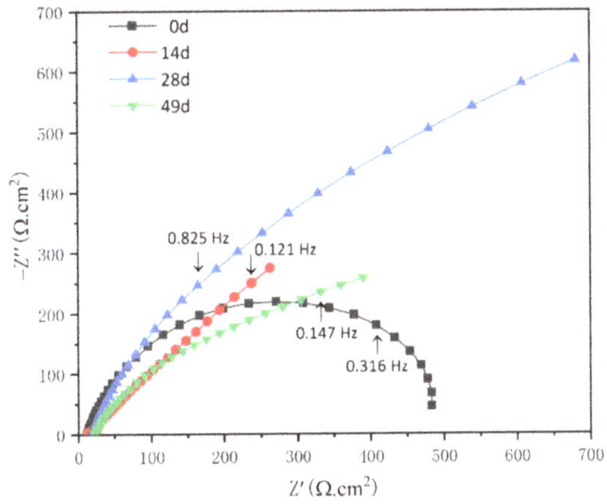

Figure 4. Nyquist diagrams of the EH47 sample immersed in seawater for different durations. Z′—real part of electrochemical impedance; Z″—imaginary part of electrochemical impedance.

Figure 5 shows the Bode diagrams of the samples immersed in seawater for different durations. Figure 5a shows the phase angle versus frequency. It can be seen that the phase angle of 28 days immersion is close to the maximum value. Figure 5b shows that the impedance modulus value increases with the decrease in frequency in the frequency range from 10^{-1} Hz to 10 Hz, and the maximum impedance is obtained at day 28 of immersion. Figure 5 further demonstrates that the corrosion resistance of the day 28 immersion sample is the best.

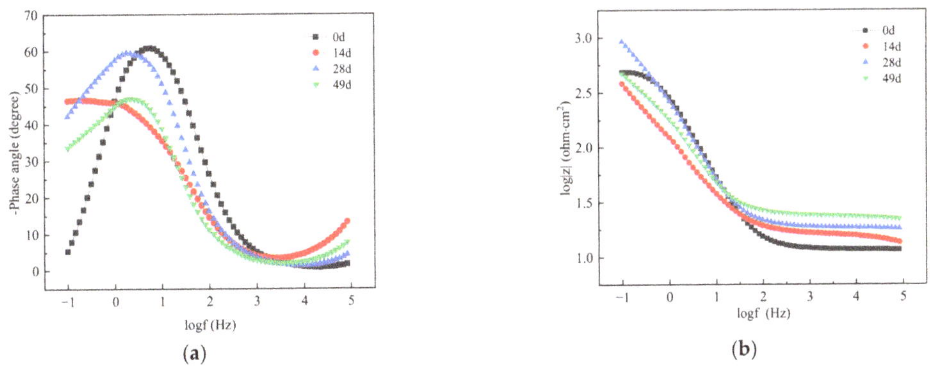

Figure 5. Bode diagrams of EH47 samples immersed in seawater for different durations. (**a**) Bode diagram of phase angle versus frequency. (**b**) Bode diagram of Z versus frequency.

The equivalent circuit diagram used for fitting the impedance data is shown in Figure 6. In this circuit, R_s represents the solution resistance, R_{ct} is the charge transfer resistance of the corroded samples and R_{rust} is the resistance of the corrosion products on the sample surface.

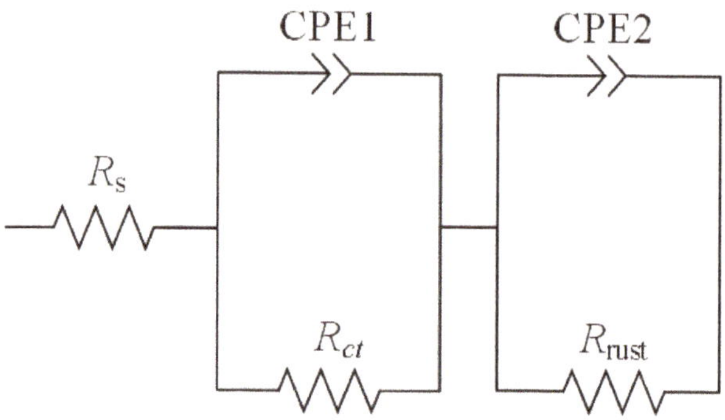

Figure 6. The equivalent circuit used for fitting the impedance data.

CPE (constant phase element) reflects the capacitance behavior, which is a frequency-dependent capacitance expressed the following equation:

$$CPE = [Q\,(J\omega)^{n}]^{-1} \tag{1}$$

where Q is a frequency independent constant, J is the imaginary unit, ω is the angular frequency, n is an exponential term. If $n = 0$, the impedance is ideal resistance, while it is ideal capacitance if $n = 1$, and $0 < n < 1$ represents the deviation from the ideal capacitance, which is correlated with the surface roughness and defect [35,36].

The n values corresponding to the four groups of different days of immersion in this experiment are all greater than 0.6, which indicates that the diffused impedance characteristics with limited retention layer [26], and the corrosion layer generated on the sample surface after corrosion reaction has produced a barrier effect on the solution, meanwhile, the diffusion reaction of ions contained in the solution to the matrix is limited. Y_0 is a parameter with dimensions of $\Omega^{-1}cm^{-2}{\cdot}s^{-n}$, *CPE*1 represents the capacitance in parallel with R_{ct} and *CPE*2 represents the capacitance in parallel with R_{rust}. Meanwhile, n_1 is the exponential term correlated with R_{ct}, and n_2 is the exponential term correlated with R_{rust}.

The inhibition efficiencies of the inhibitor from EIS are calculated using the following equation:

$$\text{IE\%} = \frac{\text{Rct} - R^0_{ct}}{R_{ct}} \times 100 \tag{2}$$

where R_{ct} and R^0_{ct} are the charge transfer resistance in the presence and absence of inhibitor, respectively.

The data was fitted by Zview (Zview3.1, San Francisco, CA, USA). The fitting results are listed in Table 2.

Table 2. Fitting parameters of EIS of different EH47 samples.

t(d)	R_s/ ($\Omega{\cdot}cm^2$)	R_{ct}/ ($\Omega{\cdot}cm^2$)	*CPE*1 $Y_0/{\cdot}\Omega^{-1}cm^{-2}{\cdot}s^{-n}$	n_1	R_{rust}/ ($\Omega{\cdot}cm^2$)	*CPE*2 $Y_0/{\cdot}\Omega^{-1}cm^{-2}{\cdot}s^{-n}$	n_2	*IE*%
0	12	12	0.0007	0.98	380	0.0017	0.75	0
14	8	8	0.0029	0.60	1623	0.0020	0.87	50
28	19	171	0.0015	0.75	2458	0.0016	0.93	92
49	23	90	0.0003	0.66	1103	0.0019	0.95	86

3.3. *The Corrosion Mechanism of EH47*

3.3.1. The Phase Composition of the Corrosion Products for EH47

In order to study the phase composition of the corrosion products of EH47, the X'Pert Powder X-ray diffractometer was used to analyze the corrosion layer of the experimental steel. The X-ray diffraction analysis results of the corrosion products of EH47 after immersion in seawater for 49 days are shown in Figure 7. It can be seen that the corrosion products formed on the surface of EH47 steel are mainly composed of FeOOH, Fe_3O_4 and Fe_2O_3 phases.

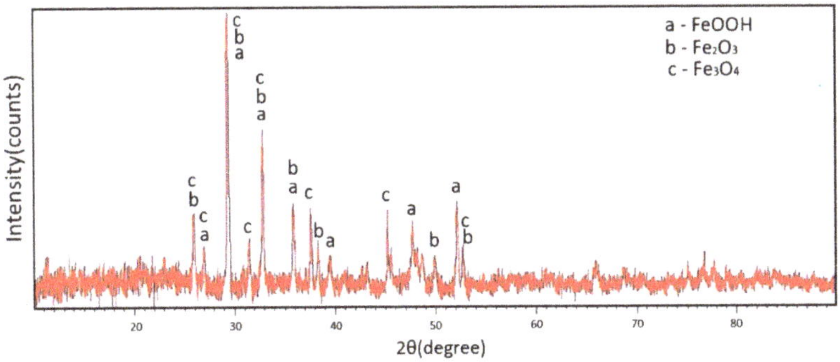

Figure 7. XRD pattern of the corrosion product of EH47 after full immersion in seawater for 49 days.

After the EH47 steel sample is immersed in seawater for 49 days, the corrosion product layer extends from the outer layer to the inner part, and the content of O element gradually decreases, while the content of Fe element gradually increases, corresponding to the results of the previous energy spectrum analysis, which reflects the characteristics of comprehensive corrosion [37]. In the early stage of corrosion, due to the existence of inclusions in EH47 steel and the inhomogeneous composition of the sample surface, numerous microcells will occur on the surface of the sample immersed in seawater. The equation of this reaction is as follows:

Anodic reaction:

$$Fe \rightarrow Fe^{2+} + 2e^{-} \quad (3)$$

Cathodic reaction:

$$O_2 + 2H_2O + 4e^{-} \rightarrow 4OH^{-} \quad (4)$$

Dissolved Fe^{2+} is deposited on the metal surface and is hydrolyzed and oxidized to Fe^{3+} and finally γ-FeOOH is formed. The corrosion reaction is as follows:

$$Fe^{2+} + H_2O \rightarrow FeOH^{+} + H^{+} \quad (5)$$

$$2FeOH^{+} + O_2 + 2e^{-} \rightarrow 2\gamma\text{-}FeOOH \quad (6)$$

With the progress of the reaction, the dissolved iron will continue to be oxidized to the unstable intermediate after deposition, and the intermediate will be further oxidized after dehydration to form β-FeOOH. With further corrosion, the corrosion layer will thicken and the dissolved oxygen will be difficult to diffuse to the surface of the steel substrate. A part of γ-FeOOH is transformed into α-FeOOH by amorphous iron hydroxide, and parts of β-FeOOH and γ-FeOOH are reduced to Fe_3O_4 due to the cathodic reaction.

Cathodic reaction:

$$6FeOOH + 2e^{-} \rightarrow 2Fe_3O_4 + 2H_2O + OH^{-} \quad (7)$$

Therefore, the corrosion products of EH47 steel after immersion in seawater for 49 days are mainly composed of FeOOH, Fe_3O_4 and Fe_2O_3.

3.3.2. The Microstructure of Corrosion Products of EH47 Formed on the Sample Surface under Different Full Immersion Corrosion Times

It can be seen that corrosion products formed on the sample surface are light yellow and loose granular clusters in a network distribution, and a small part of the corrosion products are aggregated into blocks, while a small part of the dark brown substrate is exposed after 20 days of immersion in seawater (Figure 8a). The corrosion products on the surface of the sample become large lumps, and some of the corrosion products fall off, and the exposed part of the black matrix increases as shown in Figure 8b after 40 days of immersion. As shown in Figure 8c, the color of corrosion products on the surface of the sample becomes darker after 60 days of immersion. The newly generated light yellow corrosion products appear on the original dark yellow corrosion products in irregular network blocks. The surface of the sample is almost completely covered by the newly pale yellow corrosion products, with obvious thickness differences between the upper and lower layers, as shown in Figure 8d, after 80 days of immersion.

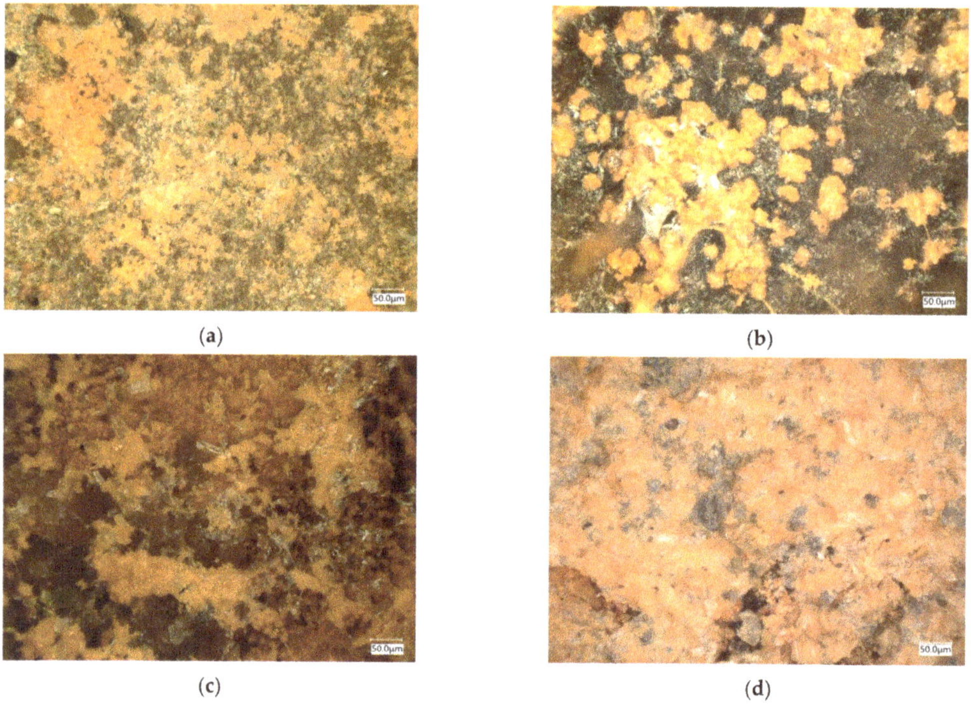

Figure 8. The OM microstructure of corrosion products of EH47 formed on the sample surface under different full immersion corrosion times: (**a**) 20 days, (**b**) 40 days, (**c**) 60 days, and (**d**) 80 days.

3.3.3. The Mechanism of Immersion Corrosion of EH47

In order to investigate the mechanism of immersion corrosion of EH47, the full immersion corrosion tests were performed in natural seawater for different total immersion durations. Furthermore, the rust layer on the surface of the sample for different total immersion durations was removed and the cross section of the sample matrix after the full immersion corrosion tests in natural seawater were measured by SEM and EDS to study the mechanism of immersion corrosion of EH47.

Figure 9 shows the SEM and EDS diagrams of the sample matrix. In order to investigate the corrosion mechanism, the sample matrix was obtained after the rust layer on the sample surface for different durations was removed.

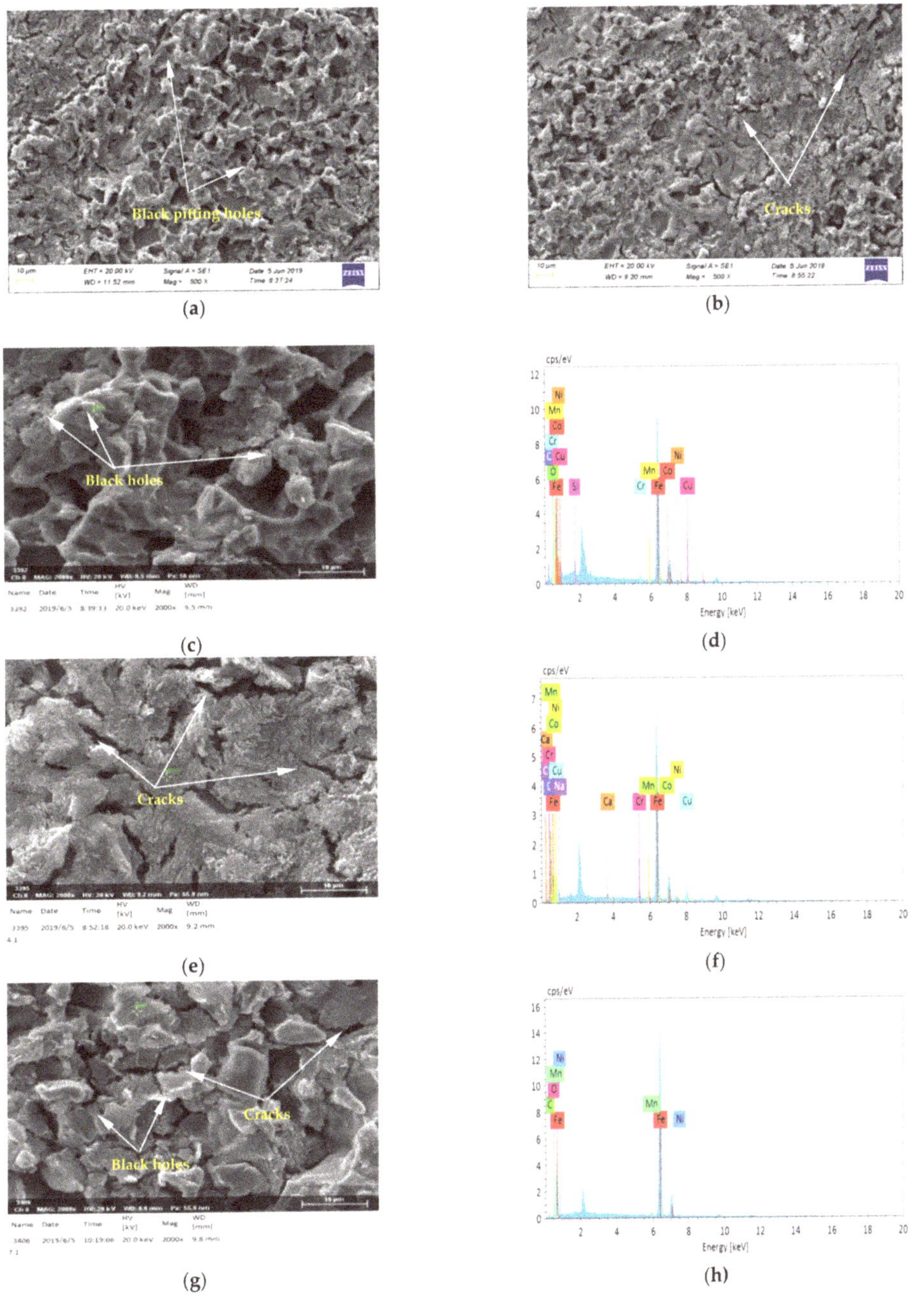

Figure 9. SEM and EDS diagram of the sample surface in which the rust layer was removed after different seawater immersion durations: (**a**) 20 day (500×); (**b**) 40 day (500×); (**c**) 20 days; (**d**) EDS (20 days); (**e**) 40 days; (**f**) EDS (40 days); (**g**) 80 days; (**h**) EDS (80 days).

Figure 9a,b show low magnification SEM images of the sample matrix. It can be seen from that the corrosion products on the surface of the substrate after 20 days of immersion show massive accumulation. Figure 9b shows the rust layer on the surface of the sample after 40days of immersion. It can be seen from the figure that the corrosion products on the surface of the matrix are of poor continuity and there are many black cracks and black holes on the corrosion product film.

Figure 9c,e,g show high magnification SEM images of the sample matrix. Figure 9c presents that the continuity of the corrosion products is poor, and the black pitting holes in the corrosion product film indicate that pitting corrosion has occurred. The EDS test results on the surface of the substrate are shown in Figure 9d. After 20 days of immersion, the surface of the substrate is mainly Fe, O, C, Mn and Cu, with the highest content being that of Fe, and a small amount of Cr, Co and Ni are also detected. It can be seen from Figure 9e, which indicates the corrosion mechanism here is mainly crevice corrosion, that there are many black cracks on the matrix of corrosion products after 40 days of immersion. Crevice corrosion is a strong localized corrosion of a metal surface immersed in seawater (or other corrosive media), often occurring in crevices. The principle of crevice corrosion is similar to pitting corrosion. The anode is in the crevice, and the cathode is in the large area outside the crevice, thus forming the corrosion battery. Cl^- can be enriched and H^+ can be formed in the crevices, reducing the pH value. The crevice corrosion is mainly due to the existence of gaps, leading to dielectric inhomogeneity caused by the media. The EDS test results on the surface of the substrate are shown in Figure 9f. After 40 days of immersion, the surface of the substrate is mainly Fe, O, C, Mn and Cu, and a small amount of Cr, Co, Ni and Na can also be detected. Figure 9g shows the SEM of the sample substrate after 80 days of immersion. It can be seen that the corrosion products on the surface of the matrix are of poor continuity, where black holes and fine cracks appear in the corrosion products. This shows that pitting and crevice corrosion occur with the increase in corrosion time. The EDS test results on the surface of the substrate are shown in Figure 9h. After 80 days of immersion, Fe, O, C and Mn are the main parts of the surface of the substrate. A small amount of Cr, Co and Ni are detected because the EDS energy spectrum just hits the matrix, and is not affected by the corrosion products, and the seawater composition is not detected.

Figure 10 shows the SEM and EDS of the cross section of the sample matrix after immersion in natural seawater for different total immersion durations, which removed the rust layer on the surface of the sample. As shown in Figure 10a, a continuous inner rust layer was formed after 20 days of corrosion. The uniformity of the rust layer is poor, and some areas show pitting morphology. The EDS test is carried out on the cross section of the matrix shown in Figure 10b. The test results show that little Cl^- is detected in the inner area of the rust layer, which indicates that the content of Cl^- in the rust layer near the substrate is very low, the rust layer has a protective effect, and the content of Ca^{2+} is high. Figure 10c shows the SEM of the cross section of the sample matrix after 40 corrosion days. It can be seen that the continuous inner rust layer has been formed on the sample surface and the uniformity of the rust layer is poor. Meanwhile, the flake corrosion products and obvious pitting pits can be observed in some areas. The EDS test results on the cross section of the matrix are shown in Figure 10d. The EDS results show that Ca^{2+} content detected inside the rust layer is high, which may be caused by the deposition of salt in seawater on the surface of the corrosion products. The content of Cl^- compared with that of 20 corrosion days is increased. Figure 10e shows that the rust layer on the surface of the sample is still discontinuous after 80 days of corrosion, and there are obvious cracks separating the rust layer, which is caused by crevice corrosion. The EDS test results are shown in Figure 10f. Ca^{2+} and O^{2+} are detected in the inner area of rust layer after 80 days of corrosion, and Cl^- is also detected, which is lower than that of 40 days of immersion and higher than that of 20 days of immersion.

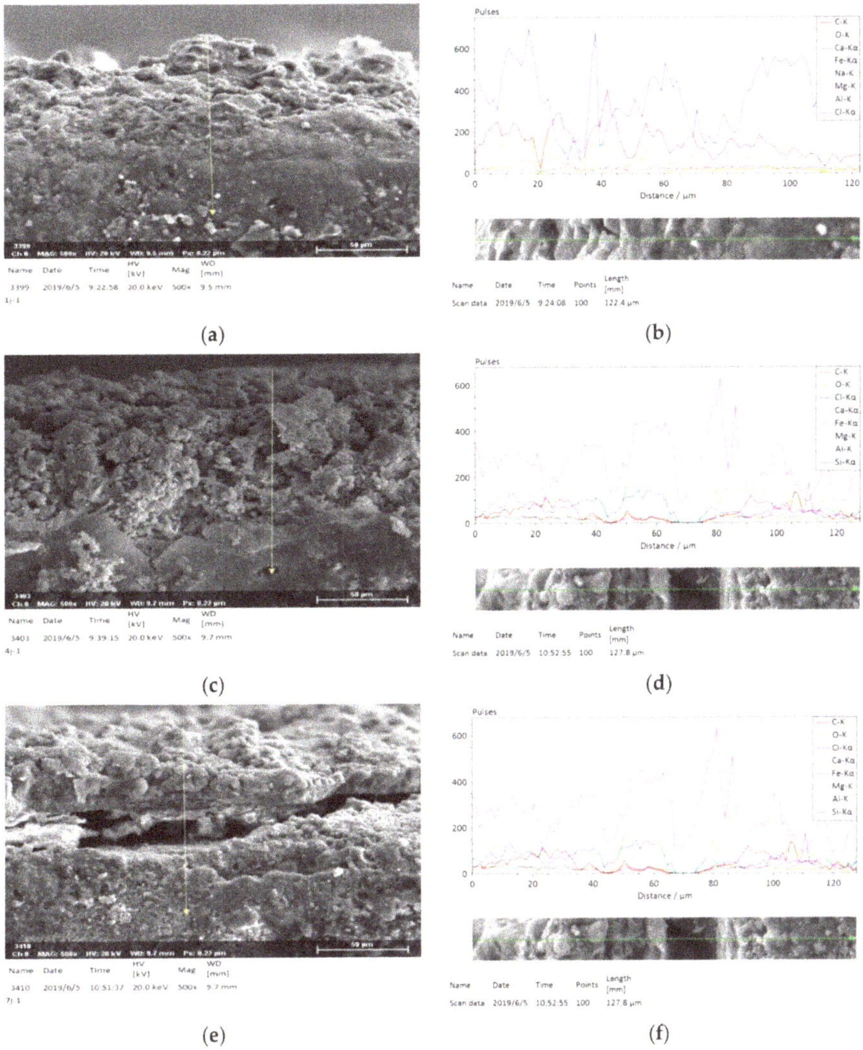

Figure 10. SEM and EDS diagram of a cross-section of sample substrate after different seawater immersion durations: (**a**) 20 days, (**b**) EDS (20 days), (**c**) 40 days, (**d**) EDS (40 days), (**e**) 80 days, (**f**) EDS (80 days).

SEM and EDS were used to observe and analyze the rust layer on the surface of the sample after 20 days of total immersion corrosion in seawater. it is found that the corrosion products are mainly laminated with poor continuity. There are black holes in the corrosion product film, which indicates that pitting corrosion is the main corrosion mechanism. After 40 full days of immersion, the corrosion products are laminated and reticulated, and there are still black holes in the corrosion products. The samples show slight cracks, which indicates that the corrosion mechanism is mainly pitting corrosion, accompanied by a small amount of crevice corrosion. After 80 days of corrosion, obvious cracks appear on the surface of the sample, which is the starting point of crevice corrosion. Local corrosion, such as pitting and crevice corrosion, occurs easily in seawater due to the local failure of passivation. Through the EDS analysis of the cross section of the matrix, it can be seen that Ca^{2+} appears on the surface of the matrix, and the content of Ca^{2+} is high, which may be caused by salt deposition on the surface of the corrosion products in seawater. At the same time, a very small amount of Cl^- appears, indicating that the rust layer still has a protective effect, preventing Cl^- penetration, and the corrosion layer is relatively stable.

4. Conclusions

In this paper, electrochemical corrosion tests and the full immersion corrosion experiments were performed in seawater at room temperature to investigate the electrochemical corrosion behavior and the corrosion mechanism of high-strength EH47. The main conclusions obtained by the analyses are as follows:

- With an increase in immersion time, the open circuit potential of EH47 decreases and then increases, with the minimum at 28 days of immersion.
- With an extension in immersion time, the corrosion current density (I_{corr}) of EH47 steel first decreases and then increases, with the minimum at 28 days of immersion. Furthermore, the arc radius of the 28 days of immersion sample is the largest, which indicates that the impedance is the largest, the corrosion rate is the smallest and the corrosion resistance is good.
- The corrosion products of EH47 steel after immersion in seawater are mainly composed of FeOOH, Fe_3O_4 and Fe_2O_3 phases.
- The corrosion mechanism of EH47 steel in the initial stage of corrosion is mainly pitting corrosion, and furthermore, the corrosion mechanism is mainly pitting corrosion with a small amount of crevice corrosion with increased corrosion time.

Author Contributions: Conceptualization, H.Z. and L.Y.; methodology, H.Z.; software, Y.Z. and F.A.; validation, Z.J., L.Y. and H.Z.; formal analysis, H.Z.; investigation, Y.Z., Y.L. and F.A.; writing—original draft preparation, H.Z. and Y.Z.; writing—review and editing, H.Z., Z.J. and H.L.; project administration, F.A. and Y.L. All authors have read and agreed to the published version of the manuscript.

Funding: This research was funded by the State Key Laboratory of Metal Material for Marine Equipment and Application—University of Science and Technology Liaoning co-project (No. SKLMEA-USTL 2017010) and (No. HGSKL-USTLN(2019)05).

Data Availability Statement: Data presented in this article are available at request from the corresponding author.

Conflicts of Interest: The authors declare no conflict of interest.

References

1. Xiao, Q.Y.; Wu, X.G. *Corrosion and Protection Technology of Ships*; Harbin Engineering University Press: Harbin, China, 2012; pp. 102–133.
2. Ping, Y. Research on the evaluation method and requirements of the corrosion state of ship structure. *Ships* **2010**, *4*, 16–19.
3. Sun, B.L.; Cao, J.; Si, T.Z.; Zhang, Q.A. Electrochemical corrosion behavior of rusted NSB steel in Simulated Seawater. *Anhui Univ. Technol.* **2013**, *30*, 124–128.
4. Zou, Y.; Wang, J.; Zheng, Y.Y. Electrochemical techniques for determining corrosion rate of rusted steel in seawater. *Corros. Sci.* **2011**, *53*, 208–216. [CrossRef]
5. Wang, H.W.; Zhou, P.; Huang, S.W.; Chi, Y. Corrosion mechanism of low alloy steel in NaCl solution with CO_2 and H_2S. *Int. J. Electrochem. Sci.* **2016**, *11*, 1293–1309.
6. Han, E.H.; Chen, J.M.; Su, Y.J. *Corrosion and Protection of Marine Engineering Materials and Structures*; Chemical Industry Press: Beijing, China, 2017; pp. 304–382.
7. Prakashaiah, B.G.; Vinaya Kumara, D.; Anup Pandith, A.; Nityananda Shetty, A.; Amitha Rani, B.E. Corrosion inhibition of 2024-T3 aluminum alloy in 3.5% NaCl by thiosemicarbazone derivatives. *Corros. Sci.* **2018**, *136*, 326–338. [CrossRef]
8. Lebrini, M.; Lagrenée, M.; Vezin, H.; Gengembre, L.; Bentiss, F. Electrochemical and quantum chemical studies of new thiadiazole derivatives adsorption on mild steel in normal hydrochloric acid medium. *Corros. Sci.* **2005**, *47*, 485–505. [CrossRef]
9. Chen, X.Q.; Chang, W.S.; Chen, D.B. Study on the law and characteristics of pitting corrosion induced by inclusions in carbon steel. *J. Nav. Acad. Eng.* **2004**, *16*, 30–36.
10. Sazou, D.; Pavlidou, M.; Pagitsas, M. Potential oscillations induced by localized corrosion of the passivity on iron in halide-containing sulfuric acid media as a probe for a comparative study of the halide effect. *J. Electroanal. Chem.* **2012**, *675*, 54–67. [CrossRef]

11. Usman, B.J.; Umoren, S.A.; Gasem, Z.M. Inhibition of API 5L X60 steel corrosion in CO_2 -saturated 3.5% NaCl solution by tannic acid and synergistic effect of KI additive. *J. Mol. Liq.* **2017**, *237*, 146–156. [CrossRef]
12. Deyab, M.; Mohamed, N.H.; Moustafa, Y. Corrosion protection of petroleum pipelines in NaCl solution by microcrystalline waxes from waste materials: Electrochemical studies. *Corros. Sci.* **2017**, *122*, 74–79. [CrossRef]
13. Yang, S.C.; Bi, H.J.; Niu, S.R.; Jin, S.T.; Shen, K. Effect of temperature and Cl~-mass fraction on pitting corrosion resistance of 304 stainless steel. *J. Lanzhou Univ. Technol.* **2010**, *36*, 5–9.
14. Xia, J.M.; Li, Z.Y.; Jiang, J.C.; Wang, X.Q.; Zhang, X.D. Effect of Flow Rates on erosion corrosion behavior of hull steel in real seawater. *Int. J. Electrochem. Sci.* **2021**, *16*, 1–13.
15. Jia, C.Y.; Niu, J.Q.; Liu, D.S. Research on marine atmospheric corrosion resistance of ship plate steel. *Shanghai Met.* **2013**, *35*, 26–30.
16. Melchers, R.E.; Jeffrey, R. Early corrosion of mild steel in seawater. *Corros. Sci.* **2005**, *47*, 1678–1693. [CrossRef]
17. Soares, C.G.; Garbatov Zayed, Y. Effect of environmental factors on steel plate corrosion under marine immersion conditions. *Corros. Eng. Sci. Technol.* **2011**, *46*, 524–541. [CrossRef]
18. Panayotova, M.; Garbatov, Y.; Guedes, S.C. Factors Influencing Atmospheric Corrosion and Corrosion in Closed Spaces of Marine Steel Structural. In Proceedings of the 7th International Conference on Marine Science and Technology, Varna, Bulgaria, 7–9 October 2004; Union of Scientists of Varna: Varna, Bulgaria, 2004.
19. Li, G.M.; Chang, W.S.; Chen, X.Q. Study on simulated pitting corrosion test method of carbon steel and low alloy steel. *Mater. Dev. Appl.* **2008**, *3*, 40–42.
20. Gan, Y.; Li, Y.; Lin, H.C. Laboratory simulation of local corrosion process of low alloy steel in seawater. *J. Corros. Protec.* **2001**, *21*, 82–87.
21. Wang, J.M.; Chen, X.Q.; Li, G.M. Comparison of pitting corrosion resistance of two kinds of marine low alloy steels. *J. Corros. Protec.* **2005**, *25*, 356–360.
22. Cao, G.L.; Li, G.M.; Chen, S.; Chang, W.S.; Chen, X.Q. Comparison of pitting corrosion resistance of Ni and Cr in typical seawater corrosion resistant steels. *Acta Metal. Sin.* **2010**, *46*, 748–754. [CrossRef]
23. Zhang, Y.C.; Chen, X.Q.; Chen, D.W.; Li, G.M.; Pan, R.Y. Pitting induced sensitivity and mechanism of different low carbon steels. *J. Corros. Protec.* **2001**, *21*, 265–272.
24. Liu, W.; Wang, H.T.; Yang, H.Y.; Huang, G.Q. Compilation and standard interpretation of seawater corrosion test method for metallic materials. *Environ. Technol.* **2016**, *34*, 57–60.
25. Song, S.Z. *Corrosion Electrochemical Methods*; Chemical Industry Press: Beijing, China, 1988; pp. 55–70.
26. Cao, C.N.; Zhang, J.Q. *Electrochemical Impedance Spectroscopy*; Science Press: Beijing, China, 2002; pp. 46–88.
27. Yang, S.W.; Chang, T.J. *Corrosion and Protection of Materials*; Harbin Engineering University Press: Harbin, China, 2003; pp. 102–108.
28. Wu, H.; Li, Y.; Lu, Y.; Li, Z.; Cheng, X.W.; Hasan, M.; Zhang, H.M.; Jiang, Z.Y. Influences of load and microstructure on tribocorrosion behaviour of high strength hull steel in saline solution. *Tribol Lett.* **2019**, *67*, 1–12. [CrossRef]
29. Sun, B.L. Electrochemical Corrosion Characteristics of Low-Carbon Alloy Steel under NaCl Solution. Master's Thesis, Anhui University of Technology, Anhui, China, 2013.
30. Zheng, L.Y.; Cao, F.H.; Liu, W.J. Electrochemical study on failure behavior of Q235 steel in simulated natural environment. *Equip. Environ. Eng.* **2011**, *8*, 8–15.
31. Liang, C.F.; Hou, W.T. Study on atmospheric exposure corrosion of low carbon alloy steel and low alloy steel for 16 years. *J. Corros. Protec.* **2005**, *25*, 2–6.
32. Sherif El-Sayed, M.; Almajid, A.A. Electrochemical corrosion behavior of API X-70 5L grade steel in 4.0 wt.% sodium chloride solutions after different immersion periods of time. *Int. J. Electrochem. Sci.* **2015**, *10*, 34–45.
33. Han, P.; Han, P.J.; Xie, R.Z.; Bin, H.; Bai, X.H. Study of the electrochemical corrosion behaviour of X70 steel in H_2SO_4 contaminated silty soil. *Int. J. Electrochem. Sci.* **2018**, *13*, 8694–8710. [CrossRef]
34. Peng, X. Corrosion Electrochemical Behavior and Correlative Parameters of Rusty Carbon Steels in Seawater Environment. Master's Thesis, Ocean University of China, Qingdao, China, 2013.
35. Dai, N.; Zhang, L.C.; Zhang, J.X.; Chen, Q.M.; Wu, M.L. Distinction in corrosion resistance of selective laser melted Ti-6Al-4V alloy on different planes. *Corros. Sci.* **2016**, *111*, 703–720. [CrossRef]
36. Yan, Z.W.; Yuan, X.J.; Tan, Z.J.; Tang, M.Q.; Feng, Z.Q. Effect of in situ ion nitridetreatment on the corrosion behavior of titanium. *Int. J. Electrochem. Sci.* **2018**, *13*, 353–361. [CrossRef]
37. Sun, L.L. Mechanism of Q235 Steel and Stainless Steel in Seawater. Master's Thesis, Shandong University, Shandong, China, 2010.

metals MDPI

Article

Structural Characterization, Global and Local Electrochemical Activity of Electroless Ni–P-Multiwalled Carbon Nanotube Composite Coatings on Pipeline Steel

Mara Cristina Lopes de Oliveira [1], Olandir Vercino Correa [2], Rejane Maria Pereira da Silva [2], Nelson Batista de Lima [2], Jefferson Thadeu Dias de Oliveira [1], Leandro Antônio de Oliveira [1] and Renato Altobelli Antunes [1,*]

1 Center for Engineering, Modeling and Applied Social Sciences (CECS), Federal University of the ABC (UFABC), Santo André 09210-580, SP, Brazil; mara.oliveira@ufabc.edu.br (M.C.L.d.O.); jefferson.oliveirah@gmail.com (J.T.D.d.O.); antoli.leandro@gmail.com (L.A.d.O.)
2 IPEN/CNEN-SP, Av. Prof. Lineu Prestes 2242, Cidade Universitária, São Paulo 05508-000, SP, Brazil; ovcorrea@ipen.br (O.V.C.); rejanep2silva@gmail.com (R.M.P.d.S.); nblima@ipen.br (N.B.d.L.)
* Correspondence: renato.antunes@ufabc.edu.br; Tel.: +55-1-4996-8241

Abstract: In this work, composite Ni–P-multiwalled carbon nanotube films were produced by electroless deposition. The main goal was to investigate the influence of multiwalled carbon nanotube loading on the local electrochemical behavior of the composite films, as probed by scanning electrochemical microscopy (SECM). The coatings were also characterized with respect to their crystalline structure, surface, and cross-section morphologies. Adhesion strength was examined by scratch tests. The global electrochemical behavior was evaluated by potentiodynamic polarization. The local electrochemical activity was investigated by probing the Fe^{2+} oxidation in the surface generation/tip collection mode of the SECM. The results revealed that multiwalled carbon nanotubes increased the adhesion strength and reduced the electrochemical activity on the surface of the coated samples.

Keywords: Ni–P/MWCNT composite coatings; electroless deposition; corrosion; scanning electrochemical microscopy

Citation: de Oliveira, M.C.L.; Correa, O.V.; da Silva, R.M.P.; de Lima, N.B.; de Oliveira, J.T.D.; de Oliveira, L.A.; Antunes, R.A. Structural Characterization, Global and Local Electrochemical Activity of Electroless Ni–P-Multiwalled Carbon Nanotube Composite Coatings on Pipeline Steel. *Metals* **2021**, *11*, 982. https://doi.org/10.3390/met11060982

Academic Editor: Aleksander Lisiecki

Received: 27 May 2021
Accepted: 17 June 2021
Published: 20 June 2021

1. Introduction

Pipeline steels are manufactured from high-strength low-alloy steels (HSLA) to meet the strict requirements of the oil and gas industry with respect to its increasing demand for weight reduction and enhanced productivity in the transportation of their products [1,2]. The intrinsically high mechanical strength of HSLA steels accounts for their reliable operation in oilfield production [3]. Notwithstanding, the harsh environments to which they are subject pose challenging design issues related to both wear and corrosion control [4,5]. In view of the critical role played by HSLA steels in the safe operation of transmission pipelines, it is of prime importance to properly manage surface properties and to guarantee long-term operation without failure [6].

Corrosion is particularly pointed as a major cause of degradation of the load bearing capacity of HSLA transmission pipelines [7]. Coatings have been traditionally employed to protect the internal tubing walls from corrosion in the petroleum industry [8,9]. Epoxy-based organic coatings are often employed with this purpose due to their chemical inertness and strong adhesion to metallic substrates [10]. It is well-known, though, that these materials lose their barrier properties with time, allowing electrolyte penetration through pores and flaws and, ultimately, to adhesion failure [11].

Electroless nickel coatings have emerged as a viable alternative to overcome the above-mentioned limitations [12]. These coatings are based on conventional binary Ni–P films and have consolidated engineering applications in the automotive, aerospace, and food industries [13]. They owe their outstanding performance to a combination of strong adhesion in a variety of substrates, shapes and sizes, and high hardness in the annealed state, due to the precipitation of nickel phosphides promoting good wear and corrosion resistance [14,15]. In recent decades, further developments in the electroless deposition of nickel-based coatings have been driven by growing needs for optimized surface properties of metallic materials to expand their applicability to more aggressive environments. Ternary, quaternary, and composite coatings have been developed with this purpose [16–19]. Co-deposition of a variety of inorganic particles has been tested for the electroless plating of Ni–P composite coating, such as SiO_2, SiC, TiO_2, ZrO_2 and B_4C [20–24].

Carbon nanotubes (CNT) have also been employed to reinforce particles for electroless Ni–P coatings. Several authors [25,26] reported that multiwalled carbon nanotubes (MWCNT) improved the friction and wear properties of conventional electroless Ni–P. In addition to improved wear behavior, corrosion resistance is often reported as being favorably affected by incorporating CNT particles into the Ni–P matrix. The intrinsic chemical inertness and high length-to-diameter ratio of the CNTs would account for the optimized corrosion protection ability of the Ni–P–CNT coatings [27,28]. The ability of the CNT particles to block pores and cavities in the coating layer is associated with the improved barrier properties of the composite coatings [29,30].

In the present work, we expand the current knowledge related to the development of electroless Ni–P/MWCNT coatings by evaluating the local electrochemical activity using scanning electrochemical microscopy (SECM). Its use as an analytical tool to investigate local corrosion processes of a variety of metallic alloys has been reported in the literature. Both uncoated and coated alloys have been probed [31,32]. SECM can detect the local electrochemical activity associated with nucleation of pits, dissolution of metallic inclusions, and defects through coatings [33–35]. In this respect, for the first time we report the assessment of the local electrochemical activity of Ni–P/MWCNT composite coatings by SECM. Furthermore, Ni–P/MWCNT on HSLA pipeline steels is innovative. In order to support discussion on the electrochemical results, film structure and adhesion properties of the deposited layers were also evaluated.

2. Materials and Methods

2.1. Substrate and Coating Preparation

API 5L X80 pipeline steel (Usiminas, Ipatinga, MG, Brazil) was employed as substrate. Its chemical composition is shown in Table 1. Specimens were cut from the as-received plate into rectangular pieces with the final dimensions of 30 mm × 30 mm × 5 mm. Before deposition, the specimens were ground with silicon carbide waterproof paper up to grit 1200. Next, the surface was cleaned with alcohol, rinsed with deionized water, and dried with a heat gun.

Table 1. Chemical composition of the API 5L X80 steel plate (wt.%).

C	Mn	Si	P	S	Nb	Al	Cr	V	Fe
0.04	1.75	0.20	0.02	0.002	0.065	0.025	0.11	0.025	Bal.

Electroless deposition was accomplished by preparing a Ni–P plating bath consisting of nickel sulfate (30 $g.L^{-1}$), nickel hypophosphite (40 $g.L^{-1}$), sodium citrate (10 $g.L^{-1}$), acetic acid (10 $mL.L^{-1}$), lactic acid (10 $mL.L^{-1}$), and sodium hydroxide (40 $g.L^{-1}$). Sodium dodecyl sulfate (2 $g.L^{-1}$) was added to facilitate dispersion of the multiwalled carbon nanotubes in the bath. The bath was operated at 88 °C and was magnetically stirred during deposition. The pH was 4.5, adjusted with ammonium hydroxide. Multiwalled carbon nanotubes (MWCNT) were purchased from the Federal University of Minas Gerais (Brazil). Three different CNT concentrations were added to the plating bath: 0.25 $g.L^{-1}$, 0.50 $g.L^{-1}$, and 1.0 $g.L^{-1}$. These samples are designated as CNT-0.25, CNT-0.50, and CNT-1.0 throughout the text.

Before deposition, the specimens were cleaned in an alkaline solution consisting of 10 wt.% NaOH at 50 °C and activated in a 50% vol. H_2SO_4 solution at room temperature. After washing with deionized water, the specimens were immersed in the plating bath. The total deposition time was 2 h. After deposition, the specimens were annealed at 400 °C for 1 h in a tubular furnace under argon atmosphere, followed by cooling inside the furnace.

2.2. Structural, Morphological and Adhesion Characterization

The crystalline character of the Ni–P/MWCNT composite coatings was assessed by X-ray diffractometry (Rigaku DMAX-2000) in the θ-2θ configuration, employing Cu-kα radiation. The 2θ range was from 20° to 70°. The surface morphology and cross-sections of the different coatings were examined by scanning electron microscopy (SEM) coupled to an X-ray energy dispersive spectrometer (EDS) to study the elemental composition at the coating/substrate interface.

The adhesion strength of the Ni–P/MWCNT layers was evaluated through scratch tests by means of a Ducom T101 apparatus equipped with a Rockwell C-type diamond tip. The normal load was continuously increased from 1 N to 38 N at a rate of 2 $N.min^{-1}$, a scratch velocity of 0.5 $mm.s^{-1}$, and a total scratch length of 10 mm. Confocal laser scanning microscopy (Olympus, LEXT OLS4100) was employed to evaluate the penetration depth of the indenter and topographic features at the interface between the unscratched and scratched regions.

2.3. Global and Local Electrochemical Tests

Conventional electrochemical tests were carried out using an Autolab M101 potentiostat/galvanostat. A classical three-electrode cell set up was employed with a platinum wire as the auxiliary electrode, Ag/AgCl as the reference, and the coated API 5L X80 specimens as the working electrodes. The tests were performed in 3.5 wt.% NaCl solution at room temperature. Initially, the open circuit potential was monitored for 1 h. Right after, electrochemical impedance spectroscopy measurements were made at the OCP in the frequency range from 100 kHz to 10 mHz. The amplitude of the perturbation signal was ±10 mV and the acquisition rate was 10 points per decade. Next, the samples were subject to potentiodymamic polarization by sweeping the potential between −300 mV versus the OCP up to +1, 0 $V_{Ag/AgCl}$ at 1 $mV.s^{-1}$.

SECM current maps were acquired using a commercial Sensolytics system, operating in the substrate generation-tip collection mode (SG-TC). In this operation mode, electroactive species generated on the corroding surfaces are reduced or oxidized at the tip, providing the values of the current related to electrochemically active sites on the material [36,37]. The SG-TC operation mode was reported in the investigation of localized corrosion sites on the stainless steel [33,38], whereby the selective monitoring of reacting species provided from the specific reactions required the operation of the corroding system without the need to insert a redox mediator to reach electrochemical responses. In this present work, the SG-TC operation mode was used to investigate the electrochemical activity of the studied material associated with the localized production of Fe^{2+} ions above the surfaces. Thus, Fe^{2+} ions produced from the corrosion of the material is sensed in an oxidation reaction at the Pt tip, according to Equation (1).

SECM current maps were acquired using a commercial Sensolytics system, operating in the substrate generation-tip collection mode (SG-TC). The reaction shown in Equation (1), which is typical of ferrous alloys corrosion in aqueous media [33,37], was probed by biasing the tip at +600 $V_{Ag/AgCl}$.

$$Fe^{2+} \leftrightarrow Fe^{3+} + e \tag{1}$$

A glass-insulated 10 μm diameter Pt microelectrode was used as the tip, Ag/AgCl as the reference electrode, and a Pt wire as the auxiliary electrode. The specimens were mounted horizontally facing upwards. The tip was at a height of 30 μm above the substrate surface. The potentials were controlled with a bipotentiostat coupled to the SECM system.

The specimens were at the open circuit potential. The measurements were carried out in a 0.1 M NaCl solution at room temperature. The electrolyte was less concentrated than that used for the global electrochemical tests, as the results obtained in 3.5 wt.% NaCl did not allow to distinguish the local electrochemical response of each sample by SECM due to excessively high current densities measured throughout the probed area.

3. Results

3.1. X-ray Diffraction

XRD patterns of the Ni–P and Ni–P/MWCNT composite coatings are shown in Figure 1.

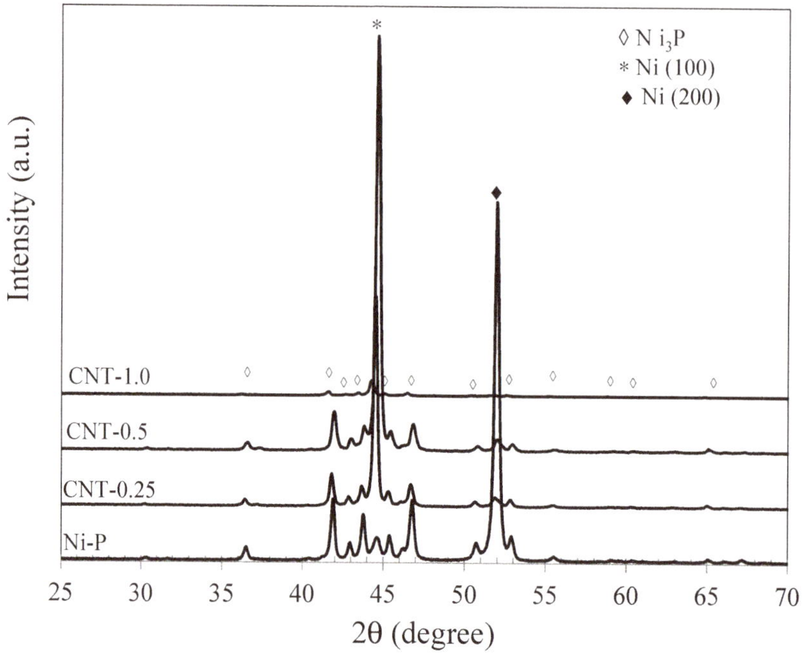

Figure 1. XRD patterns of the Ni–P and Ni–P/MWCNT coatings.

The presence of Ni_3P crystallites is clearly observed in the pattern of the conventional Ni–P film. These species are typical of electroless nickel coatings and are formed during annealing [39,40]. Strong reflections of the Ni (100) and (200) crystalline planes are also observed. The crystalline character of the composite coatings was affected by the carbon nanotubes. The strong Ni (200) reflection of the conventional Ni–P coating was markedly decreased for the Ni–P–CNT coatings, whereas the Ni (100) became the preferential orientation. Furthermore, the CNT-1.0 presented a steep reduction in the intensity of its diffraction peaks, indicating that the coating became less crystalline when CNT concentration was increased. This effect was observed by other authors [41], being attributed to a distortion of the crystalline nickel matrix by the incorporation of CNT particles.

3.2. SURFACE and Cross-Section Morphology

The morphology of the conventional Ni–P and Ni–P/MWCNT composite coatings was examined by scanning electron microscopy (SEM). Figure 2 shows the SEM micrographs (secondary electrons mode) of the top surfaces.

The binary Ni–P film displays a typical nodular morphology (Figure 2A). The nodular structure is associated with nucleation and growth of the deposit during electroless plating. A high nucleation rate is reported to enhance the number of nodules during deposition [42].

By adding the CNT filler into the plating, the nodules size became finer. The nodular morphology is less clearly perceived for CNT-0.25 (Figure 2B) due to their small size; this effect is enhanced for the coatings obtained in the baths with higher CNT loadings (Figure 2C,D). In this respect, the CNT particles tend to increase the nucleation rate during electroless deposition, as previously observed by Xu et al. [39]. As a consequence, the coating surface assumes a more compact character which, in turn, can affect the corrosion resistance in a positive way.

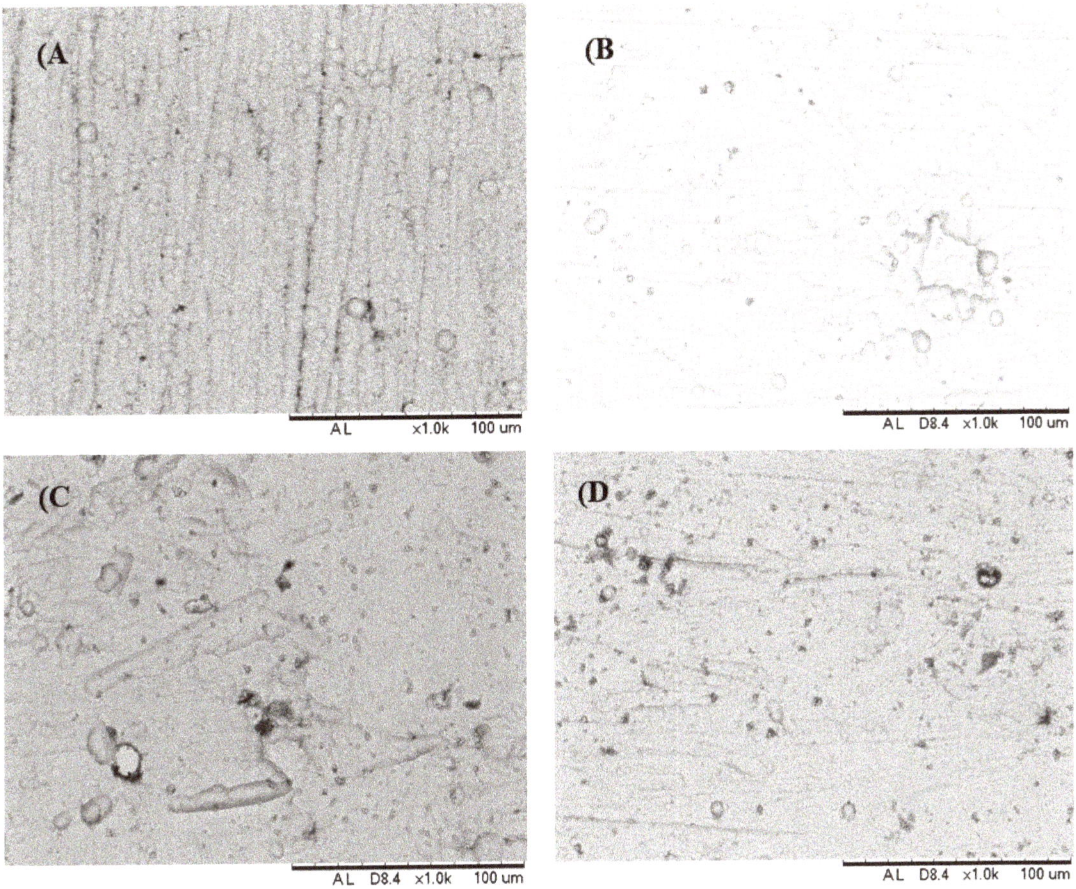

Figure 2. SEM micrographs of the top surfaces of the nickel-based coatings: (**A**) Ni–P; (**B**) CNT-0.25; (**C**) CNT-0.5; (**D**) CNT-1.0.

Coating thickness was examined by SEM/EDS analyses of the cross-sections. Figure 3 shows the results obtained for the CNT-0.25 film. The results for the other samples are very similar (not shown) and are provided as supplementary files.

The interface between film and substrate can be clearly distinguished by examining Figure 3B (Ni mapping), Figure 3C (P mapping), and Figure 3D (Fe mapping). Coating thickness was approximately 4.5 μm. The coating follows a continuous interface with the substrate with no signs of broken-off sites, suggesting it is well-adhered. Similar features were observed for the other Ni–P/MWCNT films (Supplementary Materials—Figure S1). Electroless nickel coatings are reported to present good adhesion to metallic substrates [40]. The results obtained in the present work point towards this direction.

Additional characterization of the adhesion properties of the composite coatings was undertaken by scratch tests. The results are shown in Section 3.3.

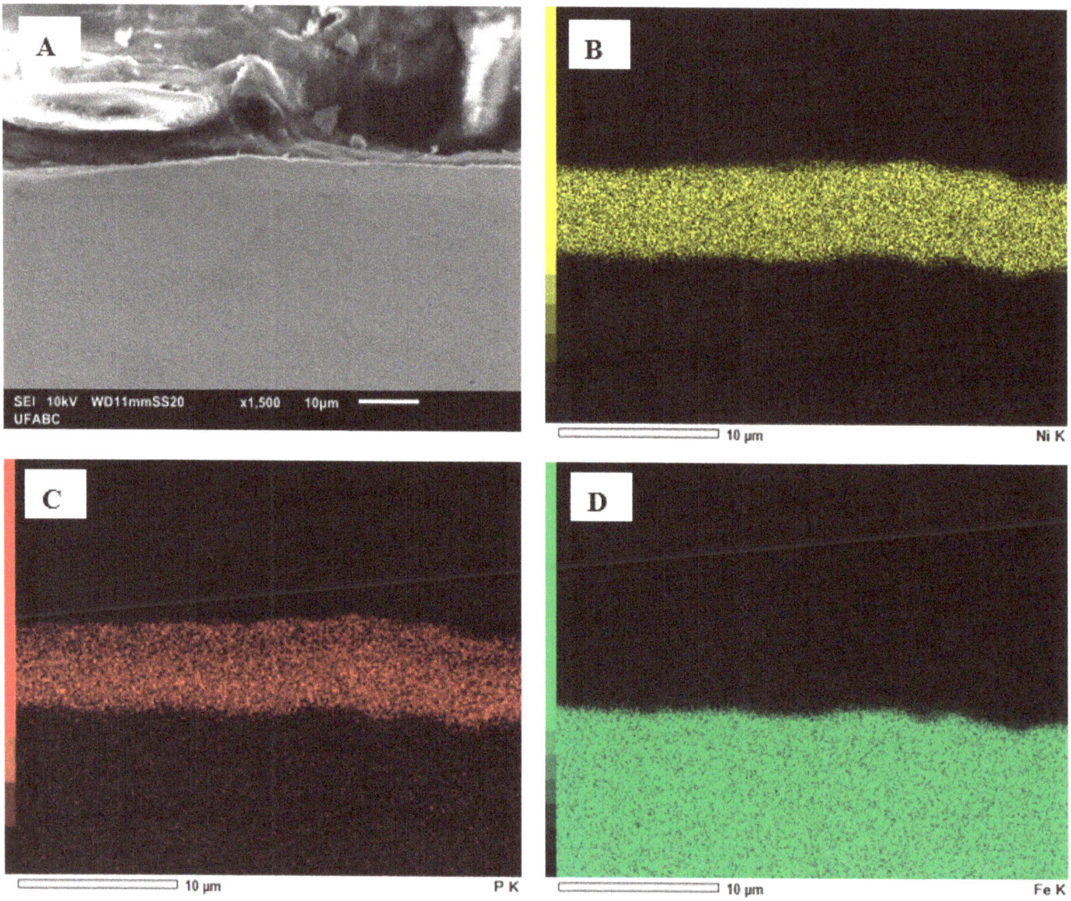

Figure 3. EDS mapping of the cross-sections for CNT-0.25: (**A**) SEM micrograph; (**B**) Ni; (**C**) P; (**D**) Fe.

3.3. Scratch Tests

Adhesion is a must-attend feature of protective coatings. In spite of its relevance, adhesion properties of Ni–P/MWCNT composite coatings are not commonly reported in the literature. The results obtained by scratch tests of the Ni–P and Ni–P/MWCNT coatings are shown in Figure 4. The CLSM 3D views of the scratched regions are shown as well as the transverse profiles along the lines marked in the 3D micrographs. The width (average of ten measurements over the scratch length) and maximum depth of the scratched regions were determined from these lines. The results are shown in Table 2. By evaluating the transverse profiles, it is evident that MWCNT loading greatly affected the scratched region. Both the width and the depth were reduced as the MWCNT loading in the plating bath increased. The shallowest scratch was measured for the CNT-1.0 sample which was also the narrowest one. The conventional Ni–P film, in turn, presented the deepest and widest scratch. This result points to the strong hardening effect of MWCNT addition into the Ni–P matrix, confirming the results obtained by other authors [43].

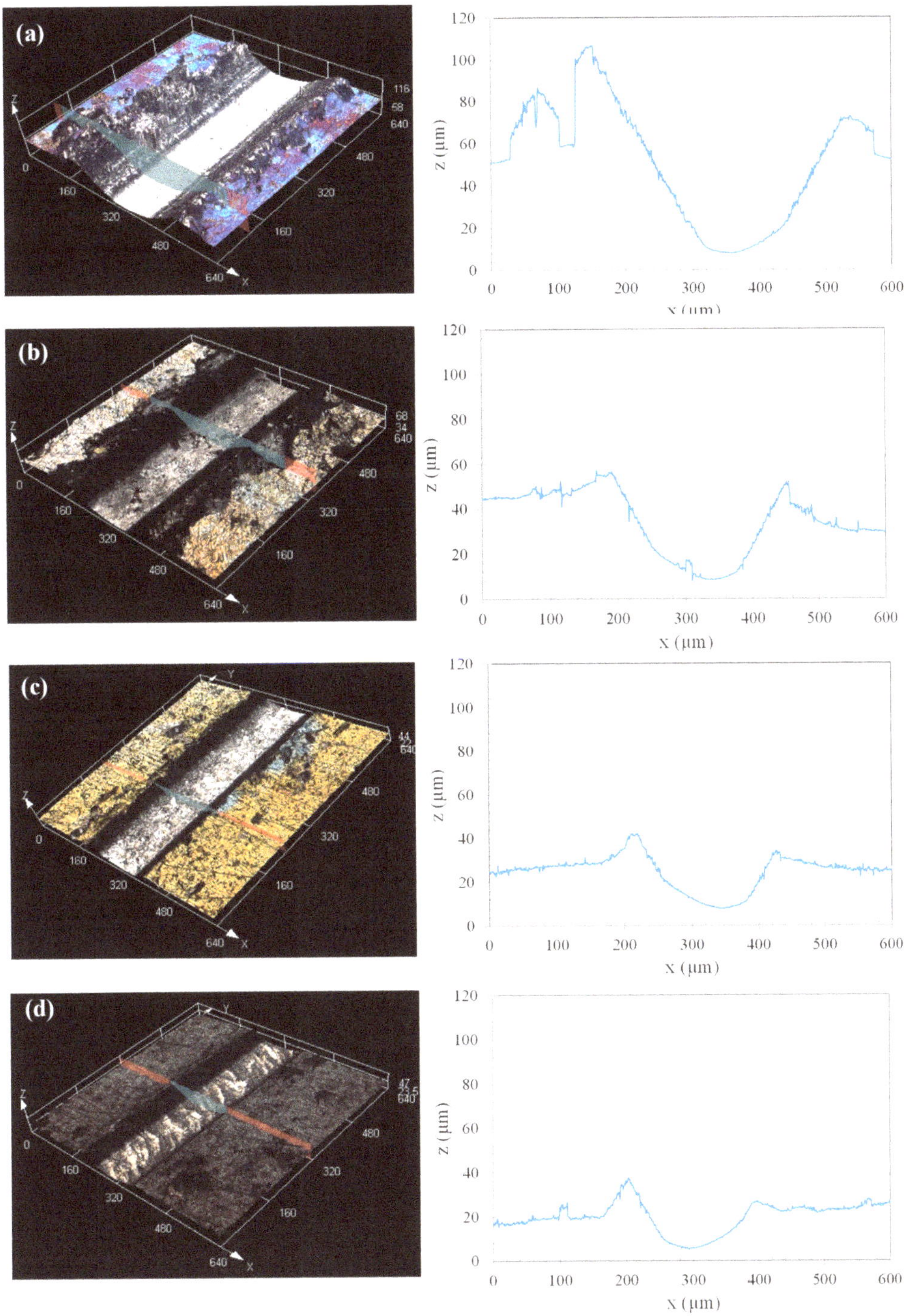

Figure 4. 3D views of the scratched region and the corresponding transverse profile along the lines marked in the micrographs: (**a**) Ni–P; (**b**) CNT-0.25; (**c**) CNT-0.5; (**d**) CNT-1.0.

Table 2. Scratch dimensions of the Ni–P and Ni–P/MWCNT coatings.

Sample	Width (μm)	Maximum Depth (μm)
Ni–P	538 ± 35	101
CB-0.25	442 ± 27	47
CB-0.50	252 ± 24	37
CB-1.0	188 ± 3	32

The critical role played by MWCNTs in the adhesion strength of the Ni–P layer to the pipeline steel substrate is also perceived from the 3D views presented in Figure 4. The conventional binary Ni–P coating was delaminated over the scratch length, as can be seen from the relatively wide dark region in the vicinity of the scratched area in Figure 4a. In spite of the reduction in both the scratch width and depth with respect to the Ni–P film, the CNT-0.25 film also presented a relatively wide dark region along the scratch, indicating that coating delamination (Figure 4b) had occurred in a similar way of that observed for the conventional unfilled Ni–P matrix. As the MWCNT loading increased, though, different features could be perceived along the scratch line. As seen in Figure 4c, the delaminated area was greatly reduced for the CNT-0.50 sample in comparison with the CNT-0.25 and Ni–P. Such a trend was also observed for the CNT-1.0 sample (Figure 4d). For this condition, the spalling area in the surroundings of the scratch scar is confined within a narrower region. Spalling avoidance is associated with a better mechanical load accommodation during scratch tests of nickel-based coatings [44]. In this respect, our results point that the CNT particles had a beneficial effect on the adhesion properties of the Ni–P/MWCNT composite coatings.

3.4. Global Electrochemical Tests

EIS results are represented as Nyquist plots, as shown in Figure 5. All samples are characterized by capacitive loops in the medium to low frequencies, whose diameter depends on the MWCNT loading in the coating. As pointed out in the literature, the corrosion resistance is associated with the diameter of the Nyquist plots, since it is associated with the polarization resistance of the electrode [45,46]. The uncoated substrate presented very low impedance values when compared to the coated samples. As a consequence, its Nyquist plot is only seen when the impedance scales are expanded, as shown in the inset of Figure 5. The impedance values were greatly enhanced for the coated samples, indicating its beneficial effect on the corrosion resistance of the steel substrate.

The Nyquist plot of the Ni–P exhibits a bigger diameter when compared to the uncoated sample, revealing the increased corrosion resistance imparted by the electrolessly deposited film. A progressive increase in the diameter of the Nyquist plot is observed by incorporating MWCTs into the Ni–P. The CNT-1.0 is the most corrosion-resistant, as suggested by its large capacitive loop.

The corrosion resistance of the Ni–P and Ni–P/MWCNT coated samples was further evaluated by potentiodynamic polarization tests after 1 h of immersion in 3.5 wt.% NaCl solution at room temperature. The results are shown in Figure 6. The uncoated API 5L X80 steel substrate was also tested for comparison purposes. The corrosion potential (E_{corr}) and corrosion current densities (i_{corr}) were determined from these curves by means of the Tafel extrapolation method. The results are displayed in Table 3, along with the protection efficiency (P%) of the different coatings, as calculated from Equation (2).

$$P\% = \left(1 - \frac{i^*_{corr}}{i^0_{corr}}\right) \times 100 \quad (2)$$

where i^*_{corr} and i^0_{corr} are the corrosion current densities of the coated and uncoated substrate, respectively.

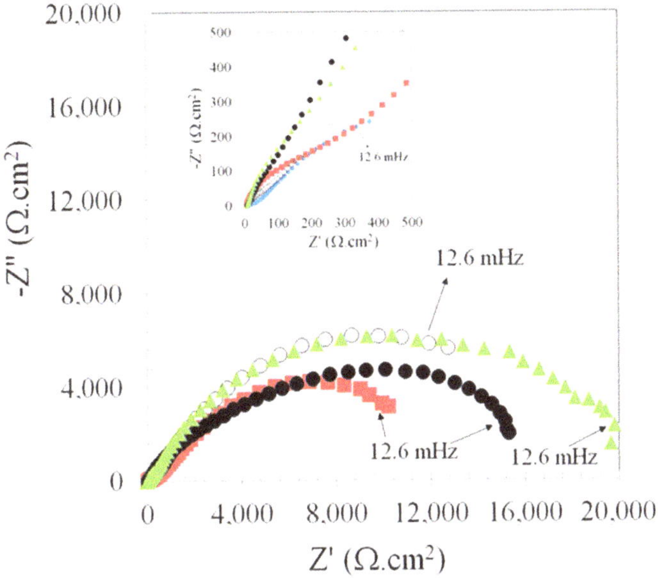

Figure 5. Nyquist plots of the Ni–P and Ni–P/MWCNT coatings after 1 h of immersion in 3.5 wt.% NaCl solution at room temperature.

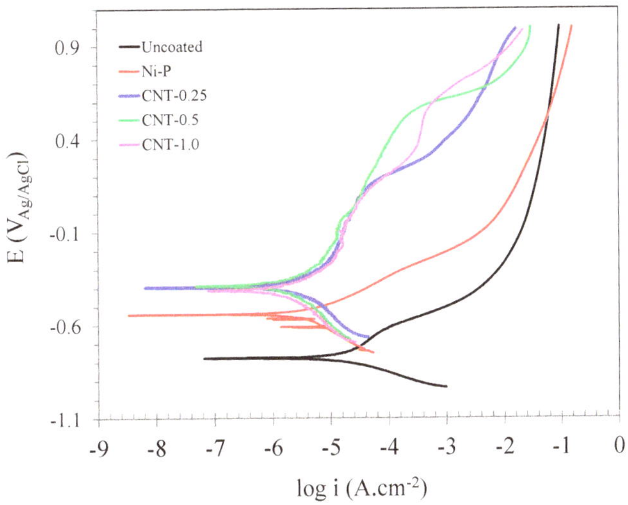

Figure 6. Potentiodynamic polarization curves of the Ni–P and Ni–P/MWCNT coatings after 1 h of immersion in 3.5 wt.% NaCl solution at room temperature.

Table 3. Corrosion parameters of the uncoated substrate, Ni–P and Ni–P/MWCNT coatings.

Coating	E_{corr} ($mV_{Ag/AgCl}$)	i_{corr} ($\mu A.cm^{-2}$)	p(%)
Uncoated	−778	20.3	—
Ni–P	−540	4.17	79.5
CNT-0.25	−394	3.28	83.8
CNT-0.5	−388	2.38	88.3
CNT-1.0	−413	2.01	90.1

The corrosion potential (E_{corr}) of the conventional Ni–P coating was shifted to the nobler direction with respect to the uncoated substrate. This trend was also observed for the composite Ni–P/MWCNT films. Notwithstanding, there is no clear tendency of increasing E_{corr} with the MWCNT loading. The electrochemical stability of the electrode surface is associated with E_{corr}, which increases as this parameter is shifted towards more anodic values [47]. The intrinsic low reactivity of carbon nanotubes [48] is likely to be responsible for the nobler E_{corr} values, compared with the uncoated substrate and the conventional Ni–P film.

All coatings markedly decreased the corrosion current density of the API 5L X80 steel substrate. The reduced i_{corr} scaled with the carbon nanotube was loaded in the plating bath. The protection efficiency of each coating is also presented in Table 3, showing the higher protectivity of the CNT-1.0 condition. Moreover, the anodic currents are lower for the composite Ni–P/MWCNT films when compared to the conventional Ni–P layer. It is also noteworthy that the polarization curves of the composite films presented a well-defined passive region that is not seen for the uncoated substrate or the Ni–P film. Both the intrinsic chemical inertia of carbon nanotubes and the possibility of blocking small pores in the Ni–P matrix were reported as the main causes of the increased corrosion resistance of electroless Ni–P–CNT composite coatings [49].

3.5. Scanning Electrochemical Microscopy (SECM)

SECM 2D maps of the uncoated API5LX80 steel, Ni–P, and composite Ni–P/MWCNT coatings are shown in Figure 7. The maps were recorded in 0.1 M NaCl at room temperature. The tip was biased at +600 $mV_{Ag/AgCl}$ to sense the formation of Fe^{2+} ions generated at the sample surface. The sample was unbiased.

During the corrosion process of this type of material, the oxidation of Fe^{2+} to Fe^{3+} occurs in the anodic sites, in which the Fe^{2+} ions are originated from the material dissolution. Thus, Fe^{2+} ions produced from the corrosion process are available on the surface. Thereby, when the Pt tip (polarized at +600 mVAg/AgCl) passes over the anodic sites, Fe^{2+} ions are oxidized to Fe^{3+}, according to Equation (1). Hence, since Fe^{2+} production is a primary feature of the anodic regions on the surface, the SECM maps shown in Figure 6 display the electrochemical activity of the studied surfaces related to possible active domains of Fe^{2+} ions. Higher oxidation current values indicate greater electrochemical activity of the surface.

The SECM maps show current spikes where the electrochemical activity is higher at the probed surface. The uncoated substrate presented the highest currents over the whole area, indicating that the electrochemical activity is more intense when compared to the coated material. There is no preferential site for current spikes, suggesting that corrosion does not occur by a localized attack, which is in agreement with the potentiodynamic polarization curve shown in Figure 6 for the bare substrate.

The currents were significantly reduced for the Ni–P film, indicating the protective character of the electrolessly deposited layer, leading to a decrease in the electrochemical activity probed by the tip near the sample surface. The cathodic values of the measured currents confirm the low activity for Fe^{2+} oxidation.

Izquierdo et al. [50] reported that the cathodic currents can be due to the fact that once the concentration of Fe^{2+} ions is low at the metallic surface, these species are likely to be easily oxidized to Fe^{3+} before diffusing to the bulk electrolyte. The cathodic current would then be probed at regions where Fe^{3+} ions are formed according to reaction (1). The presence of such regions would indicate that the film, although presenting low electrochemical activity, is prone to corrosion at its defective sites. It is possible to see in the SECM map shown in Figure 7B that the current values vary throughout the surface, suggesting that it is not homogeneous with respect to the sites where Fe^{2+} oxidation occurs.

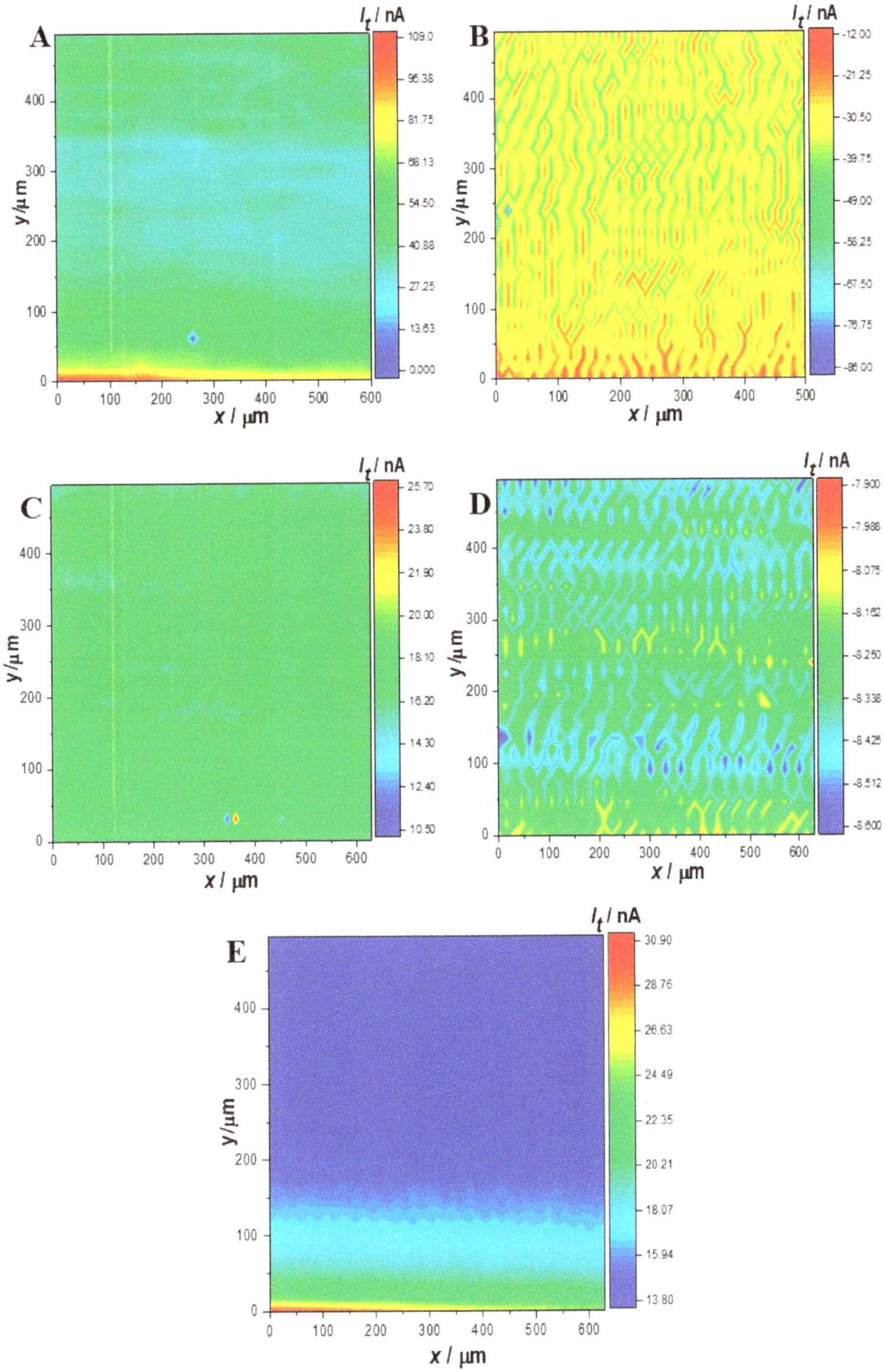

Figure 7. SECM 2D maps obtained with the tip biased at +600 $mV_{Ag/AgCl}$ and the sample at the open circuit potential: (**A**) uncoated substrate; (**B**) Ni–P layer; (**C**) CNT-0.25; (**D**) CNT-0.50; (**E**) CNT-1.0. Electrolyte: 0.1 M NaCl solution at room temperature.

The probed anodic currents are low on the surface of the CNT-0.25 sample (Figure 7C) when compared to the uncoated substrate, revealing the beneficial effect of carbon nanotubes in reducing the electrochemical activity of the metallic substrate. The current values are homogeneous over the probed area. The CNT-0.50 film, in turn, presented several current fluctuations on the SECM map and values that are slightly cathodic (Figure 7D). These low cathodic currents are likely due to the absence of the electroactive probed species (Fe^{2+} ions), indicating the lower activity of the CNT-0.50 film when compared with the CNT-0.25 film. This result agrees well with the global electrochemical behavior described in the previous section.

The currents are predominantly anodic over the surface of the CNT-1.0 sample, as shown in Figure 7E. The maximum currents are low over most part of the probed area, as indicated by the blue color scale at the right part of the SECM map. Hence, the increment of CNT loading in the coating layer has led to a decrease in the electrochemical activity. This is unequivocally perceived by comparing CNT-1.0 and CNT-0.25 samples. Notwithstanding, the difference is not so obvious when CNT-1.0 and CNT-0.50 samples are compared, since their SECM maps indicate low electrochemical activity in both cases. This result is also in agreement with the evaluation of the global electrochemical behavior the by potentiodynamic polarization curves (Figure 6 and Table 3). SECM proved to be sensitive to the electrochemical activity of the composite Ni–P–CNT films. Our results suggest that CNT-0.50 would give a suitable performance with respect to the adhesion and corrosion properties of the composite films. CNT-1.0 gave the best overall performance.

4. Discussion

From the previous sections it was possible to observe that MWCNT incorporation decreased the anodic dissolution rate of the Ni–P layer, leading to a significant reduction in the electrochemical activity, as shown in the SECM maps displayed in Figure 7. As mentioned in Section 3.4, literature reports ascribe the beneficial effect of carbon nanotubes on reducing the corrosion rate of Ni–P-coated substrates to its blocking effect with respect to intrinsic coating defects [47]. Alishahi et al. [28] reported a reduction of approximately 55% in the corrosion current density of a Ni–P/MWCNT composite film, with respect to a conventional Ni–P layer in 3.5 wt.% NaCl solution. Our results point to an even higher protection efficiency of the Ni–P/MWCNT layer. Based on these results, the proposed corrosion mechanism of the Ni–P/MWCNT coated samples is schematically illustrated in Figure 8.

Figure 8A shows the corrosion mechanism of the conventional Ni–P binary coating where anodic dissolution of the steel substrate occurs through intrinsic coating defects such as pores as cracks, releasing Fe^{2+} ions to the electrolyte. In fact, the high electrochemical activity of this sample was observed in the SECM map displayed in Figure 7B, supporting the occurrence of this mechanism. A different scenario was established after MWCNT incorporation in the Ni–P matrix. The electrochemical activity was gradually reduced, as described in Section 3.5, slowing down the anodic dissolution rate of the steel substrate. As a consequence of the blocking effect of the MWCNT particles, the release of Fe^{2+} ions was gradually reduced. This situation is illustrated in Figure 8B for the CNT-0.25 sample. As the MWCNT loading in the composite film increased, the amount of Fe^{2+} ions released to the solution was further reduced. The progressively lower electrochemical activity of the CNT-0.50 (Figure 7D) and CNT-1.0 (Figure 7E) films would be a consequence of this effect, as illustrated in Figure 8C,D, respectively.

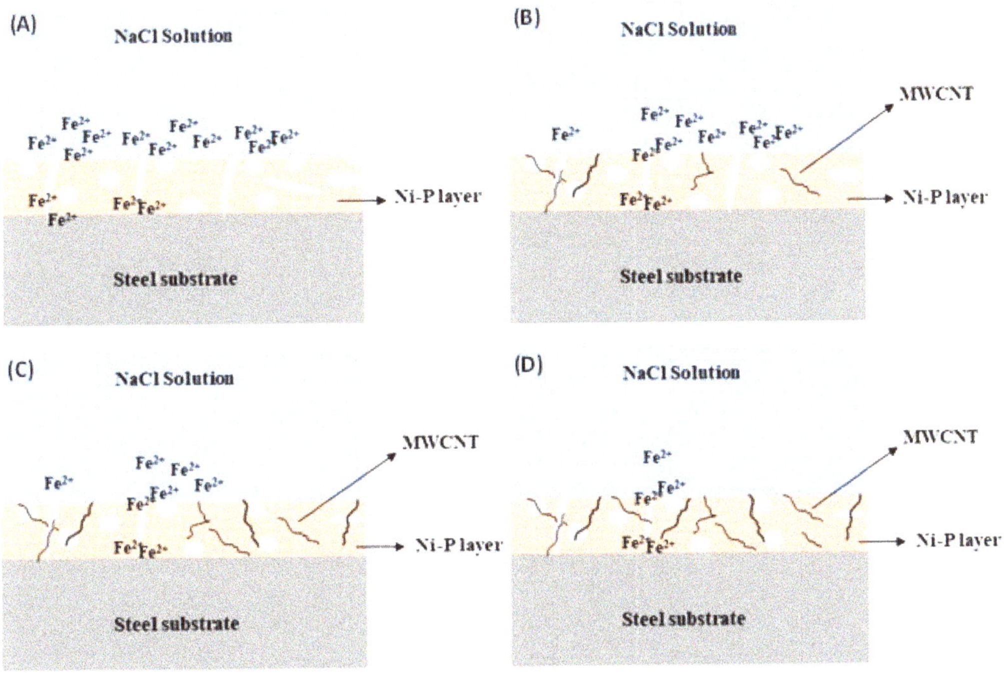

Figure 8. Schematic illustration of the corrosion mechanism of the Ni–P and Ni–P/MWCNT coatings: (**A**) Ni–P layer; (**B**) CNT-0.25; (**C**) CNT-0.50; (**D**) CNT-1.0.

5. Conclusions

Composite Ni–P/MWCNT films were successfully obtained by electroless deposition on the API 5L X80 substrate. MWCNT loading in the plating bath did not affect the thickness of the deposited layer, but reduced the size of the nodules on the coating, compared to the conventional binary Ni–P film. Adhesion strength was improved by adding the CNTs. The corrosion resistance of the composite films was increased due to the intrinsic low reactivity of the carbon nanotubes. This aspect was also evidenced by SECM analysis in the surface generation/tip collection mode (SG/TC). The SECM maps of the composite films with higher MWCNT loadings revealed low electrochemical activity for the Fe^{2+} oxidation reaction.

Supplementary Materials: The following are available online at https://www.mdpi.com/2075-4701/11/6/982/s1, Figure S1: EDS mapping of the cross-section for the CNT-0.50 film: (A) SEM micrograph; (B) Ni; (C) P; (D) Fe.; Figure S2: EDS mapping of the cross-section for the CNT-1.0 film: (A) SEM micrograph; (B) Ni; (C) P; (D) Fe.

Author Contributions: Conceptualization: M.C.L.d.O. and R.A.A.; Methodlogy: R.A.A., R.M.P.d.S. and N.B.d.L.; investigatFion: J.T.D.d.O., O.V.C. and L.A.d.O.; Writing-original draft preparation: M.C.L.d.O. and R.A.A.; Supervision: R.A.A.; funding acquisition: R.A.A. All authors have read and agreed to the published version of the manuscript.

Funding: This research was funded by CNPq (Process 470944/2013-7).

Institutional Review Board Statement: Not Applicable.

Informed Consent Statement: Not Applicable.

Data Availability Statement: The data obtained in this study are available from the corresponding.

Acknowledgments: Usiminas (Brzil) is kindly acknowledged for providing the pipeline steel used in this work. The Multiuser Experimental Facilities (UFABC) is acknowledged for the experimental support to this work.

Conflicts of Interest: The authors declare no conflict of interest.

References

1. Tanguy, B.; Luu, T.T.; Perrin, G.; Pineau, A.; Besson, J. Plastic and damage behavior of a high strength API5L X100 pipeline steel: Experiments and modeling. *Int. J. Press. Vessel. Pip.* **2005**, *85*, 322–335. [CrossRef]
2. Liu, Z.Y.; Cui, Z.Y.; Li, X.G.; Du, C.W.; Xing, Y.Y. Mechanistic aspect of stress corrosion cracking of X0 pipeline steel under non-stable cathodic polarization. *Electrochem. Commun.* **2014**, *48*, 127–129. [CrossRef]
3. Ituen, E.B.; Asuquo, J.E. Inhibition of X80 steel corrosion in oilfield acidizing environment using 3-(2-chloro-5,6-dihydrobenzol[b][1]benz 11-yl-)-N,N-dimethylpropan-1-amine and its blends. *J. King Saud Univ. Sci.* **2019**, *31*, 127–135. [CrossRef]
4. Xie, F.; Li, X.; Wang, D.; Wu, M.; Sun, D. Synergistic effect of sulphate-reducing bacteria and external tensile stress on the corrosion behavior of X80 pipeline steel in neutral soil environment. *Eng. Fail. Anal.* **2018**, *91*, 383–396. [CrossRef]
5. Islam, M.d.A.; Farhat, Z.N. Mechanical and electrochemical synergism of API X42 pipeline steel during erosion-corrosion. *J. Bio. Tribo. Corros.* **2015**, *1*, 26. [CrossRef]
6. Kong, D.; Wu, Y.; Long, D. Salt spray corrosion resistance of aluminized coatings on X70 pipeline steel by laser thermal radiation. *Rare Met. Mat. Eng.* **2014**, *43*, 2083–2088. [CrossRef]
7. Zeinoddini, M.; Mo'tamedi, M.; Zandi, A.P.; Talebi, M.; Shariati, M.; Ezzati, M. On the ratcheting of defective low-alloy, high-strength steel pipes (API-5L X80) under cyclic bending: An experimental study. *Int. J. Mech. Sci.* **2017**, *130*, 518–533. [CrossRef]
8. Yang, X.-H.; Zhu, W.-L.; Lin, Z.; Huo, J.-J. Aerodynamic evaluation of an internal epoxy coating in nature gas pipeline. *Prog. Org. Coat.* **2005**, *54*, 73–77. [CrossRef]
9. Wang, C.; Farhat, Z.; Jarjoura, G.; Hassan, M.K.; Abdullah, A.M. Indentation and bending behavior of electroless Ni-P-Ti composite coatings on pipeline steel. *Surf. Coat. Technol.* **2018**, *334*, 243–252. [CrossRef]
10. Darmiani, E.; Danaee, I.; Rashed, G.R.; Zaarei, D. Formulation and study of corrosion prevention behavior of epoxy cerium nitrate-montmorillonite nanocomposite coated carbon steel. *J. Coat. Technol. Res.* **2013**, *10*, 493–502. [CrossRef]
11. Abdou, M.I.; Ayad, M.I.; Diab, A.S.M.; Hassan, I.A.; Fadl, A.M. Influence of surface modified ilmenite/melamine formaldehyde composite on the anti-corrosion and mechanical properties of conventional polyamine cured epoxy for internal coating of gas and oil transmission pipelines. *Prog. Org. Coat.* **2017**, *113*, 1–14. [CrossRef]
12. Krishnan, K.H.; John, S.; Srinivasan, K.N.; Praveen, J.; Ganesan, M.; Kavimani, P.M. An overall aspect of electroless Ni-P depositions—A review article. *Metall. Mater. Trans. A* **2006**, *37*, 1917–1926. [CrossRef]
13. Fayyad, E.M.; Abdullah, A.M.; Hassan, M.K.; Mohamed, A.M.; Wang, C.; Jarjoura, G.; Farhat, Z. Synthesis, characterization, and application of novel Ni-P-carbon nitride nanocomposites. *Coatings* **2018**, *8*, 37. [CrossRef]
14. Oliveira, M.C.L.; Correa, O.V.; Ett, B.; Sayeg, I.J.; Lima, N.B.; Antunes, R.A. Influence of the tungsten content on surface properties of electroless Ni-W-P coatings. *Mater. Res.* **2018**, *21*, 1–13. [CrossRef]
15. Goettems, F.S.; Ferreira, J.Z. Wear behavior of electroless heat treated Ni-P coatings as alternative to electroplated hard chromium deposits. *Mater. Res.* **2017**, *20*, 1300–1308. [CrossRef]
16. Roy, S.; Sahoo, P. Optimization of electroless Ni-P-W coatings for minimum friction and wear using grey-Taguchi method. *J. Coat.* **2013**, *2013*, 608140. [CrossRef]
17. Liao, Y.; Zhang, S.T.; Dryfe, R. A study of corrosion performance of electroless Ni-P and Ni-W-P coatings on AZ91D magnesium alloy. *Materialwiss. Werkst.* **2011**, *42*, 833–837. [CrossRef]
18. Balaraju, J.N.; Selvi, V.E.; William Grips, V.K.; Rajam, K.S. Electrochemical studies on electroless ternary and quaternary Ni-P based alloys. *Electrochim. Acta* **2006**, *52*, 1064–1074. [CrossRef]
19. Cissé, M.; Abouchane, M.; Anik, T.; Himm, K.; Belakhimima, R.A.; Touhami, M.E.; Touir, R.; Amiar, A. Corrosion resistance of electroless Ni-Cu-P ternary alloy coatings in acidic and neutral corrosive mediums. *Int. J. Corros.* **2010**, *2010*, 246908. [CrossRef]
20. Sadreddini, S.; Afshar, A. The effect of heat treatment on properties of Ni-P-SiO_2 nano-composite coating. *Prot. Met. Phys. Chem. Surf.* **2016**, *52*, 492–499. [CrossRef]
21. Islam, M.; Azhar, M.R.; Khalid, Y.; Khan, R.; Abdo, H.S.; Dar, M.A.; Oloyede, O.R.; Burleigh, T.D. Electroless Ni-P/SiC nanocomposite coatings with small amounts of SiC nanoparticles for superior corrosion resistance and hardness. *J. Mater. Eng. Perform.* **2015**, *24*, 4835–4843. [CrossRef]
22. Gadhari, P.; Sahoo, P. Study of wear behavior of Ni-P-TiO_2 composite coatings by optimizing coating parameters. *Mater. Today: Proc.* **2017**, *4*, 1883–1892. [CrossRef]
23. Yang, Y.; Chen, W.; Zhou, C.; Xu, H.; Gao, W. Fabrication and characterization of electroless Ni-P-ZrO_2 nano-composite coatings. *Appl. Nanosci.* **2011**, *1*, 19–26. [CrossRef]
24. Araghi, A.; Paydar, M.H. Wear and corrosion characteristics of electroless Ni-W-P-B_4C and Ni-P-B_4C coatings. *Tribol. Mater. Surf. Interfaces* **2014**, *8*, 146–153. [CrossRef]
25. Chen, X.H.; Peng, J.C.; Li, X.Q.; Deng, F.M.; Wang, J.X.; Li, W.Z. Tribological behavior of carbon nanotubes-reinforced nickel matrix composite coatings. *J. Mater. Sci. Lett.* **2001**, *20*, 2057–2060. [CrossRef]

26. Meng, Z.-Q.; Li, X.-B.; Xiong, Y.-J.; Zhan, J. Preparation and tribological performances of Ni-P-multiwalled carbon nanotubes composite coatings. *Trans. Nonferrous Met. Soc. China* **2012**, *22*, 2719–2725. [CrossRef]
27. Zarebidaki, A.; Allahkaram, S.R. Effect of surfactant on the fabrication and characterization of Ni–P–CNT composite coating. *J. Alloy. Compd.* **2011**, *509*, 1836–1840. [CrossRef]
28. Alishahi, M.; Monirvaghefi, S.M.; Saatchi, A.; Hosseini, S.M. The effect of carbon nanotubes on the corrosion and tribological behavior of electroless Ni-P-CNT composite coating. *Appl. Surf. Sci.* **2012**, *258*, 2430–2446. [CrossRef]
29. Chen, X.H.; Chen, C.S.; Xiao, H.N.; Cheng, F.Q.; Zhang, G.; Yi, G.J. Corrosion behavior of carbon nanotubes-Ni composite coating. *Surf. Coat. Technol.* **2005**, *191*, 351–356. [CrossRef]
30. Yang, Z.; Xu, H.; Li, M.K.; Shi, Y.L.; Huang, Y.; Li, H.L. Preparation and properties of Ni/P/single-walled carbon nanotubes composite coatings by means of electroless plating. *Thin Solid Films* **2004**, *466*, 86–91. [CrossRef]
31. Marques, A.G.; Izquierdo, J.; Souto, R.M.; Simões, A.M. SECM imaging of the cut edge corrosion of galvanized steel as a function of pH. *Electrochim. Acta* **2015**, *153*, 238–245. [CrossRef]
32. Singh, A.; Ansari, K.R.; Kumar, A.; Liu, W.; Chen, S.; Lin, Y. Electrochemical, surface and quantum Chemical studies of novel imidazole derivatives as corrosion inhibitors for J55 steel in sweet corrosive environment. *J. Alloys Compd.* **2017**, *712*, 121–133. [CrossRef]
33. Yin, Y.; Niu, L.; Lu, M.; Guo, W.; Chen, S. In situ characterization of localized corrosion of stainless steel by scanning electrochemical microscope. *Appl. Surf. Sci.* **2008**, *255*, 9193–9199. [CrossRef]
34. Paik, C.H.; White, H.S.; Alkire, R.C. Scanning electrochemical microscopy detection of dissolved sulfur species from inclusions in stainless steel. *J. Electrochem. Soc.* **2000**, *147*, 4120–4124. [CrossRef]
35. Xia, D.-H.; Wang, J.; Wu, Z.; Qin, Z.; Xu, L.; Hu, W.; Behnamian, Y.; Luo, J.-L. Sensing corrosion within an artificial defect in organic coating using SECM. *Sens. Actuators B* **2019**, *280*, 235–242. [CrossRef]
36. Bastos, A.C.; Simões, A.M.; González, S.; González-García, Y.; Souto, R.M. Imaging concentration profiles of redox-active species in open-circuit corrosion processes with the scanning electrochemical microscope. *Electrochem. Commun.* **2004**, *6*, 1212–1215. [CrossRef]
37. Völker, E.; Inchauspe, C.G.; Calvo, E.J. Scanning electrochemical microscopy measurement of ferrous ion fluxes during localized corrosion of steel. *Electrochem. Commun.* **2006**, *8*, 179–183. [CrossRef]
38. González-García, Y.; Burstein, G.T.; González, S.; Souto, R.M. Imaging metastable pits on austenitic stainless steel in situ at the open circuit corrosion potential. *Electrochem. Commun.* **2004**, *6*, 637–642. [CrossRef]
39. Xu, S.; Hu, X.; Yang, Y.; Chen, Z.; Chan, Y.C. Effect of carbon nanotubes and their dispersion on electroless Ni-P under bump metallization for lead-free solder interconnection. *J. Mater. Sci. Mater. Electron.* **2014**, *25*, 2682–2691. [CrossRef]
40. Xie, G.; Liu, S.; Guo, D.; Wang, Q.; Luo, J. Investigation of the running-in process and friction coefficient under the lubrication of ionic liquid/water mixture. *Appl. Surf. Sci.* **2009**, *255*, 6408–6414. [CrossRef]
41. Li, Z.H.; Wang, X.Q.; Wang, M.; Wang, F.F.; Ge, H.L. Preparation and tribological properties of the carbon nanotubes-Ni-P composite coating. *Tribol. Int.* **2006**, *39*, 953–957. [CrossRef]
42. Balaraju, J.N.; Kalavati; Rajam, K.S. Surface morphology and structure of electroless ternary NiWP deposits with various W and P contents. *J. Alloys Compd.* **2009**, *486*, 468–473. [CrossRef]
43. Wang, Q.; Callisti, M.; Miranda, A.; McKay, B.; Deligkiozi, I.; Milickovic, T.K.; Zoikis-Karathanasis, A.; Hrissagis, K.; Magagnin, L.; Polcar, T. Evolution of structural, mechanical and tribological properties of Ni-P/MWCNT coatings as a function of annealing temperature. *Surf. Coat. Technol.* **2016**, *302*, 195–208. [CrossRef]
44. Alexis, J.; Gaussens, C.; Etcheverry, B.; Bonino, J.-P. Development of nickel phosphorus coatings containing micro particles of talc phyllosilicates. *Mater. Chem. Phys.* **2013**, *137*, 723–733. [CrossRef]
45. Hamdy, A.S.; El-Shenawy, E.; El-Bitar, T. Electrochemical impedance spectroscopy study of the corrosion behavior of some niobium bearing stainless steels in 3.5% NaCl. *Int. J. Electrochem. Sci.* **2006**, *1*, 171–180.
46. El-Sayed, M.S. Comparative study on the electrochemical corrosion behavior of iron and X-65 steel in 4.0 wt.% sodium chloride solution after different exposure intervals. *Molecules* **2014**, *19*, 9962–9974. [CrossRef]
47. Serdar, M.; Zulj, L.V.; Bjegovic, D. Long-term corrosion behaviour of stainless reinforcing steel in mortar exposed to chloride environment. *Corros. Sci.* **2013**, *69*, 149–157. [CrossRef]
48. Praveen, B.M.; Venkatesha, T.V.; Naik, Y.A.; Prashantha, K. Corrosion studies of carbon nanotubes-Zn composite coating. *Surf. Coat. Technol.* **2007**, *201*, 5836–5842. [CrossRef]
49. Yang, Z.; Xu, H.; Shi, Y.-L.; Li, M.-K.; Huang, Y.; Li, H.-L.; Huang, Y.; Li, H.-L. The fabrication and corrosion behavior of electroless Ni-P-carbon nanotube composite coatings. *Mater. Res. Bull.* **2005**, *40*, 1001–1009. [CrossRef]
50. Izquierdo, J.; Martín-Ruíz, L.; Fernández-Pérez, B.M.; Rodríguez-Raposo, R.; Santana, J.J.; Souto, R.M. Scanning microelectrochemical characterization of the effect of polarization on the localized corrosion of 304 stainless steel in chloride solution. *J. Electroanal. Chem.* **2014**, *728*, 148–157. [CrossRef]

metals MDPI

Article

Comparison of Corrosion Resistance of the AA2524-T3 and the AA2024-T3

Fernanda Martins Queiroz [1], Maysa Terada [2,*], Aline F. Santos Bugarin [3], Hercílio Gomes de Melo [4] and Isolda Costa [3]

1 Escola e Faculdade de Tecnologia SENAI Suiço-Brasileira "Paulo Ernesto Tolle", Av. Bento Branco de Andrade Filho, 379, São Paulo 04757-000, Brazil; mq_fernanda@yahoo.com.br
2 Instituto SENAI de Inovação em Manufatura Avançada e Microfabricação, Av. Bento Branco de Andrade Filho, 379, São Paulo 04757-000, Brazil
3 IPEN, Instituto de Pesquisas Energéticas e Nucleares, Av. Prof. Lineu Prestes, 2242, São Paulo 05508-000, Brazil; afbugarin@usp.br (A.F.S.B.); icosta@ipen.br (I.C.)
4 Escola Politécnica da University de São Paulo, Departament de Engenharia Metalúrgica e de Materiais, Av. Prof. Mello de Moraes, 2463, São Paulo 05508-030, Brazil; hgdemelo@usp.br
* Correspondence: maysaterada@uol.com.br

Abstract: The 2XXX Al alloys are characterized by their superior mechanical properties resulting from alloying elements and precipitation hardening treatments. The AA2524-T3 alloy was developed to replace the AA2024-T3 alloy in the aerospace industry. However, both alloys present many intermetallic particles (IMCs) in their microstructure, and this is the main reason for their high susceptibility to localized corrosion (such as pitting and stress corrosion cracking). Despite the similarities between these alloys (e.g., chemical composition and type of intermetallics) the literature comparing their properties is scarce and focuses mainly on their mechanical properties, not their corrosion resistances. In this investigation, the corrosion resistance of the AA2524-T3 alloy was compared to the AA2024-T3 alloy. The microstructure of both alloys was analyzed by Scanning Electron Microscopy before and after immersion in the test electrolyte, and the number and area fraction of intermetallics of each alloy was determined. The corrosion resistance of both alloys was monitored as a function of exposure time by electrochemical impedance spectroscopy and the results were fitted using electrical equivalent circuits. The AA2524-T3 alloy presented not only higher impedance values but also less corroded areas than the AA2024-T3 alloy.

Keywords: 2XXX aluminium alloys; intermetallics; aircraft industry; corrosion

Citation: Queiroz, F.M.; Terada, M.; Bugarin, A.F.S.; de Melo, H.G.; Costa, I. Comparison of Corrosion Resistance of the AA2524-T3 and the AA2024-T3. *Metals* **2021**, *11*, 980. https://doi.org/10.3390/met11060980

Academic Editor: Gilbert Henaff

Received: 6 May 2021
Accepted: 15 June 2021
Published: 19 June 2021

1. Introduction

The 2XXX Al alloys are largely used in the aircraft industry due to their superior mechanical properties resulting from alloying elements addition, mainly Cu and Mg, and precipitation hardening treatments. The AA2524-T3 alloy is a relatively new alloy (formerly known as C188-T3) developed by ALCOA to replace the AA2024-T3 alloy in fuselage skins. It is currently used in the Boeing 777 aircraft [1,2]. It presents high damage-tolerance and excellent fatigue properties [1]. However, both the AA2024-T3 and AA2524-T3 alloys possess a large amount of intermetallic particles (IMCs) in their microstructure, and this is the main reason for their high susceptibility to localized corrosion such as pitting and stress corrosion cracking [3–8]. Localized corrosion in aluminium alloys uses to be caused by microgalvanic effects between IMCs and the matrix [7–10]. This generally results in pitting corrosion, which is considered to be one of the main microstructure dependent damage mechanisms of high strength aluminium alloys [7]. Particles constituted by the alloying elements present different electrochemical activity from the matrix and, therefore, represent preferential sites for pitting nucleation. In potential-controlled conditions, pitting is characterized by two types of events, metastable and stable, which occur before and after the pitting potential, respectively [11].

According to Birbilis and Buchheit [12], three main types of IMCs particles can be found in high strength Al alloys: precipitates, dispersoids, and constituent particles. The first ones are formed by nucleation and growth from a supersaturated solution and their diameters are around few tenths to hundreds of nanometers. They contribute to an increase in the mechanical resistance of the alloy when they are coherent and homogeneously dispersed in the matrix [13]. It is reported that they can initiate intergranular corrosion and stress corrosion cracking when clustered near grain boundaries [12,14,15]. While dispersoids are typically bigger than the precipitates, they are still within the submicrometric to micrometric range. Their main role is to control recrystallisation and grain size [12,13]. Since they are mostly passive in environments where Al alloys are usually used and are uniformly distributed, they rarely increase corrosion susceptibility [12]. Finally, the constituent particles, in the present work denominated intermetallics (IMCs), are the largest ones. Its size range is from a few tenths of a micrometer to up to 10–20 micrometers. They tend to break down during mechanical treatments [12] and are frequently noticed in clusters [16]. They are enriched in different alloying elements and there is a universal agreement that pitting corrosion of high strength Al alloys is associated with these particles, mainly in chloride containing media.

In 2XXX series aluminium alloys, θ′ (Al_2Cu) and S (Al_2CuMg) are two of the most important types of coarse IMCs particles. The S phase is associated with pitting nucleation [17] and some mechanisms have been proposed in the literature [18]. According to Moreto, et al. [19] the AA2524-T3 aluminium alloy presents two types of coarse intermetallics particles (Al-Cu-Fe-Mn) and incoherent S (Al_2CuMg) precipitates, which are also found in AA2024-T3 alloys [9,20]. The corrosion behaviour of these intermetallic particles has been largely investigated in the literature [17,21–27]. The effect of coarse Al-Cu-Mg IMCs on the localized corrosion behaviour of the AA2024-T3 alloy has been investigated by Buchheit, et al. [17] and Zhu, et al. [28]. Both authors [17,28] have noticed that this phase is an anodic site, favouring pitting nucleation. According to the literature [17,24], at an initial stage, anodic dissolution occurs on S phase particles. During this period, the Mg is preferentially oxidized, resulting in Cu-rich remnants. This leads to potential change into the nobler direction, promoting anodic dissolution of the alloy matrix at its neighbouring boundary. Zhu, et al. [28] proposed a "cathodic corrosion" mechanism to explain the localized corrosion in 2XXX alloys, which was in disagreement with Buchheit's proposal [17]. According to Zhu, et al. [28], during the anodic reaction of S phase particles, cathodic reactions of water and/or oxygen reduction occur at the neighbouring sites of the matrix, generating hydroxyl ions (OH^-) [28]. Accordingly, local alkalization occurs around the S phase remnant and, once the local pH exceeds 9, chemical dissolution of the surrounding aluminium oxide layer takes place. Subsequently, the bare aluminium matrix would be oxidized to form a new oxide layer. On the other hand, the Al-Cu-Fe-Mn IMCs are usually cathodic sites, which are more stable and have less effect on the AA2024-T3 alloy's corrosion resistance than the Al-Cu-Mg ones [25,29–32].

Very few studies can be found concerning AA2524 [4,33–37], and only a few studies compare this alloy to the AA2024, focusing on its mechanical properties [2,38]. Although it is generally accepted that the localized corrosion caused by the IMCs is the nucleation site for other types of corrosion in high strength Al alloys [23,24], the role of the IMCs on these alloys' corrosion initiation has not been well established yet. This is mainly due to their small size, their chemical and microstructural complexity, and the dependence of reactivity to the environment [24]. This paper aims to evaluate the corrosion resistances of both AA2024 and AA2524, to compare the results, and to correlate them with their microstructure.

2. Materials and Methods

The materials studied in this work were the AA2524-T3 and the AA2024-T3 alloys produced by ALCOA, both provided as 1.6 mm sheets. Their chemical compositions are shown in Table 1 and their main mechanical and corrosion properties are shown in Table 2.

The AA2524-T3 alloy has lower amounts of Fe and Si, which are considered as deleterious to the corrosion resistance, and a narrower range for the main alloying elements Cu and Mg [32].

Table 1. Analyzed and nominal chemical composition (wt %) of the AA2024-T3 and AA2524-T3 alloys [39].

Elements	AA2024-T3	AA2524-T3
Cu	4.06 (3.8–4.9)	3.84 (4.0–4.5)
Mg	1.77 (1.2–1.8)	1.31 (1.2–1.6)
Mn	0.63 (0.3–0.9)	0.56 (0.45–0.7)
Si	0.11 (0.5 max)	0.04 (0.06 max)
Fe	0.14 (0.5 max)	0.06 (0.12 max)
Ti	−(0.15 max)	0.029 (0.10 max)
Zn	0.02 (0.25 max)	0.01 (0.15 máx)
Al	Balance	Balance

Table 2. Mechanical and corrosion properties of the AA2024-T3 and AA2524-T3 alloys.

	Tensile Strength (MPa)	Yield Strength (MPa)	Elongation (%)	Electrical Conductivity % IACS	Open Circuit Potential (V_{SCE})
AA2024 [40]	435	290	10–15	30	−0.722 [41]
AA2524 [42]	445.4	340.3	19	34	−0.590 [43]

Samples of 400 mm^2 were cut from the sheets. Their surfaces were prepared by grinding them with silicon carbide paper up to # 4000, followed by polishing with a diamond paste (up to 0.25 μm) to a mirror surface, degreasing in ethyl alcohol, washing with deionized water, and drying under a hot air stream. The corrosion resistance of both aluminium alloys was monitored in 0.01 mol L^{-1} NaCl as a function of exposure time by immersion tests and electrochemical impedance spectroscopy (EIS).

Experiments were performed using a Solartron 1287 electrochemical interface coupled to a Solartron 1260 frequency response analyser (FRA), which was used to obtain the electrochemical impedance spectroscopy (EIS) diagrams. The perturbation range was 20 mV (rms). The acquisition rate was 10 points per decade in a frequency range from 10 mHz to 100 kHz for all samples. The electrochemical tests were performed using a three-electrode setup configuration with the aluminium alloys as working electrodes and an exposed area of 1.0 cm^2 (Figure 1). The reference electrode used was an Ag/AgCl, 3 mol L^{-1} KCl, and a platinum wire was employed as an auxiliary electrode. The electrochemical experiments were carried out in naturally aerated solution at room temperature. Each test was performed six times to ensure their reproducibility.

The microstructures of the aluminium alloys were analyzed by a Field Emission Gun Scanning Electron Microscope FEI Quanta 650 (FEG–SEM) prior and after different exposure periods to the test electrolyte.

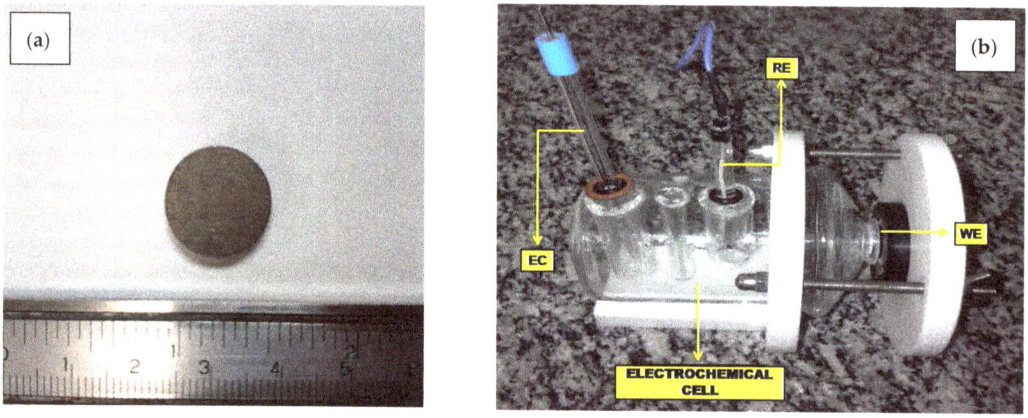

Figure 1. (**a**) Specimen of AA2524 in as-received condition and (**b**) three-electrode setup configuration cell. The reference electrode (RE) was an Ag/AgCl, 3 mol L^{-1} KCl, and a platinum wire was employed as an auxiliary electrode (EC). The sample was the working electrode (WE).

3. Results and Discussion

The surface of both the AA2024-T3 and AA2524-T3 alloys were observed by SEM after 2 h exposure to the 0.01 mol.L^{-1} NaCl solution. A general view of the microstructures of the AA2024-T3 and AA2524-T3 alloys are shown in Figure 2a,b, respectively (both in an as-received condition). The two main types of coarse IMCs particles (Al-Cu-Mg (Figure 2c,d) and Al-Cu-Mn-Fe-(Si) (Figure 2e,f)) were identified by EDX, and they are in agreement with the literature [9,10,19]. However, in the AA2524-T3 alloy, the Al-Cu-Mn-Fe-(Si) particles were present in a lower amount and with different morphologies from those found in the AA2024-T3 alloy, whereas the Al-Cu-Mg are more numerous, smaller, and more uniformly distributed. The AA2024-T3 alloy is characterized by a smaller number of IMCs, which, however, are generally larger than those found in the AA2024-T3 alloy (Figure 3). Consequently, the area fraction of the AA2024-T3 IMC's is higher than the AA2524-T3 (Table 3).

Table 3. Comparison of IMCs area fraction of AA2024 and AA2524 alloys.

Material	IMCs Area Fraction (%)
AA2024-T3	2.52 ± 0.42
AA2524-T3	1.64 ± 0.14

The general analysis of the micrographs (Figure 4) show that the corrosive attack is more evenly distributed on the AA2524-T3 alloy (Figure 4b), whereas for the AA2024-T3 alloy (Figure 4a) it is mainly concentrated around larger IMCs, including the dark areas around the bigger Al-Cu-Mg precipitates (indicated by arrows). On both alloys, the Al-Cu-Mn-Fe precipitates showed a rather random behaviour with some of them remaining intact after 2 h of immersion, likely due to a protective oxide film that hinder the cathodic reaction, as suggested by Birbilis, et al. [12] (Figure 4a,b).

After 7 h of testing of the AA2024-T3 alloy, the corrosive attack was mainly concentrated at the interface between the matrix and the larger intermetallic particles. No localized corrosion at the grain boundaries was detected. (Figure 4c). Figure 4d shows that most of the smaller precipitates had been removed from the surface of the AA2524-T3 sample by the corrosive attack, resulting in a "cleaner" surface in comparison with 2 h of test, even though some local activity remained around large IMCs. On the other hand, the corrosive attack of the AA2024-T3 continued to be mainly concentrated at the interface between the matrix and the larger intermetallic particles (Figure 4c), which, as previously showed, were more numerous in this material. For longer immersion times the "cleaning" effect on the AA2524-T3 surface is even more pronounced than that of the AA2024-T3 (Figure 4f,e, respectively).

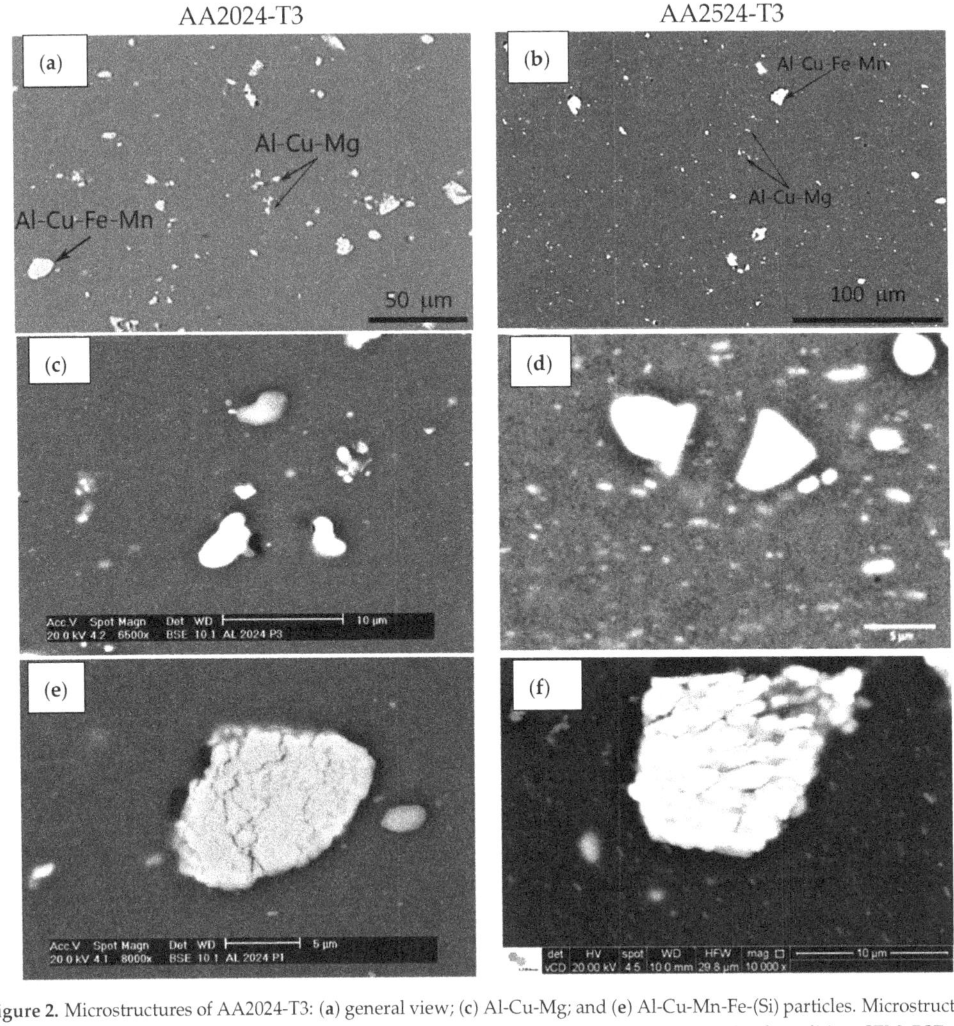

Figure 2. Microstructures of AA2024-T3: (**a**) general view; (**c**) Al-Cu-Mg; and (**e**) Al-Cu-Mn-Fe-(Si) particles. Microstructures of AA2524-T3: (**b**) general view; (**d**) Al-Cu-Mg; and (**f**) Al-Cu-Mn-Fe-(Si) particles. As-received condition. SEM–BSE.

After 11 h of exposure to the test electrolyte, most of the small precipitates were removed from the surface of both alloys (Figure 4e,f). Additionally, Figure 4g,h show the selective attack near the IMCs for longer exposure periods.

The results showed that IMCs' area fraction of the AA2524-T3 is lower in comparison with the AA2024-T3. More than 94% of these IMCs of the first alloy are smaller than 10 µm, and only 82% of the latter, confirming the theory that the corrosion attack on the AA2524-T3 alloy surface was faster due to its larger number of small particles.

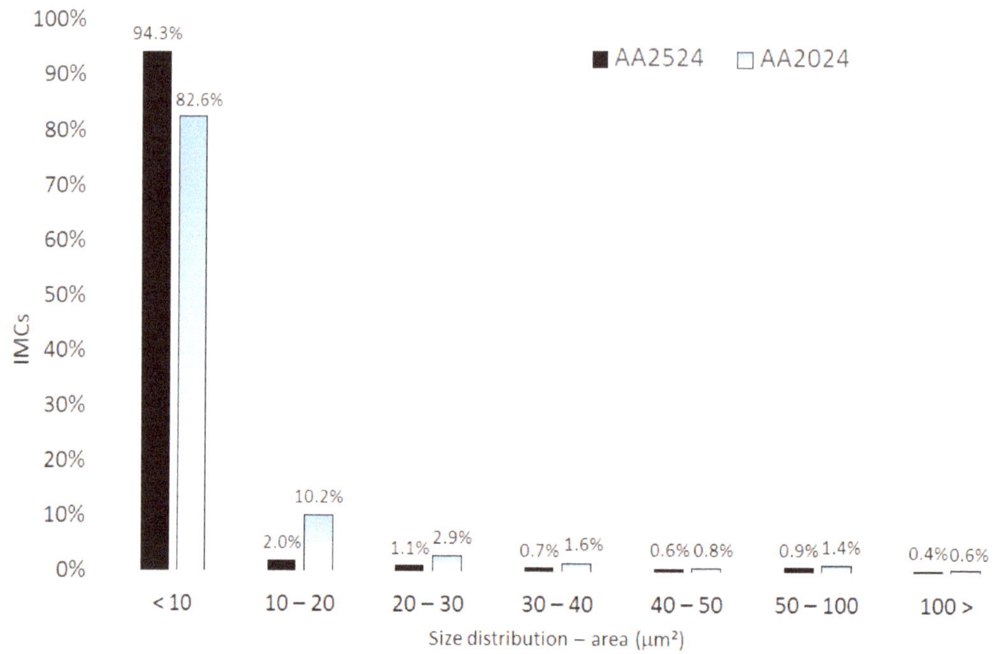

Figure 3. Frequency (%) and size distribution (μm^2) of IMCs for as-received samples. More than 94% of the AA2524-T3 alloy and 82% of the AA2024-T3 alloy IMCs are smaller than 10 μm^2.

The corrosion resistance of both alloys was also evaluated by EIS with immersion time and the results are shown in Figures 5 and 6. It is clearly seen that the impedance modulus associated with the AA2524-T3 alloy was higher than that of the AA2024-T3 during the whole test period. Independently of the alloy brand, EIS diagrams are characterized by a broad high frequency depressed capacitive loop followed by a low frequency capacitive loop that evolves to a diffusion-controlled process after 11 h of exposure to the test electrolyte.

The EIS results for both alloys were fitted with the two/three-time constants equivalent electric circuits (EEC) shown in Figure 7. These ECC have been established in the literature on aluminium alloys EIS fittings and were also tested in previous works from our research group [9,44–46]. Table 4 presents the parameters obtained from the equivalent electrical circuit fitted to the AA2024-T3 and AA2524-T3 EIS data. The EEC of Figure 7a was used to fit the data acquired from the first hour until 9 h of immersion, whereas the diagrams obtained from 11 h until the end of the test were fitted with the EEC of Figure 7b. In this latter case, the low frequency *R//CPE* element was substituted for a single *CPE*, meant to simulate a diffusion-controlled process. According to Campestrini, et al. [47], when the resistance associated with the diffusion becomes very large the *R//CPE* element may be replaced by a simple *CPE* element with the exponent "α" value of 0.5, which, in an ideal situation, is represented by a Warburg element. In the proposed circuits the pair $R_{ox}//CPE_{ox}$ stands for the capacitance of the oxide film in parallel with conductive pathways associated with defective sites created by the IMCs, which leads to the unprotected metal surface. $R_{ct}//CPE_{dl}$ refers to the charge transfer resistance coupled to the charging of the double layer, whereas $CPE_{cor}//R_{cor}$ is ascribed to the low frequency corrosion processes that kinetically control the alloys' deterioration, which gradually evolves to a diffusion-controlled process and is likely associated with the oxygen reduction reaction taking place in nobler IMCs.

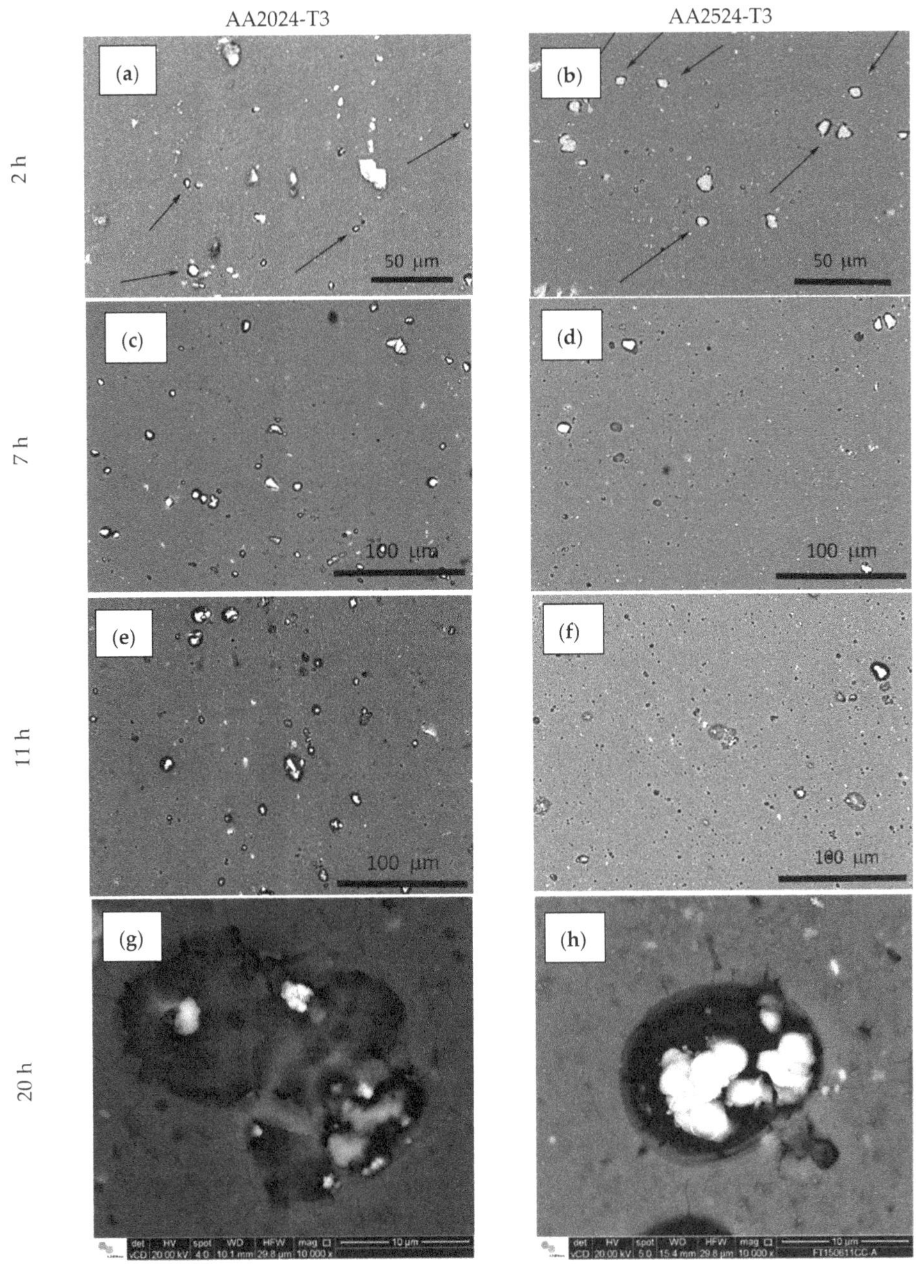

Figure 4. Microstructures of AA2024-T3 after (**a**) 2 h, (**c**) 7 h, (**e**) 11 h, and (**g**) 20 h. AA2524-T3 after (**b**) 2 h, (**d**) 7 h, (**f**) 11 h, and (**h**) 20 h of exposure to the 0.01 mol.L^{-1} NaCl solution. SEM–BSE. The micrographs show the evolution of the localized corrosion at the interface IMC/matrix (indicated by arrows).

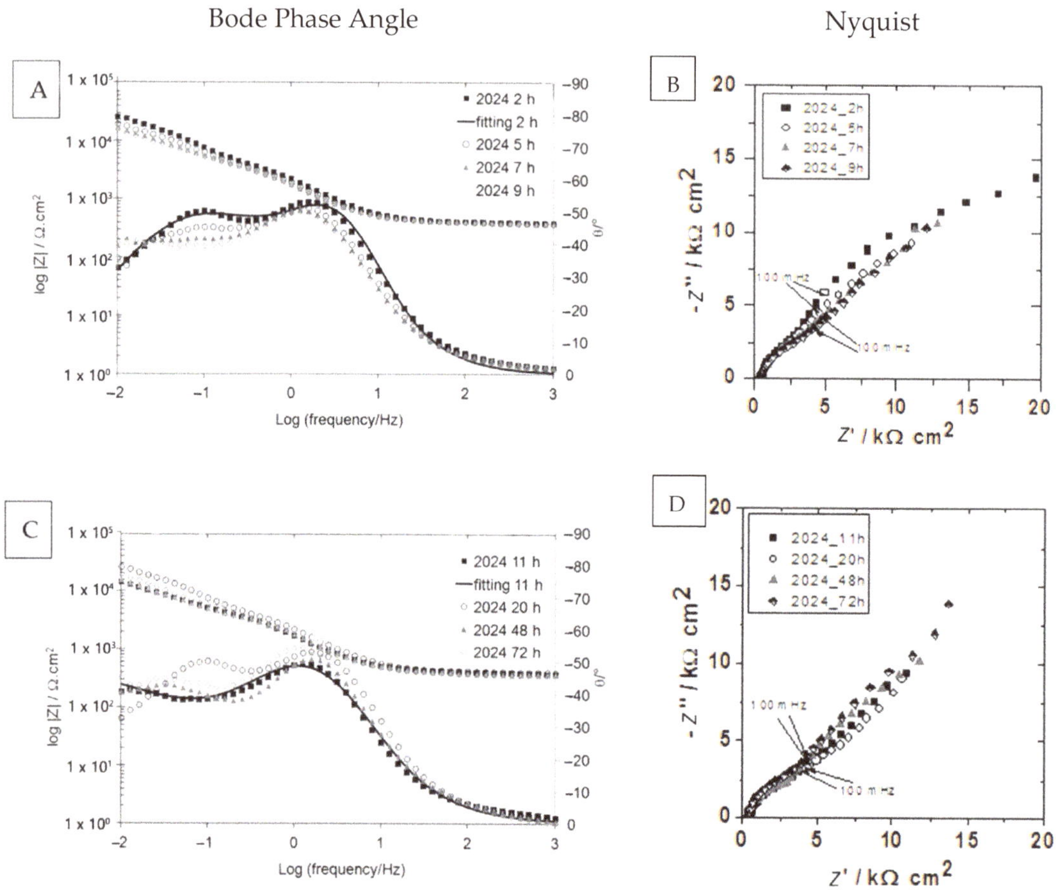

Figure 5. Evolution of EIS results for the AA2024-T3 as a function time of exposure to 0.01 mol L^{-1} NaCl electrolyte. (**A**) Bode phase angle and (**B**) Nyquist diagram. Measurements were taken after 2, 5, 7 and 9 h of the experiment. (**C**) Bode phase angle and (**D**) Nyquist diagram. Measurements were taken after 11, 20, 48 and 72 h of the experiment.

Table 4 shows the evolution of the $R_{ox}//CPE_{ox}$ values. For both alloys, CPE_{ox} increased with exposure time. SEM analysis has shown that as the corrosive attack proceeded, holes were formed in the samples surface as a result of the dissolution/detachment of the IMCs particles increasing the effective area exposed to the electrolyte [10]. If pits did not progress in these regions, a fresh oxide layer was formed. However, it should be more defective. Both facts contribute to capacitance increase. In addition, from the beginning of the exposure period, and during the whole test, CPE_{ox} was smaller for AA2524-T3, indicating the presence of a thicker oxide layer in this alloy, in agreement with the results from previous studies in the literature [46,48].

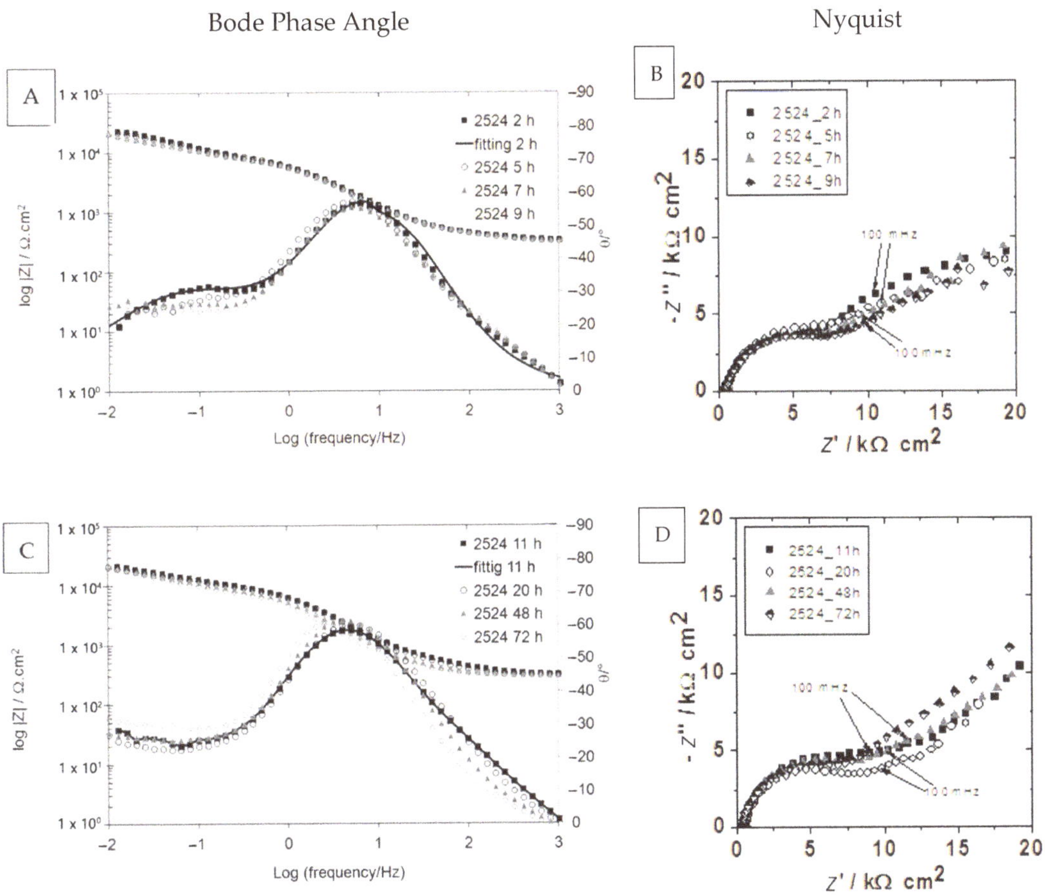

Figure 6. Evolution of EIS results for the AA2524-T3 as a function time of exposure to 0.01 mol L^{-1} NaCl electrolyte. (**A**) Bode phase angle and (**B**) Nyquist diagram. Measurements taken after 2, 5, 7, and 9 h of the experiment. (**C**) Bode phase angle and (**D**) Nyquist diagram. Measurements taken after 11, 20, 48, and 72 h of the experiment.

Concerning R_{ox}, Table 4, it also increased with immersion time, indicating that electrochemically active sites on both alloys' surfaces become less numerous and/or were less easily reachable. As already demonstrated in the SEM analysis (Figure 4), the selective attack near the IMCs results in surface "cleaning", leading to a lower number of active sites. In addition, deposition of corrosion products was frequently found above large IMCs. These features would hinder the access of aggressive species to the matrix surface and can contribute to the R_{ox} increase. Accordingly, as the surface of AA2524-T3 progressively became "cleaner" than AA2024-T3, the rate of the R_{ox} increase was much faster for the former alloy. Therefore, it is proposed that the attack of the small particles and their removal from surface are the main reasons for the R_{ox} increase, as this reduces the number of defective sites where electrochemical reactions could take place. However, blocking of active site by aluminium hydroxide precipitation may also have contributed to the R_{ox} increase.

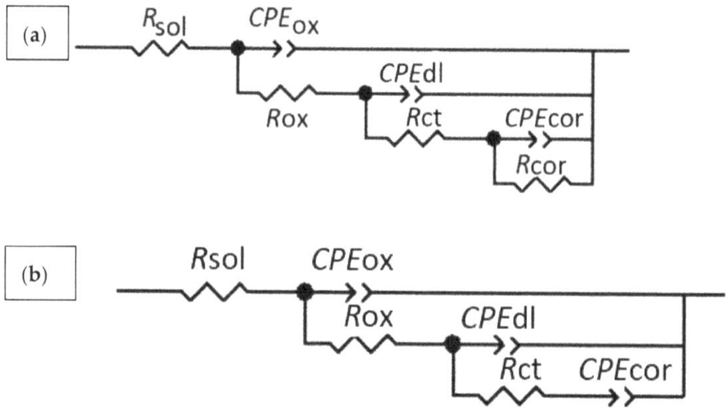

Figure 7. Equivalent electrical circuits (EEC) proposed to fit the data of the AA2024-T3 and AA2524-T3 (**a**) until 9 h and (**b**) from 11 h of immersion in a 0.01 mol L^{-1} NaCl solution.

As mentioned earlier, the $R_{ct}//CPE_{dl}$ pair is associated with the charge transfer reactions (R_{ct}) occurring at the interface of the matrix at the defective sites of the oxide (near active IMCs). The evolution for the components R_{ct} and CPE_{dl} with exposure time to the 0.01 mol L^{-1} NaCl electrolyte is shown in Table 4. For AA2524-T3, R_{ct} behaviour was significantly oscillating during the whole immersion period, which must has been a consequence of the detachment and outbreak of new IMCs as corrosion proceeded. For this sample, CPE_{dl} remained almost constant during the first 15 h, indicating that the electrochemically active surface area remained almost unchanged. However, from this period onwards the decrease in CPE_{dl} might be attributed to the formation of corrosion products on some active areas, clearly indicated on the AA2524-T3 surface, that partially blocked these sites.

Conversely, for AA2024-T3, after up to 40 h of testing, R_{ct} decreased continuously. This tendency may be explained by the enrichment in the more noble components of the remaining active IMCs. Therefore, even though in smaller number, due to detachment, enhanced galvanic activity would be expected between the matrix and the Cu-enriched IMCs. For longer immersion periods, an increase of R_{ct} was observed. This can be ascribed to the precipitation of corrosion products above the IMCs and their vicinity, which provided a barrier between the electrolyte and the metallic surface. Accordingly, CPE_{dl} initially increased, indicating that the electrochemically active surface augmented. However, for longer exposure times, it started to decrease due to active surface area reduction as corrosion products precipitated.

Finally, the low frequency time constant, corresponding to the rate-controlling step of the corrosion process, was mainly represented by CPE_{cor} and its "α_{cor}", and R_{cor} were important only up to nine hours of exposure. During the initial immersion period, α_{cor} values ranged between 0.69 and 0.83, suggesting a mixed controlled process for both alloys, corresponding to the period where the EIS diagrams were fitted with the EEC of Figure 6. For longer immersion times, "α_{cor}" varied between 0.4 and 0.6, showing the main contribution of diffusive processes likely through the precipitated corrosion products.

Higher capacitance values were associated with the AA2024-T3 alloy comparatively to the AA2524-T3 (Table 4) after the first hour of immersion, although there was much larger variation in the CPE_{cor} values for this latter alloy, likely due to the stronger contribution of partially soluble corrosion products on active sites at its surface which were periodically removed.

The electrochemical results demonstrated that the corrosion resistance of the AA2524-T3 was higher than the AA2024-T3.

Table 4. Parameters obtained from the equivalent electrical circuit fitted to the AA2024-T3 and AA2524-T3 EIS data.

	2 h		5 h		7 h		9 h		11 h		20 h		48 h		72 h	
	2524	**2024**	**2524**	**2024**	**2524**	**2024**	**2524**	**2024**	**2524**	**2024**	**2524**	**2024**	**2524**	**2024**	**2524**	**2024**
Rsol $\Omega \cdot cm^2$	346.40	394.20	342.10	388.70	334.90	382.80	330	380.40	328.40	381.20	325.10	373.80	316.90	362.30	318.80	376.90
Error (%)	0.40	0.27	0.50	0.22	0.40	0.16	0.32	0.16	0.34	0.27	0.34	0.25	0.37	0.37	0.32	0.30
*CPE*ox $\mu F \cdot cm^{-2} \cdot s^{(\alpha-1)}$	6.27	16.60	5.99	16.30	6.07	14.90	6.39	15.30	6.84	19.20	8.58	20.80	13.80	34.70	18.00	43.30
Error (%)	2.44	8.14	2.36	5.84	1.75	3.88	1.28	3.52	1.29	4.57	1.27	3.41	1.39	4.28	1.44	3.63
Rox $\Omega \cdot cm^2$	327.10	57.15	351.60	56.04	413.10	57.71	479.10	64.06	541.80	79.83	752.30	107	954.50	210.20	761.40	239.70
Error (%)	6.18	9.32	4.53	5.17	3.78	3.02	3.15	2.77	3.37	3.78	5.21	3.33	7.58	9.46	7.78	10.02
*CPE*dL $\mu F \cdot cm^{-2} \cdot s^{(\alpha-1)}$	14.80	58.90	16.40	75.40	16.60	89.30	16.10	91.50	14.40	83.50	11.50	80.80	8.85	44.00	13.10	50.20
Error (%)	3.23	3.39	3.26	2.19	3.13	1.40	2.53	1.35	2.84	2.06	4.65	2.14	6.20	6.93	4.73	6.19
α dL	0.86	0.89	0.88	0.88	0.88	0.87	0.87	0.87	0.89	0.89	0.90	0.90	0.97	0.99	0.94	0.96
Error (%)	0.81	0.66	0.81	0.49	0.83	0.35	0.71	0.36	0.78	0.50	1.33	0.57	1.74	1.79	1.40	1.69
Rct $k\Omega \cdot cm^2$	7.37	4.66	7.73	4.37	6.98	4.28	7.27	4.06	7.88	3.16	6.26	3.48	5.43	2.47	6.54	3.50
Error (%)	2.09	2.94	2.89	2.78	2.61	2.20	2.01	2.11	2.88	2.61	3.79	3.43	4.30	5.09	3.03	5.48
*CPE*cor $\mu F \cdot cm^{-2} \cdot s^{(\alpha-1)}$	174.00	180.04	187.70	241.00	239.59	313.00	274.00	358.00	204.00	341.00	283.00	327.00	199.00	352.41	275.00	295.00
Error (%)	1.95	1.00	3.31	1.28	3.05	1.38	2.88	1.44	3.07	1.16	4.46	1.87	2.65	1.75	2.71	1.98
α cor	0.76	0.81	0.63	0.74	0.65	0.69	0.64	0.64	0.45	0.54	0.46	0.50	0.42	0.59	0.57	0.60
Error (%)	1.77	1.00	2.78	1.42	3.26	1.51	2.72	1.43	2.25	0.79	3.84	1.26	1.70	1.33	2.16	1.56
Rcor $k\Omega \cdot cm^2$	22.46	30.15	27.51	27.98	27.91	33.66	26.06	54.26								
Error (%)	3.36	1.93	5.35	3.70	8.76	5.49	6.94	7.53								

4. Conclusions

In this study, the corrosion resistances of the AA2024-T3 and the AA2524-T3 alloys were compared. The differences between these alloys are the number and the precipitates' area fraction. The AA2024-T3 alloy has a smaller number of IMCs, which, however, are larger than those found in the AA2524-T3 alloy. Therefore, the IMCs area fraction of the former is higher than the latter. The SEM results suggest that the rapid removal of active sites from the AA2524-T3 surface (smaller IMCs) promotes a cleaner surface.

The hypothesis that defective areas prone to electrochemical reactions were removed from AA2524-T3 was confirmed by the EEC fitting to EIS results, as AA2524 R_{ox} increases and CPE_{ox} decreases with immersion time. Additionally, the corrosion process of AA2524, represented by CPE_{cor}, "α_{cor}" and R_{cor}, are significant only up to nine hours of exposure and confirm the higher contribution of active sites of the surface, caused by partially soluble corrosion products.

The electrochemical tests also showed that the localized corrosion resistance of the AA2524-T3 was higher than the AA2024-T3 during the whole test period. This was indicated by the higher impedance values associated with the first alloy and confirmed by surface observation after various immersion periods in the tested electrolyte. On the other hand, SEM analysis showed that the corrosion attack on the AA2524-T3 alloy surface was faster due to its larger number of small particles.

Author Contributions: Conceptualization, I.C. and H.G.d.M.; methodology, F.M.Q. and M.T.; validation, I.C. and H.G.d.M.; formal analysis, F.M.Q. and M.T.; investigation, F.M.Q., M.T. and A.F.S.B.; writing—original draft preparation, M.T. and A.F.S.B.; writing—review and editing, M.T. and A.F.S.B.; visualization, I.C. and H.G.d.M.; supervision, I.C.; project administration, I.C.; funding acquisition, I.C. All authors have read and agreed to the published version of the manuscript.

Funding: This research was funded by CNPq, grants number 400895/2014-5, 400781/2013-1 and 158764/2014-5 and CAPES, grant number 1536157.

Data Availability Statement: Not applicable.

Acknowledgments: The authors acknowledge CNPq and CAPES.

Conflicts of Interest: The authors declare no conflict of interest.

References

1. Chen, Y.Q.; Pan, S.P.; Zhou, M.Z.; Yi, D.Q.; Xu, D.Z.; Xu, Y.F. Effects of inclusions, grain boundaries and grain orientations on the fatigue crack initiation and propagation behavior of 2524-T3 Al alloy. *Mater. Sci. Eng. A* **2013**, *580*, 150–158. [CrossRef]
2. Golden, P. A comparison of fatigue crack formation at holes in 2024-T3 and 2524-T3 aluminum alloy specimens. *Int. J. Fatigue* **1999**, *21*, 211–219. [CrossRef]
3. Bonzom, R.; Oltra, R. Droplet cell investigation of intergranular corrosion on AA2024. *Electrochem. Commun.* **2017**, *81*, 84–87. [CrossRef]
4. Moreto, J.A.; Broday, E.E.; Rossino, L.S.; Fernandes, J.C.S.; Bose Filho, W.W. Effect of localized corrosion on fatigue–crack growth in 2524-T3 and 2198-T851 aluminum alloys used as aircraft materials. *J. Mater. Eng. Perform.* **2018**, *27*, 1917–1926. [CrossRef]
5. Queiroz, F.M.; Donatus, U.; Prada Ramirez, O.M.; de Sousa Araujo, J.V.; Gonçalves de Viveiros, B.V.; Lamaka, S.; Zheludkevich, M.; Masoumi, M.; Vivier, V.; Costa, I.; et al. Effect of unequal levels of deformation and fragmentation on the electrochemical response of friction stir welded AA2024-T3 alloy. *Electrochim. Acta* **2019**, *313*, 271–281. [CrossRef]
6. Cabrini, M.; Bocchi, S.; D'Urso, G.; Giardini, C.; Lorenzi, S.; Testa, C.; Pastore, T. Stress corrosion cracking of friction stir-welded AA-2024 T3 alloy. *Materials* **2020**, *13*, 2610. [CrossRef] [PubMed]
7. Nisancioglu, K. Corrosion of aluminium alloys. In Proceedings of the 3rd International Conference on Aluminium Alloys, Trondheim, Norway, 25–29 October 1992; p. 239.
8. Milagre, M.X.; Donatus, U.; Machado, C.S.C.; Araujo, J.V.S.; da Silva, R.M.P.; de Viveiros, B.V.G.; Astarita, A.; Costa, I. Comparison of the corrosion resistance of an Al–Cu alloy and an Al–Cu–Li alloy. *Corros. Eng. Sci. Technol.* **2019**, *54*, 402–412. [CrossRef]
9. Queiroz, F.M.; Magnani, M.; Costa, I.; de Melo, H.G. Investigation of the corrosion behaviour of AA 2024-T3 in low concentrated chloride media. *Corros. Sci.* **2008**, *50*, 2646–2657. [CrossRef]
10. Queiroz, F.M. Estudo Do Comportamento de Corrosão Dos Intermetálicos Presentes Na Liga Aa2024-T3, Por Meio De Técnicas De Microscopia Associadas A Técnicas Eletroquímicas. Ph.D. Thesis, Universidade de São Paulo, São Paulo, Brazil, 2008.
11. Murer, N.; Buchheit, R.G. Stochastic modeling of pitting corrosion in aluminum alloys. *Corros. Sci.* **2013**, *69*, 139–148. [CrossRef]

12. Birbilis, N.; Cavanaugh, M.K.; Buchheit, R.G. Electrochemical behavior and localized corrosion associated with Al7Cu2Fe particles in aluminum alloy 7075-T651. *Corros. Sci.* **2006**, *48*, 4202–4215. [CrossRef]
13. Bousquet, E. Durabilité des Assemblages Soudès par Friction Stir Welding (FSW). Ph.D. Thesis, L'Université de Bordeaux, Bordeaux, France, 2011.
14. Jariyaboon, M.; Davenport, A.J.; Ambat, R.; Connolly, B.J.; Williams, S.W.; Price, D.A. The effect of welding parameters on the corrosion behaviour of friction stir welded AA2024-T351. *Corros. Sci.* **2007**, *49*, 877–909. [CrossRef]
15. Paglia, C.S.; Buchheit, R.G. A look in the corrosion of aluminum alloy friction stir welds. *Scr. Mater.* **2008**, *58*, 383–387. [CrossRef]
16. Zhou, X.; Luo, C.; Hashimoto, T.; Hughes, A.E.; Thompson, G.E. Study of localized corrosion in AA2024 aluminium alloy using electron tomography. *Corros. Sci.* **2012**, *58*, 299–306. [CrossRef]
17. Buchheit, R.G.; Grant, R.P.; Hlava, P.F.; Mckenzie, B.; Zender, G.L. Local dissolution phenomena associated with s phase (al2cumg) particles in aluminum alloy 2024-T3. *J. Electrochem. Soc.* **1997**, *144*, 2621–2628. [CrossRef]
18. Li, J.F.; Ziqiao, Z.; Na, J.; Chengyu, T. Localized corrosion mechanism of 2×××-series Al alloy containing S(Al2CuMg) and θ′(Al2Cu) precipitates in 4.0% NaCl solution at pH 6.1. *Mater. Chem. Phys.* **2005**, *91*, 325–329. [CrossRef]
19. Moreto, J.; Gamboni, O.; Rocha, L. Corrosion behavior of Al and Al-Li alloys used as aircraft materials. *Corrosão Protecção Mater.* **2012**, *31*, 60–64.
20. Bugarin, A.F.S.; De Abreu, C..; Terada, M.; De Melo, H.G.; Costa, I. Effect of friction stir welding (FSW) on the electrochemical behavior and galvanic coupling of AA2024-T3 and AA7475-T651. *Mater. Today Commun.* **2020**, *25*, 101591. [CrossRef]
21. Buchheit, R.G. A compilation of corrosion potentials reported for intermetallic phases in aluminum alloys. *J. Electrochem. Soc.* **1995**, *142*, 3994–3996. [CrossRef]
22. Buchheit, R.; Boger, R.; Carroll, M.; Leard, R.; Paglia, C.; Searles, J. The electrochemistry of intermetallic particles and localized corrosion in Al alloys. *JOM* **2001**, *53*, 29–33. [CrossRef]
23. Blanc, C.; Lavelle, B.; Mankowski, G. The role of precipitates enriched with copper on the susceptibility to pitting corrosion of the 2024 aluminium alloy. *Corros. Sci.* **1997**, *39*, 495–510. [CrossRef]
24. Blanc, C.; Gastaud, S.; Mankowski, G. Mechanistic studies of the corrosion of 2024 aluminum alloy in nitrate solutions. *J. Electrochem. Soc.* **2003**, *150*, B396. [CrossRef]
25. Campestrini, P.; Terryn, H.; Hovestad, A.; de Wit, J.H.W. Formation of a cerium-based conversion coating on AA2024: Relationship with the microstructure. *Surf. Coat. Technol.* **2004**, *176*, 365–381. [CrossRef]
26. Leard, R.R.; Buchheit, R.G. Electrochemical characterization of copper-bearing intermetallic compounds and localized corrosion of Al-Cu-Mg-Mn alloy 2024. *Mater. Sci. Forum* **2002**, *396–402*, 1491–1496. [CrossRef]
27. Obispo, H.M.; Murr, L.E.; Arrowood, R.M.; Trillo, E.A. Copper deposition during the corrosion of aluminum alloy 2024 in sodium chloride solutions. *J. Mater. Sci.* **2000**, *35*, 3479–3495. [CrossRef]
28. Zhu, D.; van Ooij, W.J. Corrosion protection of AA 2024-T3 by bis-[3-(triethoxysilyl) propyl]tetrasulfide in sodium chloride solution. Part 2: Mechanism for corrosion protection. *Corros. Sci.* **2003**, *45*, 2177–2197. [CrossRef]
29. Schneider, O.; Ilevbare, G.O.; Scully, J.R.; Kelly, R.G. In situ confocal laser scanning microscopy of aa 2024-t3 corrosion metrology. *J. Electrochem. Soc.* **2004**, *151*, B465. [CrossRef]
30. Leblanc, P.; Frankel, G.S. A study of corrosion and pitting initiation of AA2024-T3 using atomic force microscopy. *J. Electrochem. Soc.* **2002**, *149*, B239. [CrossRef]
31. Schmutz, P.; Frankel, G.S. Characterization of AA2024-T3 by scanning kelvin probe force microscopy. *J. Electrochem. Soc.* **1998**, *145*, 2285–2295. [CrossRef]
32. Schmutz, P.; Frankel, G.S. Corrosion study of AA2024-T3 by scanning kelvin probe force microscopy and in situ atomic force microscopy scratching. *J. Electrochem. Soc.* **1998**, *145*, 2295–2306. [CrossRef]
33. Li, Z.; Chen, L.J. Feature recognition of corrosion pit for pre-corroded AA 2524 and statistical analysis. In *Advanced Materials Research*; Trans Tech Publications Ltd.: Stafa-Zurich, Switzerland, 2014; Volume 906, pp. 263–267.
34. Costenaro, H.; Lanzutti, A.; Paint, Y.; Fedrizzi, L.; Terada, M.; de Melo, H.G.; Olivier, M.-G. Corrosion resistance of 2524 Al alloy anodized in tartaric-sulphuric acid at different voltages and protected with a TEOS-GPTMS hybrid sol-gel coating. *Surf. Coat. Technol.* **2017**, *324*. [CrossRef]
35. Guadagnin, H.C. Corrosion Resistance Study of AA2524 Anodized in Sulphuric-Tartaric Acid and Sealed with Hybrid Coatings. Ph.D. Thesis, Universidade de São Paulo, State of São Paulo, Brazil, 2017.
36. Liu, C.; Liu, Y.; Li, S.; Ma, L.; Zhao, X.; Wang, Q. Effect of creep aging forming on the fatigue crack growth of an AA2524 alloy. *Mater. Sci. Eng. A* **2018**, *725*, 375–381. [CrossRef]
37. Terada, M.; Queiroz, F.M.; Costenaro, H.; Olivier, M.; Costa, I.; Melo, H.G. The Effect of Cerium (III) on the Corrosion Protection Properties of the Film Formed on the Aa2524-T3 Alloy by Hydrothermal Treatments. In Proceedings of the European Corrosion Congress, Montpellier, France, 11–15 September 2016.
38. Yang, B.; Yan, J.; Sutton, M.A.; Reynolds, A.P. Banded microstructure in AA2024-T351 and AA2524-T351 aluminum friction stir welds. *Mater. Sci. Eng. A* **2004**, *364*, 55–65. [CrossRef]
39. The Aluminium Association. *International Alloy. Designations and Chemical Composition Limits for Wrought Aluminum and Wrought Aluminum Alloys With Support for On-Line Access From:s Aluminum Extruders Council Use of the Information*; The Aluminium Association: Arlington County, VA, USA, 2015.

40. ASM Handbook Committee. *ASM Handbook Volume 2 Properties and Selection: Nonferrous Alloys and Special-Purpose Materials*; ASM International: Cleveland, OH, USA, 1993.
41. Fontana, Á. Utilização de Polianilina Como Revestimento Protetor Contra Corrosão Das Ligas de Alumínio 2014 F, 2024 T3 e 7075 O. Master's Degree, Universidade de São Paulo, São Paulo, Brazil, 2008.
42. Moreto, J.A.; Gambonf, O.; Rucherf, C.O.F.T.; Romagnoli, F.; Moreira, M.F.; Beneduce, F.; Filho, W.W.B. Corrosion and fatigue behavior of new Al alloys. *Procedia Eng.* **2011**, *10*, 1521–1526. [CrossRef]
43. Moreto, J.A. Estudo da Corrosão e Corrosão-Fadiga em Ligas de Al e Al–Li de Alta Resistência para Aplicação Aeronáutica. Ph.D. Thesis, Universidade de São Paulo, State of São Paulo, Brazil, 2012.
44. Moreto, J.A.; dos Santos, M.S.; Ferreira, M.O.A.; Carvalho, G.S.; Gelamo, R.V.; Aoki, I.V.; Taryba, M.; Bose Filho, W.W.; Fernandes, J.C.S. Corrosion and corrosion-fatigue synergism on the base metal and nugget zone of the 2524-T3 Al alloy joined by FSW process. *Corros. Sci.* **2021**, *182*, 109253. [CrossRef]
45. Moreto, J.A.; Rossino, L.S.; Filho, W.W.B.; Marino, C.E.B.; Da Conceição Ferreira, M.; Taryba, M.; Fernandes, J.C.S. On the global and localised corrosion behaviour of the AA2524-T3 aluminium alloy used as aircraft fuselage skin. *Mater. Res.* **2019**, *22*. [CrossRef]
46. Queiroz, F.M.; Bugarin, A.F.S.; Hammel, N.P.; Capelossi, V.R.; Terada, M.; Costa, I. EIS behavior of anodized and primer coated AA2198–T851 compared to AA2024–T3 exposed to salt spray CASS test. *Surf. Interface Anal.* **2016**, *48*. [CrossRef]
47. Campestrini, P.; Van Westing, E.P.M.; De Wit, J.H.W. Influence of surface preparation on performance of chromate conversion coatings on Alclad 2024 aluminium alloy–Part II: EIS investigation. *Electrochim. Acta* **2001**, *46*, 2631–2647. [CrossRef]
48. Capelossi, V.R.; Poelman, M.; Recloux, I.; Hernandez, R.P.B.; de Melo, H.G.; Olivier, M.G. Corrosion protection of clad 2024 aluminum alloy anodized in tartaric-sulfuric acid bath and protected with hybrid sol–gel coating. *Electrochim. Acta* **2014**, *124*, 69–79. [CrossRef]

metals

MDPI

Article

Effect of Molybdenum Content on the Corrosion and Microstructure of Low-Ni, Co-Free Maraging Steels

Asiful H. Seikh [1], Hossam Halfa [2] and Mahmoud S. Soliman [3,*]

1 Center of Excellence for Research in Engineering Materials, Deanship of Scientific Research, King Saud University, P.O. Box 800, Riyadh 11421, Saudi Arabia; aseikh@ksu.edu.sa

2 Steel Technology Department, Central Metallurgical R&D Institute (CMRDI), Helawn 11421, Egypt; hosamhalfa@gmail.com

3 Department of Mechanical Engineering, College of Engineering, King Saud University, P.O. Box 800, Riyadh 11421, Saudi Arabia

* Correspondence: solimanm@ksu.edu.sa; Tel.: +966-591959335

Abstract: Molybdenum (Mo) is an important alloying element in maraging steels. In this study, we altered the Mo concentration during the production of four cobalt-free maraging steels using an electroslag refining process. The microstructure of the four forged maraging steels was evaluated to examine phase contents by optical microscopy, scanning electron microscopy (SEM), and X-ray diffraction (XRD) analysis. Additionally, we assessed the corrosion resistance of the newly developed alloys in 3.5% NaCl solution and 1 M H_2SO4 solution through potentiodynamic polarization and electrochemical impedance spectroscopy (EIS) techniques. Furthermore, we performed SEM and energy-dispersive spectroscopy (EDS) analysis after corrosion to assess changes in microstructure and Raman spectroscopy to identify the presence of phases on the electrode surface. The microstructural analysis shows that the formation of retained austenite increases with increasing Mo concentrations. It is found from corrosion study that increasing Mo concentration up to 4.6% increased the corrosion resistance of the steel. However, further increase in Mo concentration reduces the corrosion resistance.

Keywords: molybdenum; ESR; potentiodynamic polarization; EIS; Raman spectroscopy

Citation: Seikh, A.H.; Halfa, H.; Soliman, M.S. Effect of Molybdenum Content on the Corrosion and Microstructure of Low-Ni, Co-Free Maraging Steels. *Metals* **2021**, *11*, 852. https://doi.org/10.3390/met11060852

Academic Editors: Santiago Fajardo and Renato Altobelli Antunes

Received: 11 April 2021
Accepted: 16 May 2021
Published: 21 May 2021

1. Introduction

Maraging steels are developed to combine superior high strength (yield strength commercially ranging from 1030 to 3450 MPa) and fracture toughness [1]. The small amount of carbon makes maraging steels different from regular steels; that is, they are reinforced by intermetallic compounds derived from supersaturated martensite during the age hardening of the malleable low-carbon iron (Fe)–nickel (Ni) lath martensitic matrix [2].

Recently, super-high-strength maraging steels have been broadly utilized in applications including the manufacture of airplanes, missiles, rocket motor cases, and gas turbines [3–6]. Maraging steels are normally low-carbon, Fe–Ni-based alloys with substantial amounts of cobalt (Co) and molybdenum (Mo) along with a small percentage of titanium (Ti) and aluminum (Al). However, the material can be altered by changing its composition according to the application [7,8]. Due to the low carbon content, maraging steels mostly exhibit a high degree of machinability [9]. Manufacture of such steels has been performed by adding or completely removing certain alloying elements to create maraging steel with good mechanical and corrosion-resistance properties.

Previous studies report that the durability of Fe–18% Ni matrix alloyed by Ti significantly decreases at higher strength levels due to the formation of low-temperature, Ti-rich, metastable NiTi (B_2) or Ni_3Ti during the aging process [10–12]. Similarly, higher Ti levels can decrease durability via grain-boundary precipitation of the TiC or TiCN, except in cases where the carbon content is kept low and thermomechanical processing is profoundly controlled. Therefore, this suggests that recently created maraging steels likely have low Ti content. Mo is another alloying component that is valuable for both strength and durability. Previous studies showed that Mo-rich zones precipitate during the aging of maraging steel and play an exceptionally critical role in maintaining the strength of steel [13–16]. The role of Mo in these processes has been evaluated in ranges of 3–5% based on its inhibition of embrittlement due to intergranular isolation of contaminants in grain boundaries and as a potential precipitation hardener [17].

The corrosion and corrosion resistance of maraging steels in sulfuric acid solution have been examined previously [18–21], revealing their elevated corrosion resistance in acidic solution relative to tempered martensite alloy steels. In the present study, we incorporated varied contents of Mo into four maraging steels using an electroslag refining (ESR) process in order to evaluate its effect on the corrosion-resistance and microstructure properties of the resulting steels. These four new low-Ni, Co-free maraging steels containing 0, 2.9, 4.6, and 9.8 wt% Mo were investigated to determine their corrosion behavior in 1 M H_2SO_4 and 3.5% NaCl solution using potentiodynamic polarization and electrochemical impedance spectroscopy (EIS) techniques.

2. Materials and Methods

2.1. Materials

In this approach, investigated steel specimens were prepared through open-air melting followed by remelting in electroslag furnace. The consumable electrodes for ESR were made in a 20 kg air induction furnace with estimated amounts of scrap and ferroalloys. The pouring temperature was about 1600 °C, and the metal was cast into 40 mm diameter and 600 mm high vertical chill molds. These were remelted through ESR process with and without inoculation of 0.05–0.07% Ti. The process was carried out in a water-cooled steel mold of 80 mm diameter connected to the negative end of a DC power source. The applied current and voltage were about 730 A and 25 ± 2 V, respectively, with mold water flow rate of 30 L/min and base-plate water flow of 20 L/min. After ESR, cooled ingots were taken out from the mold and homogenized in a muffle furnace at 975 °C for 8–9 h. To eliminate the probability of the presence of any cavities or pipes or any other internal defects, about 20 mm and 10 mm lengths were cut from the bottom and top of the ingots, respectively. Each ESR ingot was 20 kg in weight. The ESR ingots underwent forging. The chemical compositions of the four produced steels are given in Table 1. The chemical compositions are similar except for Mo content, which was changed from 0 to 10%. The steels are identified as M0, M2.9, M4.6, and M9.8 with 0.0, 2.9%, 4.6%, and 9.8% Mo, respectively (Table 1).

Table 1. Chemical composition (wt%) of produced steels.

Composition (%) / Sample	C	Mn	Si	S	P	Ni	Cr	Mo	Ti	Al	Fe
M0	0.026	0.10	0.082	0.009	0.012	10.9	4.6	0.0	1.19	0.082	Bal.
M2.9	0.029	0.09	0.08	0.011	0.014	11.05	4.8	2.9	1.22	0.078	Bal.
M4.6	0.027	0.11	0.078	0.012	0.011	11.1	4.7	4.6	1.21	0.085	Bal.
M9.8	0.025	0.12	0.084	0.015	0.009	10.8	4.75	9.8	1.24	0.081	Bal.

2.2. Optical Microscopic Observation

Initially, specimens were cut into pieces ($1 \times 1 \times 0.5$ cm^3) from the sheets, and two samples of each grade were used for the tests. The specimens were properly polished with different grades of emery paper (120, 180, 500, and 1000) and cloth polishers. After mechanical polishing, the samples were etched in 10% ammonium persulfate and 10% nital solutions. Microstructural analysis was then performed using a Leica optical microscope (Leica, Wetzlar, Germany).

2.3. X-ray Diffraction (XRD)

The crystalline structure and phases of the forged steels were characterized by XRD using a Rigaku Ultima III X-ray Diffractometer (Rigaku, Tokyo, Japan) to record diffraction traces from monochromatic Cu Kα radiation applied at room temperature and at a scan rate of 2°/min. The crystallographic planes of X-ray diffraction were obtained from the inbuilt software of the X-ray machine.

2.4. Potentiodynamic Polarization

Potentiodynamic polarization was performed at a scan rate of 1 mV/s using a potentiostat (PC/750; Gamry Instruments, Warminster, PA, USA). Samples were degreased with acetone and ethanol solution after polishing, and a standard three-electrode system was used to measure the i_{corr} and E_{corr} of the samples. The reference electrode was a saturated calomel electrode, with graphite used as the counter electrode (the sample was used as the working electrode). NaCl solution (3.5 wt%) and 1 M H_2SO_4 solution were used as the electrolytes at room temperature. The corrosive mediums of 3.5% NaCl and 1.0 M sulfuric acid solution were prepared from analytical-reagent-grade sodium chloride, sulfuric acid, and double-distilled water. Both the NaCl solution and H_2SO_4 solution were kept in open-air condition. All the experiments were carried out at room temperature in the open atmosphere, under unstirred conditions. The surface area of samples exposed to the solution differed from sample to sample; however, during plotting, this was standardized according to the same sample area. Tables 2 and 3 present the i_{corr} and E_{corr} values obtained by the Tafel extrapolation.

Table 2. E_{corr} and i_{corr} values of maraging steel in 3.5% NaCl solution.

Sample	i_{corr} (A/cm^2)	E_{corr} (VSCE)
M0	60	−0.553
M2.9	4	−0.356
M4.6	2	−0.400
M9.8	202	−0.329

Table 3. E_{corr} and i_{corr} values of maraging steel in 1 M H_2SO_4 solution.

Sample	i_{corr} (A/cm^2)	E_{corr} (VSCE)
M0	100	−0.615
M2.9	200	−0.518
M4.6	51	−0.318
M9.8	602	−0.416

2.5. Electrochemical Impedance Spectroscopy (EIS)

EIS was performed using the potentiostat (Gamry Instruments) in a three-electrode system and at an amplitude range of 10 mV rms and a frequency interval of 0.1–30 kHz. These experiments were also done in the same solutions, i.e., 3.5% NaCl solution and 1.0 M H_2SO_4 solution. Here also all experiments were done at room temperature where solutions were kept in open air and unstirred condition. At each frequency, the absolute impedance and a Nyquist plot were obtained. The obtained impedance data were fit to an appropriate equivalent electrical circuit, with the fitting performed using a simplex fit mode and using both the real and imaginary components of the data. The impedance data were interpreted based on an equivalent electric circuit, and Nyquist plots were obtained at the open-circuit potential for all samples at different solutions.

2.6. Raman Spectroscopy

Raman estimations were performed using a RAMAN HR800 confocal micro-Raman spectrometer (Horiba Scientific, North Edison, NJ, USA) with backscattered calculation through a 10× (NA = 0.25) magnifying lens objective. An argon laser transmitting at a frequency of 514.5 nm was used as the excitation source, and the quantity of grating in the Raman spectrometer was 1800 grooves/mm. The Raman band of a silicon wafer at 520 cm^{-1} was utilized to align the spectrometer. Raman spectra were recorded in the broad range of 200–2000 cm^{-1}. Corrosion products formed on the specimens were characterized with the help of Raman spectrometer using an excitation laser of 514.5 nm wavelength coupled with an Olympus optical microscope.

2.7. Scanning Electron Microscopy (SEM) and Energy-Dispersive Spectroscopy (EDS)

In this present work, scanning electron microscopy (SEM) was used to study the surface morphology of the maraging steel with different wt% of Mo, and EDS was used to determine the composition of the corrosion products on the surface of maraging steel. All the studies were carried out with Model JEOL JSM-6360. In this present study, 200× magnification was used. All the SEM images before and after corrosion are presented in results section.

3. Results and Discussion

3.1. Optical Microscopic Observation

Figure 1 shows the optical micrographs of the maraging steel produced by ESR and with different Mo contents. The microstructure displayed a lamellar morphology with packets of martensite within prior austenite grains. As the Mo concentration increased, the martensite packets became visible both due to the special etching along their boundaries and the martensite packets inside an austenite grain that failed to extend beyond the austenite grain boundary. The microstructure of M9.8 differed from the other samples according to the existence of interlath austenite, which was not fully resolved. However, we speculated that this interlath structure affected the corrosion-resistance properties.

These findings suggested that increasing the concentration of the alloying element (Mo) resulted in an increased tendency to form retained austenite. Compared with conventional casting methods, ESR enables lower local solidification time, allowing the formation of a very fine and well-distributed microstructure not present in other production methods used to produce maraging steel.

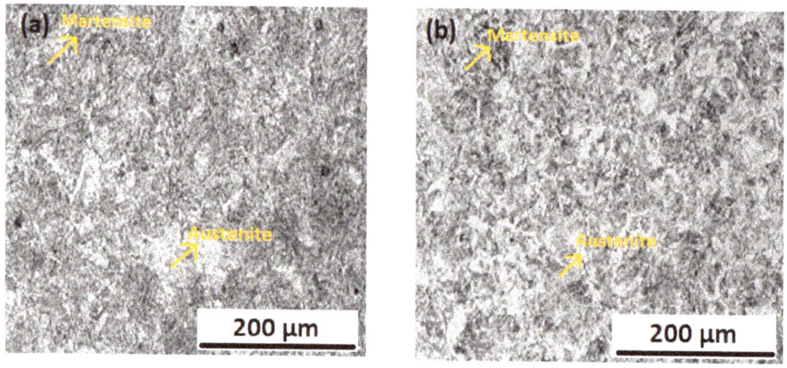

Figure 1. *Cont.*

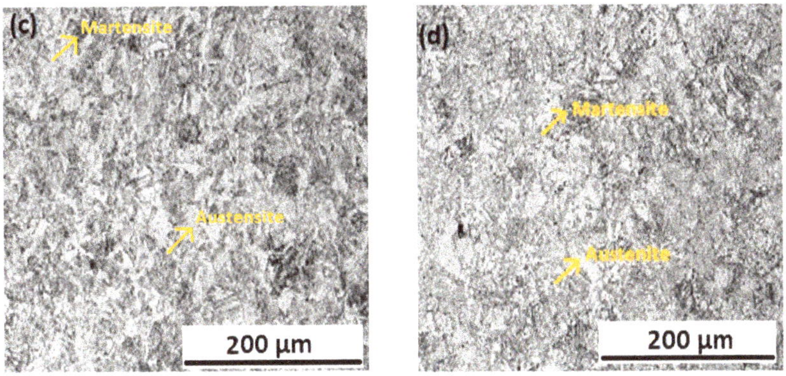

Figure 1. Microstructure analyses of (**a**) M0, (**b**) M2.9, (**c**) M4.6, and (**d**) M9.8.

3.2. XRD Analysis

Figure 2 shows the XRD patterns for the as-received and differently heat-treated maraging-steel samples produced using ESR. The diffraction peaks α(110), α(200), and α(211) of the martensitic phase corresponded to diffraction angles of 44.5°, 65°, and 82.2°, respectively. Additionally, there was a weak diffraction peak (ϒ(111)) for the austenite phase at a diffraction angle of 43.5°. The samples mainly comprised a martensitic phase and a small amount of retained austenite phase, which arose from the microsegregation of solute elements (particularly Ni) at cellular boundaries during solidification. The enrichment of Ni stabilized the retained austenite, thereby allowing the detection of the austenite phase. Moreover, the diffraction peak intensity for the austenite phase in the sample became higher as Mo concentration increased. In the M9.8 sample, we observed reversion of the martensitic phase into the austenite phase. The austenite in maraging steel comprises retained austenite, and the reverted austenite forms mostly during the aging process by a diffusion-controlled reaction for overaging conditions.

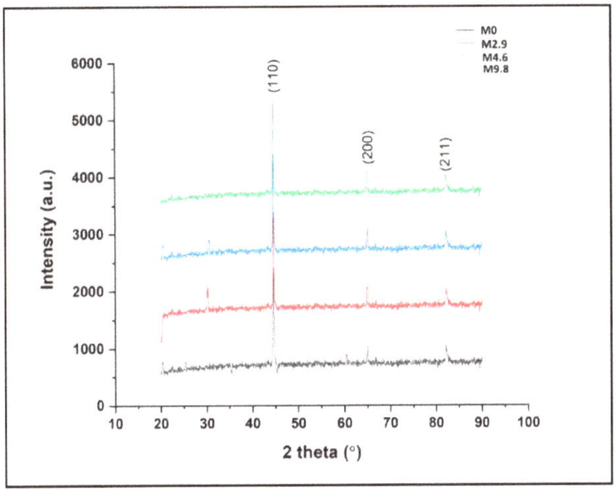

Figure 2. XRD analysis of maraging steel with different molybdenum concentrations.

It is clear from Figure 2 that the amount of retained austenite in ESR remelted steels depends mainly on the chemical composition of investigated steels. An increase in the amount of alloying element, i.e., Mo, is accompanied by an increase in the tendency to form retained austenite.

3.3. Potentiodynamic Polarization Analysis

We then assessed the effect of Mo on the corrosion-resistance properties of the developed steel. We found that this component improved the passive behavior of the protective films. Previous studies report that Mo improves chromium enrichment in the film without being incorporated [22,23]. Additionally, studies suggest that Mo in the compound breaks down into Mo particles, speeding the repassivation rate (anodic inhibitor) [24–26]. Furthermore, Mo expands the stability of the inner layers of oxide film [26] and diminishes the unfavorable action of sulfides on pitting resistance [27]. Figure 3 shows the potentiodynamic polarization curve of maraging steel in 3.5% NaCl solution.

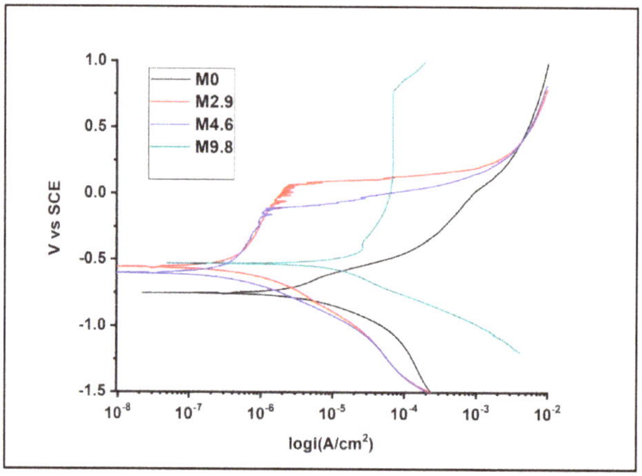

Figure 3. Potentiodynamic polarization curve of maraging steel in 3.5% NaCl solution.

In 3.5% NaCl solution, as seen in Table 2, the i_{corr} value of the maraging steel gradually increased along with increases in the Mo content in the alloy, with similar results observed in 1 M H_2SO_4 solution. In both cases, with 9.8% Mo content, the M9.8 sample showed an increase in the i_{corr} value. By studying the X-ray patterns, it can be seen that retained austenite of experiment steels increases with the increase in Mo content. On the other hand, the microstructure in Figure 1 shows that even at high magnification, retained austenite is not recognized in the studied steels except in specimen M9.8. In samples with lower Mo contents, austenite presents as lath austenite; on the other hand, the specimen containing 9.8 wt% Mo retained austenite aggregate in separate grains. So, the high corrosion of sample M9.8 may be attributed to the large amount of retained austenite and its morphology. Previous studies reported that Mo alone does not directly affect corrosion resistance but rather indirectly influence this property [28–32].

Figure 4 and Table 3 present the results of the potentiodynamic polarization study of maraging steel in 1 M H_2SO_4 solution. Interestingly, the rate of corrosion for M9.8 was worse in 1 M H_2SO_4 solution as compared with 3.5% NaCl solution, possibly due to the NaCl solution allowing Mo to produce a more protective layer.

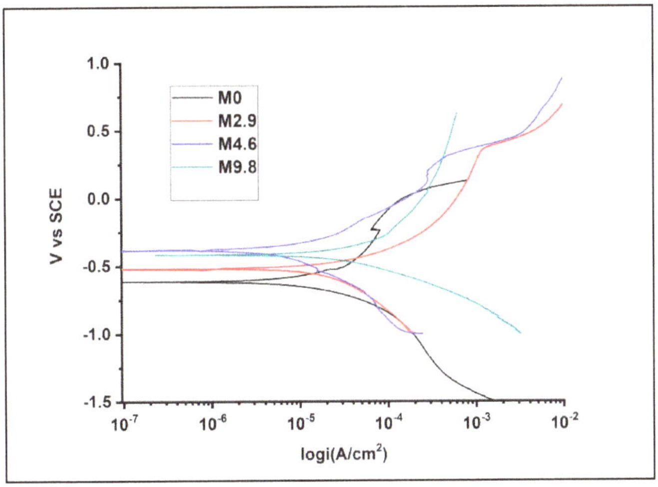

Figure 4. Potentiodynamic polarization curve of maraging steel in 1 M H_2SO_4 solution.

3.4. EIS Analysis

We then used EIS to examine the corrosion resistance of the ESR steel samples in 1 M H_2SO_4 solution to determine their mechanistic and kinetic parameters and compare them with the results of potentiodynamic polarization. The Nyquist plots (Figures 5 and 6) were examined by fitting the trial results to an equivalent circuit model (Figure 7). The circuit comprises a parallel combination of a consistent phase element (CPE, Q), the charge transfer resistance (R_{ct}) relating to the corrosion response at the metal/electrolyte interface, and the solution resistance (Rs) between the working and reference electrode [33,34]. To diminish the effects of surface anomalies in the metal, CPE (Q) was brought into the circuit instead of a pure double-layer capacitance (C_{dl}), allowing a more precise fit [35,36]. Among these parameters, R_{ct} is the factor that determines the corrosion resistance of composites. Because R_{ct} is conversely relative to i_{corr}, a higher R_{ct} value correlates to a lower i_{corr} value. The impedance of CPE can be presented as follows:

$$ZCPE = 1/Y0(j\omega)n, \quad (1)$$

where Y0 is the CPE constant, n is the exponent (phase shift), ω is the angular frequency, and j is the imaginary unit. CPE can represent resistance, capacitance, and inductance according to the estimations of n [37]. In all analyses, the estimated n ranged from 0.8 to 1.0, suggesting a capacitive response of CPE. From this circuit, we determined R_s, R_{ct}, and C_{dl} (Tables 4 and 5).

The Nyquist plots indicated that the impedance response of the samples gradually increased along with increases in Mo concentration. However, the smallest capacitive loop in a high-frequency region was observed in the M9.8 sample in both 3.5% NaCl and 1 M H_2SO_4 solutions, indicating that this sample showed the highest corrosion rate. Tables 4 and 5 present all EIS parameters for all of the maraging steels. We found that the R_{ct} value decreased gradually for samples up to M4.6, with the highest value observed for M9.8, confirming the lowest corrosion-resistance property in M9.8 sample as Rct is inversely proportional to the i_{corr} value as mentioned earlier. Furthermore, this was supported by the potentiodynamic polarization results for M9.8. Comparing the diameters of the Nyquist plots, the curves in 3.5% NaCl solution showed increased diameters relative to those in 1 M H_2SO_4 solution, verifying the potentiodynamic results.

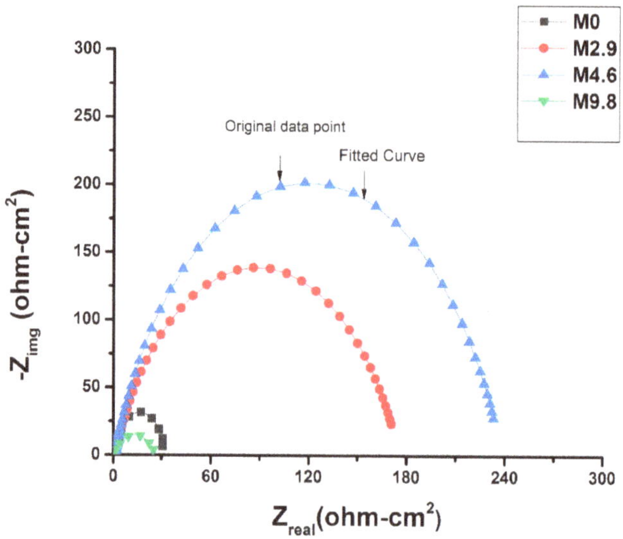

Figure 5. Nyquist plot curves of maraging steel with different molybdenum (Mo) concentrations in 3.5% NaCl solution.

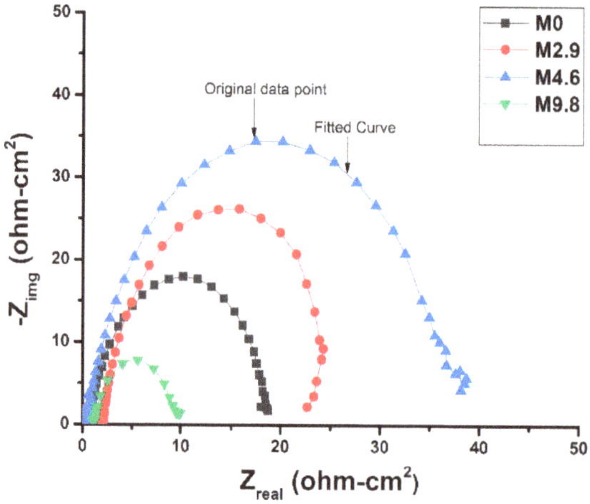

Figure 6. Nyquist plot curves of maraging steel with different Mo concentrations in 1 M H_2SO_4 solution.

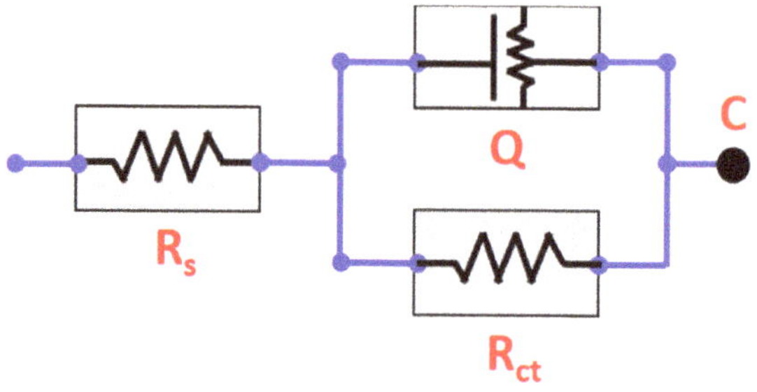

Figure 7. Equivalent circuit diagram.

Table 4. Electrochemical impedance parameters of maraging-steel samples in 3.5% NaCl solution.

Sample	R_{ct} (ohm-cm^2)	n	R_s (ohm-cm^2)	C_{dl} (mMho)
M0	12.8	0.85	1.3	3
M2.9	9.2	0.86	1.7	6
M4.6	4.5	0.84	1.5	29
M9.8	30	0.85	1.2	2

Table 5. Electrochemical impedance parameters of maraging-steel samples in 1 M H_2SO_4 solution.

Sample	R_{ct} (ohm-cm^2)	n	R_s (ohm-cm^2)	C_{dl} (mMho)
M0	14.6	0.77	1.4	3
M2.9	8.7	0.86	1.2	3
M4.6	5.4	0.83	1.45	5
M9.8	32.2	0.76	1	1

3.5. Scanning Electron Microscopy (SEM) Analysis

Figures 8 and 9 show SEM images of the samples after corrosion in 3.5% NaCl solution and 1 M H_2SO_4 solution, respectively. It is well known that Mo affects the pitting resistance of the maraging steel by reducing pitting on the sample surface. Chloride is primarily responsible for surface pitting on materials. However, in the present study, the presence of more salt resulted in decreased corrosion resistance in NaCl solution relative to H_2SO_4 solution. Moreover, Mo increased the stability of the inner layers of the oxide film, resulting in minimal effect on the E_{corr} value in both solutions. Furthermore, the surface of the M9.8 sample showed significant effects from both solutions relative to the other samples as a result of the decreased corrosion resistance associated with Mo.

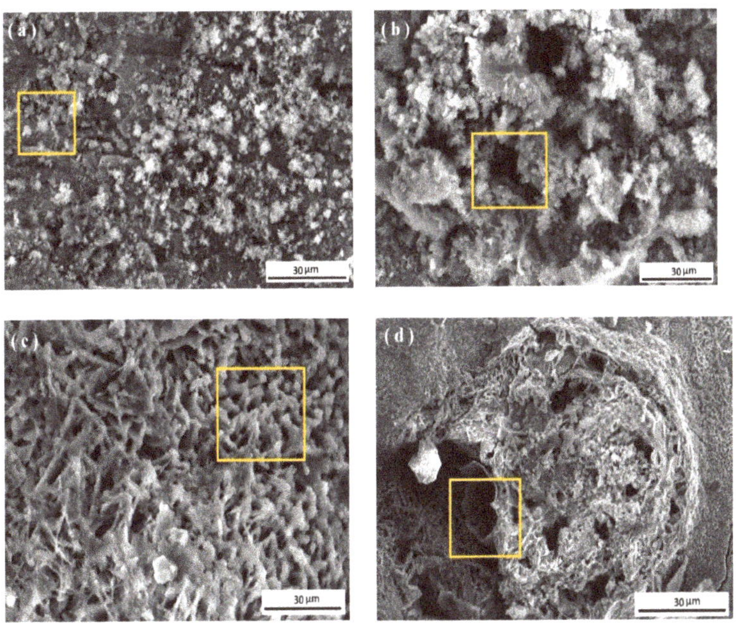

Figure 8. Scanning electron microscope (SEM) micrographs of maraging-steel samples: (**a**) M0, (**b**) M2.9, (**c**) M4.6, and (**d**) M9.8 in 3.5% NaCl solution.

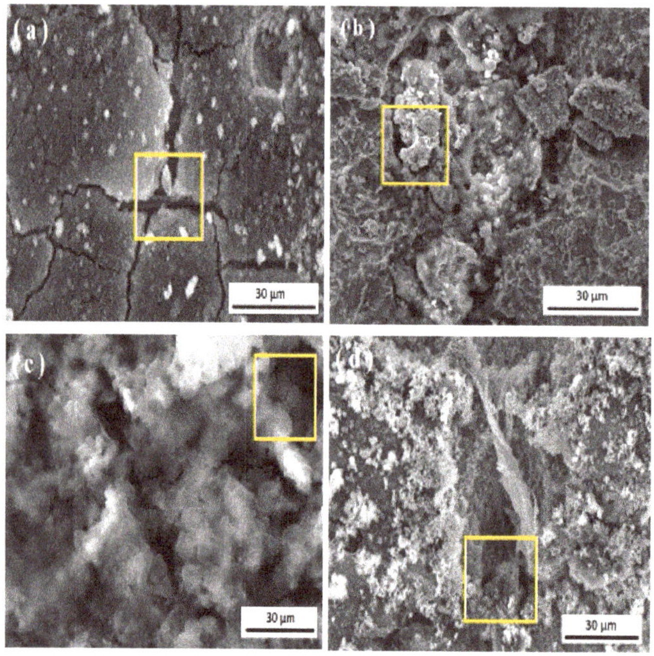

Figure 9. SEM micrographs of maraging-steel samples: (**a**) M0, (**b**) M2.9, (**c**) M4.6, and (**d**) M9.8 in 1 M H_2SO_4 solution.

3.6. Energy-Dispersive X-ray Spectroscopy (EDS) Analysis

Figure 10 shows the results of EDS analysis of the as-received and heat-treated maraging-steel samples in 3.5% NaCl solution. The EDS was completed on the yellow boxed portion as given in the SEM images in Figure 8. The EDS analysis data are given in Table 6 in wt%. The results showed that the Mo peak increased from samples M0 to M9.8, with peaks for Ti, C, Ni, and Fe also observed. The chloride peak was higher in the M9.8 sample relative to others, which is likely related to its decreased corrosion resistance and the associated increased number of chloride ions on the M9.8 surface.

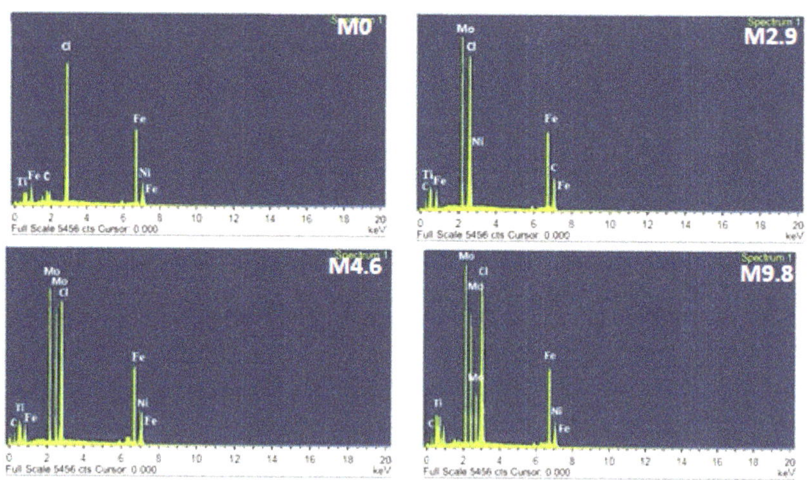

Figure 10. EDS spectra of the maraging steel after immersion in 3.5% NaCl solution.

Table 6. EDS analysis data in 3.5% NaCl solution.

Sample	Fe	Mo	Cl	Ni	Ti	C	Total
M0	55.8	0	10.9	15.3	8.8	9.2	100
M2.9	40.7	3.1	18.5	17.7	9.7	10.3	100
M4.6	38.9	4.8	25.3	13.8	10.8	6.4	100
M9.8	35.6	9.9	32.6	11.8	5.3	4.8	100

Figure 11 shows the results of EDS analysis of the as-received and heat-treated maraging-steel samples in 1 M H_2SO_4 solution. The EDS was completed on the yellow boxed portion as given in the SEM images in Figure 10. The EDS analysis data are given in Table 7 in wt%. Similarly, peaks for Ti, C, Fe, and Ni were observed following their precipitation as carbides. Additionally, we observed an increase in the intensity of the Mo peak from samples M0–M9.8, with peaks for sulfur and oxygen ions also observed as the surface reacted with the SO_4 ions present in the solution. Given the unfavorability of sulfur ions on pitting resistance, we found greater corrosive effects on the surface of the maraging-steel sample in H_2SO_4 solution.

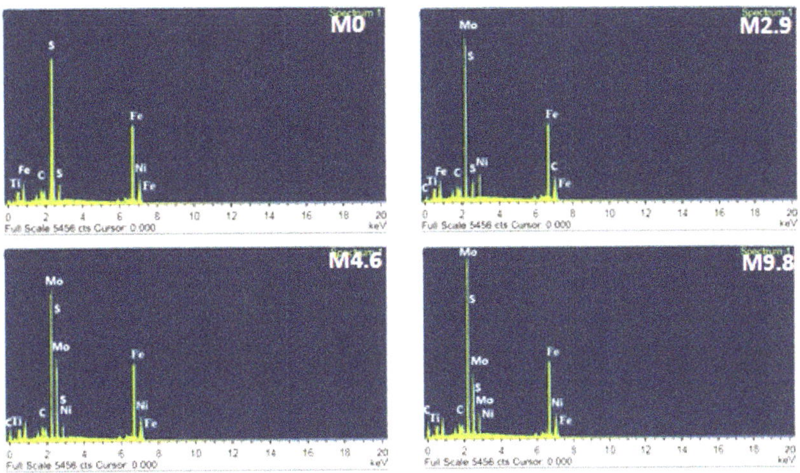

Figure 11. EDS spectra of the maraging steel after immersion in 1.0 M H_2SO_4 solution.

Table 7. EDS analysis data in H_2SO_4 solution.

Sample	Fe	Mo	S	Ni	Ti	C	Total
M0	57.5	0	20.2	10.3	5.8	6.2	100
M2.9	42.3	2.8	25.5	11.8	9.4	8.2	100
M4.6	36.9	4.5	30.9	12.6	7.2	7.9	100
M9.8	35.2	9.8	35.6	9.4	4.2	5.8	100

3.7. Raman Spectroscopy Analysis

Raman spectroscopy analyses of the corroded surfaces of the samples following potentiodynamic polarization are shown in Figures 12 and 13, for both solutions. Standard normal variate (SNV) model is an effective procedure to make the output data more comparable. In this method, the spectrum mean subtraction and standard spectrum deviation procedure is used. As long as the original scale of the spectra is not interesting, this is an efficient way of removing constant baseline effects and scaling differences from spectra. Previous studies reported that increases in the proportion of α-FeOOH and gamma * (total mass of γ-FeOOH, β-FeOOH, and magnetite) in the corrosion products suggest a decline in the corrosion rate [38,39]. In the present study, we identified α-FeOOH (280 cm^{-1}) as a significant phase of the corrosion product present on a superficial level, and it was more prominent in the M9.8 sample relative to the others. This suggested that a higher level of α-FeOOH in the M9.8 sample promoted an increase in the corrosion rate. Additionally, higher amounts of α-FeOOH and gamma * (Figure 13) were observed in 1 M H_2SO_4 solution, indicating that the maraging steel showed a higher degree of corrosion in the presence of H_2SO_4 solution as compared to the NaCl solution. In the case of sulfate solution, a phase shift occurs at 310 cm^{-1} and a new peak is seen at 1000 cm^{-1}. More γ-FeOOH occurs in sulfate solution at 80, 1320, and 1410 cm^{-1}. For the same composition, only one peak at 1380 cm^{-1} occurs for the sample in chloride solution. The β-FeOOH phase occurs at 480 cm^{-1} only in the case of sulfate solution. No phase shift occurs for the δ-FeOOH phase, and it shows a peak at 400 cm^{-1}. Only an extra peak at 610 cm^{-1} occurs in sulfate solution for δ- FeOOH.

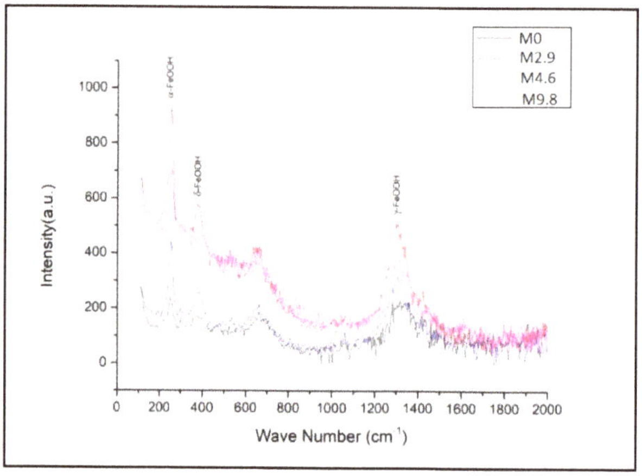

Figure 12. Raman spectra of maraging steel with different Mo concentrations in 3.5% NaCl solution.

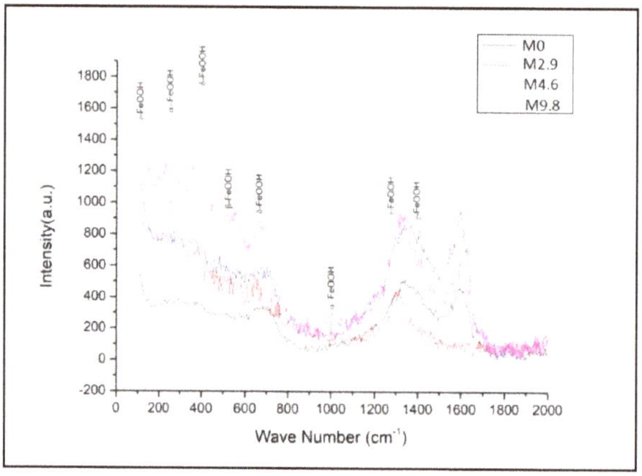

Figure 13. Raman spectra of maraging steel with different Mo concentrations in 1 M H_2SO_4 solution.

4. Conclusions

Mo plays an important role as an alloying element by increasing the strength and toughness of maraging steel. Moreover, the tendency to form retained austenite increases along with increasing Mo concentrations. In the present study, we found that in both 3.5% NaCl solution and 1 M H_2SO_4 solution, the generated maraging steel exhibited improvements in corrosion resistance according to increasing Mo content up to 4.6%. However, corrosion resistance decreased at 9.8% Mo due to increased embrittlement in the alloy. These results were confirmed by the Nyquist plots, showing the same trends in corrosion behavior. Furthermore, Raman spectroscopy verified decreases in the corrosion rate due to the presence of higher fractions of corrosion products (α-FeOOH) in the M9.8 (9.8% Mo) sample.

Author Contributions: Conceptualization, A.H.S., H.H., and M.S.S.; data curation, A.H.S., H.H., and M.S.S.; formal analysis, A.H.S., H.H., and M.S.S.; funding acquisition, M.S.S.; investigation, A.H.S. and M.S.S.; methodology, A.H.S., H.H., and M.S.S.; resources, H.H.; supervision, A.H.S., H.H., and M.S.S.; writing—original draft, A.H.S., H.H., and M.S.S.; writing—review and editing, A.H.S., H.H., and M.S.S. All authors have read and agreed to the published version of the manuscript.

Funding: This project was supported by the NSTIP Strategic Technologies Program, grant number (12-ADV2465-02), Kingdom of Saudi Arabia.

Data Availability Statement: The experimental datasets obtained from this research work and then the analysed results during the current study are available from the corresponding author on reasonable request.

Acknowledgments: This project was supported by the NSTIP Strategic Technologies Program, grant number (12-ADV2465-02), Kingdom of Saudi Arabia.

Conflicts of Interest: The authors declare no conflict of interest.

References

1. An, J.; Meng, F.; Lv, X.; Liu, H.; Gao, X.; Wang, Y.; Lu, Y. Improvement of Mechanical Properties of Stainless Maraging Steel Laser Weldments by Post-Weld Ageing Treatments. *Mater. Des.* **2012**, *40*, 276–284. [CrossRef]
2. Li, Y.; Yan, W.; Cotton, J.D.; Ryan, G.J.; Shen, Y.; Wang, W.; Shan, Y.; Yang, K. A New 1.9 GPaMaraging Stainless Steel Strengthened by Multiple Precipitating Species. *Mater. Des.* **2015**, *82*, 56–63. [CrossRef]
3. Rezek, J.; Klein, I.E.; Yahalom, J. Structure and Corrosion Resistance of Oxides Grown on Maraging Steel in Steam at Elevated Temperatures. *Appl. Surf. Sci.* **1997**, *108*, 59–165. [CrossRef]
4. Ohue, Y.; Matsumoto, K. Sliding-Rolling Contact Fatigue and Wear of Maraging Steel Roller with Ion-Nitriding and Fine Particle Shot-Peening. *Wear* **2007**, *263*, 782–789. [CrossRef]
5. Wang, W.; Yan, W.; Duan, Q.; Shan, Y.; Zhang, Z.; Yang, K. Study on Fatigue Property of a New 2.8 GPa Grade Maraging Steel. *Mater. Sci. Eng. A* **2010**, *527*, 3057–3063. [CrossRef]
6. Poornima, T.; Nayak, J.; Shetty, A.N. Corrosion of Aged and Annealed 18 Ni 250 Grade Maraging Steel in Phosphoric Acid Medium. *Int. J. Electrochem. Sci.* **2010**, *5*, 56–71.
7. Decker, R.F.; Floreen, S. *International Symposium on Maraging Steels Recent Development and Applications*; Pheonix, AZ, USA, 1998; Volume 1, pp. 1–88.
8. Stiller, K.; Danoix, F.; Bostel, A. Investigation of Precipitation in a New Maraging Stainless Steel. *Appl. Surf. Sci.* **1996**, *94–95*, 326–333. [CrossRef]
9. Klobčar, D.; Tušek, J.; Taljat, B.; Kosec, L.; Pleterski, M. Aging of Maraging Steel Welds During Aluminium Alloy Die Casting. *Comput. Mater. Sci.* **2008**, *44*, 515–522. [CrossRef]
10. Viswanathan, U.K.; Dey, G.K.; Asundi, M.K. Precipitation hardening in 350 grade maraging steel. *Metall. Trans. A* **1993**, *24*, 2429–2442. [CrossRef]
11. Vanderwalker, D.M. The precipitation sequence of Ni3Ti in Co-free maraging steel. *Metall. Trans. A* **1987**, *18*, 1191–1194. [CrossRef]
12. Sha, W.; Ye, A.; Malinov, S.; Wilson, E.A. Microstructure and Mechanical properties of low nickel maraging steel. *Mater. Sci. Eng. A* **2012**, *536*, 129–135. [CrossRef]
13. Peters, D.T.; Cupp, R.C. Kinetics of aging reactions in 18% Ni maraging steels. *Trans. AIME* **1966**, *236*, 1420.
14. Tokunaga, Y.; Takagi, S. Bull. *Jpn. Inst. Met.* **1982**, *21*, 234. [CrossRef]
15. Miner, R.E.; Jackson, J.K.; Gibbons, D.F. Internal Friction in 18 percent Ni Maraging steels. *Trans. TMS-AIME* **1966**, *236*, 1565–1570.
16. Asayama, Y.J. Aging Behavior and Role of Mo Rich Zone on the Notch Toughness in 18%Ni Maraging Steels. *Jpn. Inst. Metals* **1985**, *49*, 972–980. [CrossRef]
17. Hazarabedian, M.S.; Quadir, M.Z.; Iannuzzi, M. Characterization of Intergranular Phases in Precipitation Hardening Ni Alloy UNS N07725. *Mater. Charact.* **2020**, *171*, 110770. [CrossRef]
18. Poornima, T.; Jagannatha, N.; Shetty, A.N. Studies on Corrosion of Annealed and Aged 18 Ni 250 Grade Maraging Steel in Sulphuric Acid Medium. *Port. Electrochim. Acta* **2010**, *28*, 173–188. [CrossRef]

19. Sanatkumar, B.S.; Nayak, J.; Shetty, A.N. Corrosion Behavior of 18% Ni M250 Grade Maraging Steel under Weld-Aged Condition in Sulfuric Acid Medium. *Chem. Eng. Commun.* **2012**, *199*, 1610–1625. [CrossRef]
20. Singh, A.K.; Shukla, S.K.; Quraishi, M.A.; Ebenso, E.E. Investigation of Adsorption Characteristics of N,N′-[(Methylimino)Dimethylidyne]di-2,4-Xylidine as Corrosion Inhibitor at Mild Steel/Sulphuric Acid Interface. *J. Taiwan Inst. Chem. Eng.* **2012**, *43*, 463–472. [CrossRef]
21. Poornima, T.; JagannathaNayak, A. NityanandaShetty,3,4Dimethoxybenzaldehydethiosemicarbazone as Corrosion Inhibitor for Aged 18 Ni 250 Grade Maraging Steel in 0.5 M Sulfuric Acid. *J. Appl. Electrochem.* **2011**, *41*, 223–233. [CrossRef]
22. Galvele, J.R.; Lumsden, J.B.; Staehle, R.W. Effect of Molybdenum on the Pitting Potential of High Purity 18% Cr Ferritic Stainless Steels. *J. Electrochem. Soc.* **1978**, *125*, 1204. [CrossRef]
23. Yaniv, A.E.; Lumsden, J.B.; Staehle, R.W. The Composition of Passive Films on Ferritic Stainless. *Electrochem. Soc.* **1977**, *124*, 490–496. [CrossRef]
24. Ogawa, H.; Omata, H.; Itoh, I.; Okada, H. Auger electron spectroscopic and electrochemical analysis of the effect of alloying elements on the passivation behavior of stainless steels. *Corrosion* **1978**, *34*, 53. [CrossRef]
25. Sugimoto, K.; Sawada, Y. The Role of Alloyed Molybdenum in Austenitic Stainless Steels in the Inhibition of Pitting in Neutral Halide Solutions. *Corrosion* **1976**, *32*, 347–352. [CrossRef]
26. da Cunha Belo, M.; Rondot, B.; Pons, F.; Le Hericy, J.; Langerous, J.P. *Advances in Electronics and Electron Physics*; Elsevier B.V.: Amsterdam, The Netherlands, 1977; Volume 61.
27. Kolotyrkin, Y.M.; Freinman Korroziya, L.I. I Zashchitaot Korroziy, Protection of equipments from corrosion and salt deposition in geothermal systems, Izd. *Winiti Moscow* **1978**, *5*, 5.
28. Floreen, S.G.R. The Strength and Toughness of Maraging Steels. *Trans. As. Med.* **1964**, *57*, 714.
29. Miller, G.P.; Mitchell, W.L. Structure and Hardening Mechanisms of 18% Nickel-Cobalt-Molybdenum Maraging Steels. *J. Iron Steel Inst.* **1965**, *203*, 899–904.
30. Baker, A.J.; Swann, P.R. Metallography of Maraging Steels. *J. Appl. Phys.* **1963**, *34*, 2505.
31. Hättestrand, M.; Nilsson, J.-O.; Stiller, K.; Liu, P. Precipitation hardening in a 12%Cr–9%Ni–4%Mo–2%Cu stainless steel. *Acta Mater.* **2004**, *52*, 1023–1037. [CrossRef]
32. Peters, D.T. Precipitate Reversion in 18% Ni-Co-Mo Steels. *Trans. AIME* **1968**, *61*, 62.
33. Sherif, E.-S.M. Corrosion Behavior of Duplex Stainless Steel Alloy Cathodically Modified with Minor Ruthenium Additions in Concentrated Sulfuric Acid Solutions. *Int. J. Electron. Sci.* **2011**, *6*, 2284–2298.
34. Singh, A.K.; Shukla, S.K.; Singh, M.; Quraishi, M.A. Inhibitive Effect of Ceftazidime on Corrosion of Mild Steel in Hydrochloric Acid Solution. *Mater. Chem. Phys.* **2011**, *129*, 68–76. [CrossRef]
35. El Azhar, M.; Mernari, B.; Traisnel, M.; Bentiss, F.; Lagrenée, M. Corrosion Inhibition of Mild Steel by the New Class of Inhibitors [2,5-Bis(n-Pyridyl)-1,3,4-Thiadiazoles] in Acidic Media. *Corros. Sci.* **2001**, *43*, 2229–2238. [CrossRef]
36. Seikh, A.H.; Halfa, H.; Baig, M.; Khan, S.M.A. Microstructure Characterization and Corrosion Resistance Behavior of New Cobalt-Free Maraging Steel Produced Through ESR Techniques. *J. Mater. Eng. Perform.* **2017**, *26*, 1589–1597. [CrossRef]
37. Macdonald, J.R.; Johnson, W.B. *Theory in Impedance Spectroscopy*; Wiley: New York, NY, USA, 1987.
38. Kamimura, T.; Hara, S.; Miyuki, H.; Yamashita, M.; Uchida, H. Composition and protective ability of rust layer formed on weathering steel exposed to various environments. *Corros. Sci.* **2006**, *48*, 2799–2812. [CrossRef]
39. Yamashita, M.; Miyuki, H.; Matsuda, Y.; Nagano, H.; Misawa, T. The long term growth of the protective rust layer formed on weathering steel by atmospheric corrosion during a quarter of a century. *Corros. Sci.* **1994**, *36*, 283–299. [CrossRef]

www.ingramcontent.com/pod-product-compliance
Lightning Source LLC
Chambersburg PA
CBHW061447210926
51850CB00008B/1635
* 9 7 8 3 0 3 6 5 7 9 8 0 1 *

MDPI
St. Alban-Anlage 66
4052 Basel
Switzerland
Tel. +41 61 683 77 34
Fax +41 61 302 89 18
www.mdpi.com

Metals Editorial Office
E-mail: metals@mdpi.com
www.mdpi.com/journal/metals